BASIC EXPERIMENTAL STRATEGIES and DATA ANALYSIS for SCIENCE and ENGINEERING

John Lawson

Brigham Young University
Provo, Utah, USA

John Erjavec

University of North Dakota
Grand Forks, USA

 CRC Press
Taylor & Francis Group
Boca Raton   London   New York

CRC Press is an imprint of the
Taylor & Francis Group, an **informa** business
A CHAPMAN & HALL BOOK

CRC Press
Taylor & Francis Group
6000 Broken Sound Parkway NW, Suite 300
Boca Raton, FL 33487-2742

First issued in paperback 2020

© 2017 by Taylor & Francis Group, LLC
CRC Press is an imprint of Taylor & Francis Group, an Informa business

No claim to original U.S. Government works

ISBN 13: 978-0-367-57408-6 (pbk)
ISBN 13: 978-1-4665-1217-7 (hbk)

Visit the Taylor & Francis Web site at
http://www.taylorandfrancis.com

and the CRC Press Web site at
http://www.crcpress.com

John Lawson — *Dedicated to the memory of my late Professor Melvin Carter who inspired my lifelong interest in the design and analysis of experiments.*

John Erjavec — *Dedicated to the memory of my late Professor George E. P. Box who motivated my doctoral research and started me on the path of using statistics as a powerful tool in engineering.*

BASIC
EXPERIMENTAL
STRATEGIES
and DATA ANALYSIS
for SCIENCE and
ENGINEERING

Contents

12 Mixture Experiments   317

13 Practical Suggestions for Successful Experimentation   359

Appendix A   363

Appendix B   369

Check Figures for Selected Exercises   411

Bibliography   413

Index   417

List of Figures

List of Tables

Preface

In today's competitive business environment, many companies are seeking ways to increase their efficiency in tasks such as manufacturing process troubleshooting, new process development, and new product development. New process and product development include prototype testing, development of new component and system design configurations, raw material selection, and determination of appropriate component tolerances. Because deterministic relationships for solving these problems are often not known, industrial research programs or technical investigations usually involve trial and error.

Every technical investigation involving experimentation embodies a strategy (either consciously or unconsciously) for deciding what experiments to perform, when to quit, and how to interpret the data. There are as many different strategies as there are investigators. Some are good and some not so good. This text presents several statistically derived strategies which give greater efficiency than the more intuitive approaches. That is, the use of statistical experimentation strategies will generally get the investigator to his goal in the shortest time (i.e., running the fewest experiments), give the greatest degree of reliability to his conclusions, and keep the risk of overlooking something of practical importance to a minimum.

These statistical experimentation strategies were first developed for use in agricultural research, and they continue to be used successfully on a wide scale in that type of research. The methods have also been commonly used in chemical manufacturing and in electronic circuit design in Japan. However, for the most part these techniques have not been widely used in industry. The reasons for this lack of use are partly historical and cultural and partly due to misconceptions. Historically, statistical experimentation strategies are a relatively recent tool, having been developed for industrial use only since about 1940. Many business managers are not aware of the methods since they are not as glamorous or attention getting as factory automation, artificial intelligence, or robotics. Also, most engineers have had little or no training in the use of these methods. Culturally, industrial managers are often rewarded for working around problems or obtaining quick fixes. This does not favor use of statistical experimental strategies which seek understanding and permanent solutions to recurring problems. Finally, statistical experimentation strategies are often thought to be complicated, lengthy, and costly, and, therefore, reserved for "special" problems where they are often misused due to lack of experience. However, as this text shows, statistical experimentation strategies are actually simple to use, and are neither costly nor lengthy. Herein lies a straightforward presentation of the information necessary for engineers to begin applying these techniques. Companies that utilize these strategies as standard operating procedures can expect large cost reductions in manufacturing, improved product quality, and reduced lead time for the introduction of new products and/or manufacturing methods.

This text is a technical manual which presents methods an engineer or scientist can use to plan an experimental program in two basic situations: (1) screening out which variables are important from a multitude of possible candidates, and (2) optimizing with respect to the relatively few important variables to find the combination of variable settings that gives the best response (or the best compromise among several responses). Enough basic statistics will be presented so that the novice can perform the analysis and interpretation of experimental data, as well as appreciate the reasoning behind the statistical methods. The book illustrates how data analysis can be accomplished with common spreadsheet programs such as Excel© or the open source programs Libre Office Calc or Open Office Calc. Some examples in the book show the use of the statistical software such as the commercial program Minitab© or the open source program R, but the common spreadsheet is capable of handling the majority of the types of analysis presented.

In the past, the material in Chapters 1, 3, 4, 7, and 9 (with light emphasis on the least squares method in Chapter 8) has been covered in a three-day workshop. A prerequisite for course participants has been a familiarity with basic statistical concepts and methods covered in Chapter 2. Chapter 2 is included in the text for review purposes and for setting the stage for the statistical analysis tools used in the following chapters. Course participants should study, read, and review Chapter 2 prior to attending a workshop. Normal workshops have consisted of morning and afternoon lectures followed by in-class simulation problems wherein participants work in groups and report their results. Other workshop formats have also been used. With one or two half-day sessions per week, a wider selection of topics has been successfully covered in five- to ten-week periods. The most successful workshops have a follow-up meeting several weeks after the conclusion of the course. In this follow-up meeting, teams of course participants present results of experimental studies conducted on the job utilizing techniques they learned in the course.

In an academic setting, the whole book is generally covered in a one-semester course. If time is short, Chapter 12 can be omitted without loss of comprehension of the rest of the material. No prerequisites in statistics or mathematics are required, although some background in matrix algebra is helpful (but not essential) in Chapter 8. The text material is supplemented by requiring students to run experiments on their own using the various designs discussed in the course (e.g., screening, factorial, etc.) and then to analyze the results and report their conclusion. A course such as this goes a long way in reducing the serious lack of training in statistics and experimental design that most engineering graduates have.

There is a website for the book at https://jlawson.byu.edu/ that contains the appendix tables in spreadsheet format and other examples of computer code from the book.

Author Bios

John Lawson—is a professor in the Department of Statistics at Brigham Young University (BYU) where he has been for the last 30 years. He earned his his BS in statistics from BYU, an MS from Rutgers University, and a PhD degree in Applied Statistics from the Polytechnic Institute of N. Y. He worked as a statistician, and later senior biostatistician at Johnson and Johnson Corporation; as a senior statistician, and later manager of statistical services at FMC corporation prior to returning to academia. In industry, he used designed experiments and statistical analysis to help engineers and chemists on product development and manufacturing process improvements. At BYU, he teaches courses on experimental design and quality control and consults with clients through the BYU Center for Statistical Consultation and Collaborative Research.

John Erjavec—is a retired professor and chair of the Department of Chemical Engineering at the University of North Dakota (UND) where he taught for 20 years. He earned his BS in chemical engineering at Princeton University, and his MS and PhD in chemical engineering at the University of Wisconsin, Madison. He minored in statistics while getting his PhD, and his PhD research was motivated by a desire to show the power of some new statistical tools in experimental design and analysis to modeling in the sciences and engineering. The principal investigator on that project was G.E.P Box, a professor in the Department of Statistics. Dr. Erjavec worked for five years at American Cyanamid as a process analysis engineer, and followed that with ten years at FMC Corporation. His titles at FMC included senior statistician and manager, systems analysis. While at Cyanamid and FMC he taught (with John Lawson at FMC) numerous short courses in experimental design and data analysis to engineers and scientists throughout the company. At UND he continued teaching (in addition to the usual chemical engineering courses) engineering statistics to seniors in engineering and graduate students in both sciences and engineering.

Chapter 1

Strategies for Experimentation with Multiple Factors

1.1 Introduction

Advancements in knowledge through the scientific method are obtained by iterating through three basic steps: conjecture, observation, and analysis. In the conjecture step, a theory or hypothesis is formulated to explain the causes for the behavior of the system or phenomenon under study. In the observation and analysis steps, data is collected and analyzed to confirm or contradict the consequences of the proposed theory or hypothesis. The value of statistical methods in the analysis and interpretation of research data is well understood. Articles in reputable journals justify their conclusions based on the level of statistical significance. Less understood is the vital role of statistical methods in the collection of data, especially in research studies aimed at establishing cause and effect links, or determining optimum operating conditions.

One of the statistical methods used in data collection is called the statistical design of experiments, or just design for short. In practice, using a good strategy for data collection (i.e., a good design) is even more important than a statistical analysis of the results. That is because it is difficult, if not impossible, to remedy the problems inherent in a less than optimal data collection strategy, even with the most sophisticated statistical analysis. As the old adage goes: "You can't make a silk purse out of a sow's ear." On the other hand, it is difficult (but not impossible) to mangle the interpretation of a well-designed set of experiments.

These two tools—statistical design and statistical analysis—are used very fruitfully in finding empirical solutions to industrial problems and basic research. The areas of application include research, product design, process design, production troubleshooting, and production optimization. The response or objective in a research or industrial study is usually a function of many interrelated factors, so solving these problems is typically not straightforward. There are two broad categories of approaches to these problems: 1) finding solutions by invoking known theory or facts, including experience with similar problems or situations, and 2) finding solutions through trial and error or experimentation. Statistical design and analysis methods are very useful for the second approach, and they are much more effective than any traditional, nonstatistical method.

In this book we present some well-proven experimental design plans to (1) screen which variables (from a multitude of possible factors) have important effects, (2) determine optimum operating conditions with respect to a few important variables, and (3) describe variability in research data. In each case we illustrate the corresponding methods for analyzing the resulting data. The plans are tabulated in appendices to the book, and available for download from the website for this book. Most of the analysis can be conducted using spreadsheet software like Microsoft Excel© or Libre Office Calc©.

1

1.2   Some Definitions

Before beginning a discussion of statistical design and statistical analysis, it is important to ensure that we are all speaking the same language. So let us define several of the terms that we will use frequently.

- *Experiment* (also called a Run)—an action in which the experimenter changes at least one of the factors being studied and then observes the effect of his/her action(s). Note that the passive collection of historical data is not experimentation.

- *Experimental Unit*—the item under study upon which something is changed. In a chemical experiment it could be a batch of material that is made under certain conditions. The conditions would be changed from one unit (batch) to the next. In a mechanical experiment it could be a prototype model or a fabricated part.

- *Factor* (also called an Independent Variable and denoted by the symbol, X) —one of the variables under study which is being deliberately controlled at or near some target value during any given experiment. Its target value is being changed in some systematic way from run to run in order to determine what effect it has on the response(s).

- *Background Variable* (also called a Lurking Variable)—a variable of which the experimenter is unaware or cannot control, and which could have an effect on the outcome of an experiment. The effects of these lurking variables should be given a chance to "balance out" in the experimental pattern. Later in this book we will discuss techniques (specifically randomization and blocking) to help ensure that goal.

- *Response* (also called a Dependent Variable and denoted by the symbol, Y)—a characteristic of the experimental unit which is measured during and/or after each run. The value of the response depends on the settings of the factors or independent variables (X's).

- *Effect*— is the expected change in the response that results when the value of a factor is changed.

- *Experimental Design* (also called Experimental Pattern)—the collection of experiments to be run. We have also been calling this the experimental strategy.

- *Experimental Error*—the difference between any given observed response, Y, and the long run average (or "true" value of Y) at a particular set of experimental conditions. This error is a fact of life. There is variability (or imprecision) in all experiments. The fact that it is called "error" should not be construed as meaning it is due to a blunder or mistake. Experimental errors may be broadly classified into two types: bias errors and random errors. A bias error tends to remain constant or follow a consistent pattern over the course of the experimental design. Random errors, on the other hand, change value from one experiment to the next with an average value of zero. The principal tools for dealing with bias errors are blocking and randomization (of the order of the experimental runs). The tool to deal with random error is replication. All of these will be discussed in detail later.

Some of the terms are illustrated in Figure 1.1. It is a study of a chemical reaction ($A + B \to P$) consisting of nine runs to determine the effects of two factors (time and temperature) on one response (yield). One run (or experiment) consists of setting the temperature, allowing the reaction to go on for the specified time, and then measuring the yield. The design is the collection of all runs that will be (or were) made.

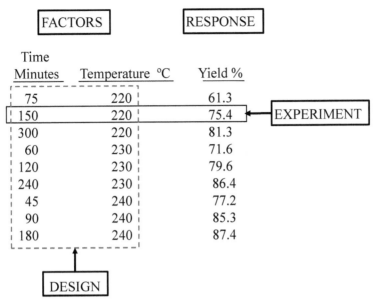

FIGURE 1.1: Example of an Experimental Study with Terminology Illustrated

1.3　Classical Versus Statistical Approaches to Experimentation

A typical strategy for a research study used by someone unfamiliar with statistical plans is the *one-at-a-time* design. In this approach, a solution is sought by methodically varying one factor at a time (usually performing numerous experiments at many different values of that factor), while all other factors are held constant at some reasonable values. This process is then repeated for each of the other factors, in turn. One-at-a-time is a very time-consuming strategy, but it has at least two good points. First, it is simple, which is not a virtue to be sneezed at. And second, it lends itself easily to a graphical display of results. People think best in terms of pictures, so this is also a very important benefit.

Unfortunately, one-at-a-time usually results in a less than optimal solution to a problem, despite the extra work and consequent expense. The reason is that one-at-a-time experimentation is only a good strategy under what we consider to be unusual circumstances. These unusual circumstances would include the following characteristics: (1) the response is a complicated function of the factors, X's (perhaps multimodal), which requires many levels of each X to elucidate, (2) the response relationship with the factors dominates the experimental error and will be recognized in the presence of this error, (3) the effect of the factor being studied (and therefore its optimum value) is not changed by the level of any of the other factors. That is, the one-at-a time strategy works only if the effects are strictly additive, and the effect of one factor does not depend on others. As was just stated, but worth repeating, these circumstances do not typically exist in the real world, and so one-at-a-time ends up being an exceedingly poor approach to problem solving.

The assumptions made by statistical designs represent a more typical set of circumstances. They are: (1) within the experimental region, the response is smooth with, at most, some curvature but no sharp kinks or inflection points, (2) experimental error is always present to "muddy the water" of experimental results, and (3) the effect of one factor can depend on the level of one or more of the other factors. In other words, there may be dependencies or interactions between factor's effects. If these assumptions hold, the classical one-at-a-time approach could do very badly. If the first assumption holds, one-at-a-time would require many more experiments to do the same job, because fewer levels of a factor are needed to fit a smooth response than to fit a complicated one. If the second and third assumption hold, one-at-a-time could lead to completely misleading conclusions.

For example, if we try to find the optimal time and temperature for the yield of a chemical reaction in one-at-a-time fashion, we could get the results shown in Figure 1.2.

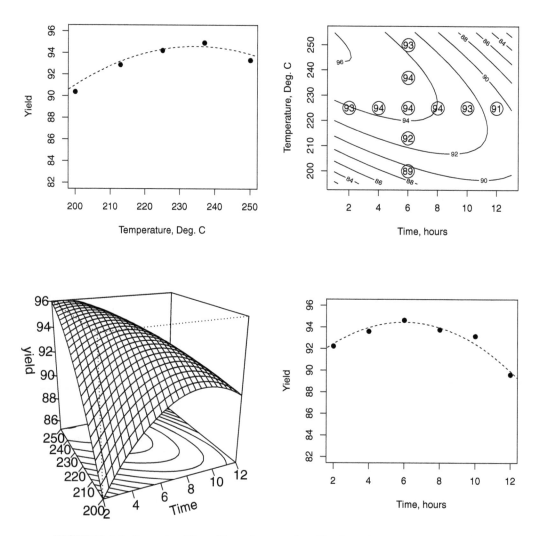

FIGURE 1.2: One-at-a-Time Experimentation Used to Optimize Process Yield

The three-dimensional surface in the lower left of the figure shows the true relationship between yield, temperature, and time. This relationship can be represented in two dimensions with the contour lines shown below the surface and enlarged in the upper right. If reaction time were methodically varied from 2 to 12 hours while temperature was held constant at 225°, the curve traced can be seen in the lower right (where points differ from the line due to experimental error). After varying reaction time, it would be concluded that the optimum time was 6 hours. Holding time constant at 6 hours and varying temperature would result in the curve shown in the upper left. This would lead to the faulty conclusion that the maximum yield of about 94% could be obtained at 6 hours and 235°. A higher yield of over 96% can be seen in the upper left of the corner of the experimental region. One-at-a-time experimentation fails because there is an interaction between time and temperature. In other words, the best time depends on the temperature (or vice versa).

An even more confusing result can occur with one-at-a-time when an experimenter is simply studying the effect of one factor holding all others constant. In Figure 1.3 it can be seen that the effect of calcination temperature on surface area looks much different depending on the level of calcination time chosen. The conclusion an experimenter would draw concerning the effect of temperature depends very much on the level at which time is held constant. It can't be repeated too often that in the real world, these interactions are the rule, not the exception.

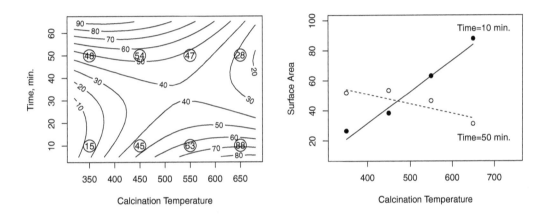

FIGURE 1.3: Effect of Calcination Temperature

The potential for failing with a one-at-a-time strategy gets even stronger in the real world where experimental errors cloud the issue. But, one-at-a-time is still far better than another common strategy used in research, namely sheer guesswork. Using this strategy, researchers assume that they know the solution to a problem, as if they had the appropriate theory or experience to guide them. Then they run a few confirmatory experiments for validation. If the confirmatory experiments produce less than the anticipated results, more guessing and confirmatory experiments are run. The result is a very haphazard approach. These poor strategies for studying many factors simultaneously in industrial research often deplete all research funds before a satisfactory solution can be reached. This, in turn, leads to less than optimal product designs, inefficient production processes, and the complete abandonment of many promising ideas.

Studying the effects of many factors simultaneously in an efficient way, while at the same time allowing valid conclusions to be drawn, is the purpose of the experimental design techniques described in this text. After a review of some basic background in probability and statistics in Chapter 2, we will begin our discussion of experimental designs in Chapter 3 with a focus on the design that will be our cornerstone—the two-level factorial design. It is one of the most basic, and yet most powerful, experimental strategies that exists.

1.4   Why Experiment? (Analysis of Historical Data)

Frequently in industrial settings, there is a wealth of data already available about the process of interest in the form of process logs and, more recently, computer databases. It is tempting, therefore, to abandon an (expensive) experimental study in favor of a (cheap) analysis of the historical data on hand. On the surface, this looks very reasonable. After all, there has certainly been variation in all of the factors of interest over the life of the process. And there is certainly much more data available from the logs than could ever be collected in an experimental program, so the precision should be excellent. Why then ever bother with experiments at the plant level? The reasons are many. All of them have to do with the inadequacies inherent in unplanned data:

- **Correlation ≠ Causation** If a significant correlation is found between the response and an independent variable (factor), it may be interesting, but it does not prove causation. For example, let us say that someone who had hip surgery years ago noticed that when their hip ached it was a sure sign that it would rain within 12 hours. It may be a perfect correlation, but it does not mean that they should bruise their hip to make it ache if their lawn needs rain. Notice that there may be value in the correlation as a predictor of what will happen. But to use it to try to control the system is very likely folly [see Box, 1966].

- **Correlated Factors** Over time, there may have been quite a bit of variation in the factors of interest, but often they move up and down together. It is then impossible to sort out which of the factors, if any, is having an impact on the response. For example, in the gasification of coal, it may be useful to know which impurities in the coal are deleterious to yield. The amounts of the impurities may change quite a bit, but if they stay in roughly the same proportions, it is impossible to sort out which impurity is the culprit, if any.

- **No Randomization → Bias** Since there is no randomization in the variations of the factors with time, the door is wide open for biases to influence the results. For example, if a plant switched from one supplier of raw material to another and noticed a deterioration in yield (i.e., the average yield for the month before the switch was better than the average yield for the month after the switch), one explanation is that the new raw material is worse than the old one. However, another explanation is that normal drifting of the process due to other things changing gave a worse yield after the switch. So the raw material source had nothing to do with the drop in yield. Note: A valid test would be to randomly switch back and forth several times between suppliers.

- **Tight Control of Factors** If an independent variable is known (or thought) to be important, then it will be controlled tightly if possible. Therefore, there will not be enough variation in that factor to get a measurable effect on the response. The few data points that may exist in which these factors do vary more widely will have occurred during plant upsets. That is not the kind of data that one wants to use to draw sound conclusions.

- **Incomplete Data** Plants are quite different from laboratories and pilot plants, in that as few variables are measured and recorded as possible, consistent with being able to control the process. Therefore, even in the case when all important variables are known, which is unusual, they are not usually recorded. And to make matters worse, in older plants where much of the data were/are recorded by hand, even the variables to be measured and recorded may be missing or suspect.

Therefore, the analysis of historical data should be done with great care. This does not mean that it is a totally worthless exercise. Some interesting correlations may emerge. These can be used for predictive purposes, or they may point to some factors that were previously thought to be unimportant that should be studied in an experimental program. But, this analysis should be undertaken with the knowledge that the chances of gleaning any worthwhile information from the data are quite low (10%). And it must be kept in mind that an analysis of historical data is NEVER a replacement for an experimental program if one needs to determine causation between the factors and the response(s). To find out what happens to a system when you interfere with it, you have to interfere with it (not just passively observe it). For a more complete discussion of this topic see [Box et al., 1978, pages 487-498].

1.5 Diagnosing the Experimental Environment

There is no single experimental design that is best in all possible cases. The best design depends very much on the environment in which the experimental program will be carried out [see Lucas, 1985]. Figure 1.4 on the following page shows different environments that will be discussed in this book. In Chapter 3, the two-level factorial experimental designs for the objective of constrained optimization will be presented. These designs provide the building block which will be used to construct designs for the other situations. Chapter 7 will present experimental designs for the objective of screening, and Chapter 9 will discuss experimental designs applicable for unconstrained optimization (response surface studies). Designs for mechanistic modeling are not discussed in this text, the reason being that they are too specific to the model being used. The characteristics that define the environment are:

- **Number of Factors** The single most important characteristic of the experimental environment is how many independent variables are to be studied. If the number is small (say three or less), then a design giving fairly complete information on all of them may be reasonable right off the bat. However, if there are many variables, it is usually more reasonable to proceed in stages—first sifting out the variables of major importance, and then following up with more effort on them.

- **Prior Knowledge** The amount of prior knowledge also shapes the experimental program to a very large degree. When the area to be studied is new, there are generally a large number of potential variables that may have an effect on the responses. However, when the area has been studied extensively in the past, the scope of the experiments is generally to further elucidate in detail the effects of a few of the key variables. If theory is available, a mechanistic model may be desirable and experiments can be set up for determining the unknown parameters in the model. The best experiments to run are specific to the model, but often experimental designs used for empirical models are good as a starting point [see Kittrel and Erjavec, 1968].

- **Cost of an Experiment** The size of a reasonable experimental program is, of course, dictated by the cost of an experimental run versus the potential benefits. The cheaper

	Present ⇓			Goal ⇓	
	0%		**Knowledge**		100%

Objective:	Preliminary Exploration	Screening Factors	Effect Estimation	Optimization	Mechanistic Modeling
No. of Factors		5 - 20	3 - 6	2 - 4	1 - 5
Model:	Variance Components	Linear	Linear + Interactions	Linear + Interactions + Quadratics	Mechanistic Model
Purpose:	Identify Sources of Variability	Identify Important Factors	Estimate Factor Effects + Interactions	Fit Empirical Model Interpolate	Estimate Parameters of Theory Extrapolate
Designs:	Nested Designs	Fractional Factorials	Two-level Factorials	RSM	Special Purpose

FIGURE 1.4: Objective of Experimentation Is to Increase Our Knowledge

an experiment is, the more thoroughly we can study the effects of the independent variables for a reasonable total cost.

- **Precision** Generally speaking, the reason for experimenting is to be able to make predictions about what will happen if you make similar actions in the future. For example, the reason for studying the effect of pH on the yield of a chemical reaction is to be able to say that the yield is 4% higher at a pH of 9 than it is at a pH of 7. The more precise you want your predictions to be, and the less precise your individual data are, the greater the number of experiments required.

- **Iteration Possible** If the duration of an experiment is relatively short, it is usually reasonable to experiment in small bite sizes and iterate toward your final goal. If, on the other hand, the time for an experiment to be completed is long (such as in stability testing), it would be necessary to initially layout a fairly extensive pattern of experiments. When it is possible to iterate, the first stage of an experimental program should usually be a set of screening experiments. At this stage, all of the factors that could conceivably be important are examined. Since the cost of looking at an extra five or ten variables is relatively low with a screening design, it is much cheaper in the long run to consider some extra variables at this stage than to find out later that you neglected an important variable. The next stage is generally a constrained optimization design in which the major variables are examined for interactions and better estimates are obtained of their linear effects. The minor variables are dropped from consideration after the screening stage (i.e., they are held at their most economical values). If necessary for further optimization, a full unconstrained optimization (response surface) design may now be run to allow for curvature in the effects of the factors on the responses.

Generally, the constrained optimization design builds upon the screening design, and the unconstrained optimization design just adds more points to the constrained optimization design, so that no points are wasted going from one stage to the next.

If it is necessary to go further and use a theoretical model, the full, unconstrained optimization design is usually a good starting point for the estimation of parameters in the model. Additional runs would have to be designed via computer. The book by Bates and Watts [2007] gives a clear description of how this can be done, but this topic is beyond the scope of this text.

1.6   Example of a Complete Experimental Program

1.6.1   Chemical Process System

An example may be useful at this point to illustrate the typical steps involved in solving a specific industrial problem experimentally. Let us say we have a chemical process under development. It consists of a batch reaction to make a product, and the objective is to find the conditions which give the best yield. The process involves charging a stirred tank reactor with solvent, catalyst, and an expensive reactant (Reactant 1). A second reactant (Reactant 2, which is inexpensive) is added slowly. Reactant 2 is added in excess to ensure that all of Reactant 1 is consumed. Some yield is lost due to the formation of by-products. After all Reactant 2 is added, the reaction is quenched by adding cold solvent, and then the product is separated by distillation.

Several variables were thought to possibly have an important influence on yield. They are listed in Table 1.1 along with a reasonable range of values for each factor.

TABLE 1.1
Variables to Be Studied for Chemical Reaction Example

Variable	Definition	Label	Range of Variables	
			Low Level$(-)$	High Level$(+)$
A	Temperature °C	X_1	75	85
B	Reactant 2, excess %	X_2	4	8
C	Time to add Reactant 2, min.	X_3	10	20
D	Agitation, RPM of stirrer	X_4	100	200
E	Solvent:Reactant 1 ratio	X_5	1:1	2:1
F	Catalyst Concentration mg/l	X_6	20	40

1.6.2   Step One: Screening

Since there were six factors to be investigated, the first step in the experimental program was to find out which variables were the most important and focus the most attention on them. It is simply too expensive to study every factor thoroughly, and it is a waste of resources to lavish attention on a variable that has a minor impact. Therefore, we start with screening experiments, which corresponds to the first step in increasing our knowledge. This step is shown in Figure 1.4 semi-graphically as the second column in the figure.

In our example, we used one of the screening designs that will be discussed in detail in Chapter 5, a twelve run Plackett—Burman design, which was copied from Table B.1-1 in

Appendix B. The design is shown in Table 1.2 along with the yield data that was collected for each run. The −'s and +'s indicate whether the low or high level of the factors were used. The main object of the experiments was to find out which variables were important, which means which variables had the greatest impact on yield when the variables were changed. The measure of that importance is the effect of the variable, which is how much the yield changed (on average) when the variable was changed from its low value to its high value. These effects are shown near the bottom of Table 1.2.

TABLE 1.2
Screening Design (Plackett—Burman Table B.1-1)

Run	A	B	C	D	E	F	— Unassigned —					Yield
No.	X_1	X_2	X_3	X_4	X_5	X_6	X_7	X_8	X_9	X_{10}	X_{11}	
1	+	+	−	+	+	+	−	−	−	+	−	62.7
2	+	−	+	+	+	−	−	−	+	−	+	74.9
3	−	+	+	+	−	−	−	+	−	+	+	44.9
4	+	+	+	−	−	−	+	−	+	+	−	72.1
5	+	+	−	−	−	+	−	+	+	−	+	61.3
6	+	−	−	−	+	−	+	+	−	+	+	54.1
7	−	−	−	+	−	+	+	−	+	+	+	43.2
8	−	−	+	−	+	+	−	+	+	+	−	79.8
9	−	+	−	+	+	−	+	+	+	−	−	8.6
10	+	−	+	+	−	+	+	+	−	−	−	84.2
11	−	+	+	−	+	+	+	−	−	−	+	77.4
12	−	−	−	−	−	−	−	−	−	−	−	10.8
Effects	24.1	-3.4	32.1	-6.1	6.8	23.9	0.9	-1.4	1.0	6.6	6.2	

For example, when the reaction temperature, Factor A, was increased from 75°C to 85°C, the yield increased by 24.1% on average. The effects are represented graphically in Figure 1.5. In this figure, effects whose bars are within the dotted lines cannot be distinguished from the variability caused by experimental error.

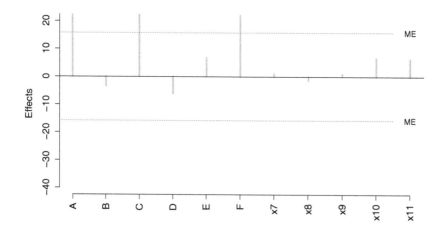

FIGURE 1.5: Factor Effects Relative to Experimental Error

Therefore only three factors were found to be significant: A, C, and F. They are the reaction temperature, the addition time of Reactant 2, and the catalyst concentration. Furthermore, they all had positive effects on the yield, which means that the higher the variable value, the higher the yield. The best yield obtained in this set of experiments can be seen to be at the high values of the three important factors (Run 10).

The other important conclusion at the end of our screening experiments was that the remaining factors were not of major importance, and therefore they were set to some reasonable values and ignored for the rest of the experimental program. This means that Factor B, the excess of Reactant 2, was set to 4% excess, Factor D, the speed of the agitator, was set to 100 rpm, and Factor E, the solvent to Reactant 1 ratio, was set to 1: 1. All were the low values of the factors, which were picked to minimize cost.

1.6.3 Step Two: Crude Optimization

Follow-up experiments were then conducted on the three important factors to determine optimum operating conditions. The strategy used was based upon a constrained optimization design called a factorial design, which makes sure we are in the appropriate region. The factorial design is outlined in the second column of Figure 1.4. An important feature of this design is that it can be augmented easily to find an unconstrained optimum, which is our ultimate goal.

The factorial design used in this study is shown in Table 1.3. It should be noted that the experiments were centered around the conditions found to be best up to that point: temperature (A) of 85°C, addition time (C) of 20 minutes, and a catalyst concentration (F) of 40 mg/l. The coded factor levels were calculated as $X_1 = (A - 85)/5$, $X_2 = (C - 20)/5$, and $X_3 = (F - 20)/5$. The resulting yield data are also given in the table. An analysis of the data, which will be discussed in much more detail in Chapter 3, showed that all the variables continued to be important in addition to some interactions. However, the main point that was learned from the data was that the response could not be described by a straight line model—a quadratic equation was needed. This can be seen directly from the data; the average yield in the center of the experimental region was 80%, while the average response at the other conditions was only 62%. This difference is extremely significant, and it means that the yield is a curved function of the three important factors. It also means that we are in the vicinity of the optimum, and can move on to the final optimization phase.

TABLE 1.3
Factorial Design (Table B.3-2, Block 1)

Run No.	A	C	F	X_1	X_3	X_6	X_1X_3	X_1X_6	X_3X_6	Average	Yield
1	80	15	30	−	−	−	+	+	+		70.2
2	90	15	30	+	−	−	−	−	+		71.1
3	80	25	30	−	+	−	−	+	−		74.6
4	90	25	30	+	+	−	+	−	−		55.3
5	80	15	50	−	−	+	+	−	−		69.5
6	90	15	50	+	−	+	−	+	−		50.1
7	80	25	50	−	+	+	−	−	+		75.5
8	90	25	50	+	+	+	+	+	+	62.0	29.8
9	85	20	40	0	0	0	0	0	0		78.8
10	85	20	40	0	0	0	0	0	0		80.3
11	85	20	40	0	0	0	0	0	0	80.0	81.0
Effects				-20.8	-6.42	-11.5	-11.7	-11.7	-0.7	18.01	

1.6.4   Step Three: Final Optimization

Curved (e.g., quadratic) functions cannot be elucidated by data from a factorial design alone, so the data were augmented by the extra runs to form a complete central composite design, which is outlined in the third column of Figure 1.4. These extra runs are shown in the bottom half of the design in Table 1.4 (called Block 2) along with the associated yield data. Notice that Block 1 consisted of the factorial design that was already in hand. This composite set of data was used to complete the optimization.

TABLE 1.4

Central Composite Design (Table B.3-2)

Run No.	A	C	F	X_1	X_3	X_6	Block	Yield
1	80	15	30	−1	−1	−1	1	70.2
2	90	15	30	1	−1	−1	1	71.1
3	80	25	30	−1	1	−1	1	74.6
4	90	25	30	1	1	−1	1	55.3
5	80	15	50	−1	−1	1	1	69.5
6	90	15	50	1	−1	1	1	50.1
7	80	25	50	−1	1	1	1	75.5
8	90	25	50	1	1	1	1	29.8
9	85	20	40	0	0	0	1	78.8
10	85	20	40	0	0	0	1	80.3
11	85	20	40	0	0	0	1	81.0
12	76.4	20	40	−1.73	0	0	2	82.6
13	93.7	20	40	1.73	0	0	2	44.6
14	85.0	11.4	40	0	−1.73	0	2	67.0
15	85.0	28.7	40	0	1.73	0	2	58.2
16	85.0	20	22.7	0	0	−1.73	2	58.2
17	85.0	20	57.3	0	0	1.73	2	58.2
18	85.0	20	40	0	0	0	2	80.9
19	85.0	20	40	0	0	0	2	77.7
20	85.0	20	40	0	0	0	2	83.5

In the analysis, a full quadratic equation was first to the twenty data points, and then the equation was used to predict the best operating conditions. The equation was fit using regression analysis, which is a tool included in spreadsheet programs as well as statistical software. Equation (1.1) is the result.

$$\text{Yield} = 80.37 - 10.67X_1 - 3.79X_3 - 5.66X_6 - 5.81X_1X_3 - 5.84X_1X_6$$
$$- 0.36X_3X_6 - 5.84X_1^2 - 7.35X_3^2 - 4.47X_6^2, \tag{1.1}$$

Analysis of this equation revealed that the predicted optimum yield occurred when catalyst concentration (F) is set at 40 mg/l. Figure 1.6 shows the predicted yield from Equation 1.1 as a function of temperature (A) and addition time of Reactant 2 (C) when catalyst concentration is set at 40 mg/l. There it can be seen that the optimum conditions are a reaction temperature of 80° C, and an addition time for Reactant 2 of about 20 minutes.

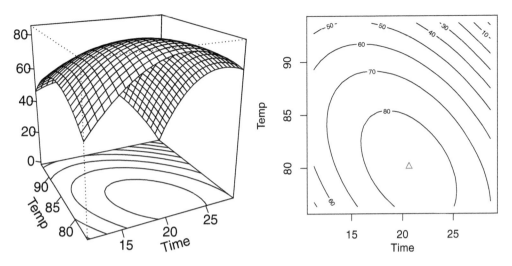

FIGURE 1.6: Graphs to Visualize Optimum Yield

1.7 Good Design Requirements

Prior to discussing good strategies for conducting experiments (beginning in Chapter 3), it is first worth discussing what "good" means. In other words, what would an ideal experimental design do for us? The following are some attributes of a good design:

- **Defined Objectives** The experimenter should clearly set forth the objectives of the study before deciding on an experimental design and proceeding with the experiments. Generally this takes the form of how many factors will be studied, and what model will be fit to the data (a simple linear model, a linear model with interactions, or a full quadratic model). In addition, the desired precision of the conclusions needs to be specified. A good design must meet all of the objectives. Once a design is selected, the experimenter can and should detail (for the sponsor of the work) not only what information will be obtained from the experimental data, but what information will not be learned, so that there are no misunderstandings.

- **Unobscured Effects** The effects of each of the factors in the experimental program should not be obscured by other variables insofar as possible. All the designs discussed in this book meet this goal, so it will not be mentioned again.

- **Free of Bias** As far as possible, the experimental results should be free of bias, conscious or unconscious. The first step in assuring this is to carefully review the experimental setup and procedure. However, some statistical tools are helpful in this regard:

  -BLOCKING (planned grouping) of runs lets us take some lurking variables into account.

  -RANDOMIZATION of the run order of experiments within each block enables us to minimize the confounding (biasing) of factor effect with background variables.

  -REPLICATION allows us to reduce the effect of experimental error by "averaging" it over replicate experiments at the same conditions. It also allows us to estimate the standard deviation of experimental error that will be used as the "yardstick" to judge the significance of factor effects.

- **Variability Estimated** To judge whether the effects of factors found in an experimental program are real or whether they could just be due to the variability in the data caused by experimental error, the experimental design should provide for estimating the variability of the results. For example, if two methods were compared with respect to their effect on purity, it appears that the two experiments in the left side of Figure 1.7 show method 2 is better. However, there is no measure of variability in purity since the experiments were not repeated. If the two experiments were repeated four additional times (as shown in the right side of Figure 1.7), the difference in average purity can be compared in light of variability caused by the inability to exactly replicate experimental results.

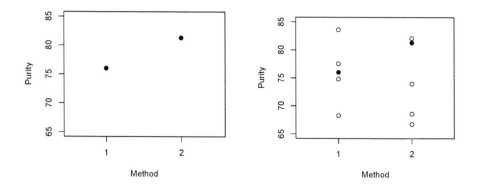

FIGURE 1.7: Variability Caused by Experimental Error

The only time replication is not needed is when there is a well known history of the process or system being studied, with quantitative estimates of the process standard deviation, σ, from process capability studies. Replication allows the variability or precision of results to be estimated, while randomization (or randomly switching between method 1 and 2) reduces the chance that the difference in average results was caused by an unanticipated change in some background variable that was not controlled. Thus randomization assures the approximate validity of results.

- **Design Precision** The precision of the total experimental program should be sufficient to meet the objectives. In other words, enough data should be taken so that effects that are large enough to have practical significance will be statistically significant. Greater precision can sometimes be achieved by refinements in experimental technique. Blocking can also help improve precision a great deal. However, the main tool at our disposal to increase precision is replication. Just how much replication is needed for a desired level of precision is discussed in Chapter 4.

1.8 Summary

In this chapter we discussed what properties are desirable in an experimental strategy. Some terms were defined to help communicate the ideas more clearly. And a few statistical tools were introduced that are used to help us get the best results we can for our efforts.

1.8.1 Important Terms and Concepts

The following is a list of the important terms and concepts covered in this chapter.

design/pattern/strategy
one-at-a-time design
experiment/run
experimental unit
factor/independent variable/X
background variable/lurking variable
response/dependent variable/Y
error
correlation vs. causation
bias
blocking
randomization
replication

Chapter 2

Statistics and Probability

2.1 Introduction

In any empirical investigation or experimentation, experimental error is a fact of life. It would be very nice if tests or experiments conducted at the same factor settings would always give the same result. In the real world this is not the case and variability in data is something that must be expected. Figure 2.1 illustrates the problem. Although experimenters may hold the controllable factors constant, differences in experimental units, unknown changes in background or lurking variables, and measurement errors will cause the response in repeat experiments at the same factor settings to vary in what seems to be a random pattern.

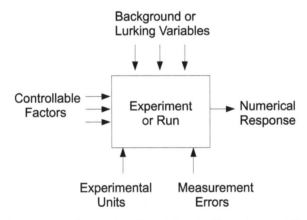

FIGURE 2.1: Causes for Variability in Experimental Data

When analyzing and interpreting data from experiments, the magnitude of factor effects must always be judged by (or compared to) the amount of variability expected from repeat experiments conducted at the same factor settings with different experimental units.

In order to quantify and visualize response variability caused by experimental error, this chapter will present some graphical and numerical methods useful for summarizing variable data. Next, some concepts from probability theory will be presented, which are helpful in understanding what type of variability in data should be expected from random sources. This understanding will enhance an experimenter's ability to correctly determine whether changes in the response resulting from factor level changes are real or due to experimental error.

2.2   Graphical and Numerical Summaries of a Single Response-Variable

2.2.1   Dot Diagrams

When there are 15 or fewer observations, a simple way to visualize the variability in the data is to construct a dot diagram. To construct a dot diagram, mark a horizontal axis with tic marks spanning the range of the data, then place a dot above the axis where each data point falls. This can be easily done with a pencil and paper, but some statistical software like (Minitab©) can generate a dot diagram for data stored in a column of the worksheet. For an example of a dot diagram, consider the bore diameters of nine consecutive parts (shown in Table 2.1) that were produced at the same conditions in a manufacturing facility.

TABLE 2.1
Bore Diameters of Nine
Parts

Part No.	Diameter
1	0.5755
2	0.5753
3	0.5756
4	0.5762
5	0.5756
6	0.5758
7	0.5758
8	0.5768
9	0.5750

The dot diagram is shown in Figure 2.2. The tic marks on the axis span the range of the data, and there is a dot placed where each response value falls on the scale. When more than one data point assumes the same value (like part numbers 3 and 5), dots are stacked on the plot. This figure graphically depicts the range of variability in the data and shows the average of the bore diameters is around 0.5757.

FIGURE 2.2: Dot Diagram of Bore Diameters

2.2.2   Sample Mean, Variance, and Standard Deviation

A numerical summary that represents the center of the data is called the average or sample mean. If the observations are labeled symbolically as Y_i for $i = 1$ to $i = n$ (where $n = 9$ in the example above), then the average ($\bar{Y}$) is calculated by summing the observations and dividing by the number of observations as shown in Equation 2.1 (which also illustrates the calculation with the data in Table 2.1.

$$\bar{Y} = \frac{\sum_{i=1}^{n} Y_i}{n}$$

$$= \frac{.5755 + .5753 + .5756 + .5762 + .5756 + .5758 + .5758 + .5768 + .5750}{9}$$

$$= 0.57573, \tag{2.1}$$

A commonly used numerical summary to represent the variability or spread in data is the sample variance, which is denoted by s^2. It is calculated with the following formula:

$$s^2 = \frac{\sum_{i=1}^{n}(Y_i - \bar{Y})^2}{n-1}, \tag{2.2}$$

This represents the average squared distance from the mean, and it is measured in squared units of the data. For example, if the bore diameters in Table 2.1 are measured in centimeters, then the sample variance s^2 of the bore diameters is measured in terms of squared centimeters. The sample standard deviation, or $s = \sqrt{s^2}$, is in the same units as the measurements Y_i, and it also measures the spread or variability in a set of data.

Common spreadsheet programs like Microsoft's Excel© or the open source Oracle Calc© or Libre Office Calc can also calculate the mean, variance, and standard deviation of a set of numbers. For example, Figure 2.3 illustrates the calculation of the mean in cell A11 and the standard deviation in cell A12.

	A	B
1	0.5755	
2	0.5753	
3	0.5756	
4	0.5762	
5	0.5756	
6	0.5758	
7	0.5758	
8	0.5768	
9	0.575	
10		
11	0.575733	
12	0.000522	

FIGURE 2.3: Calculating $\bar{Y}$ and s Using a Spreadsheet

The formula in cell A11 is =AVERAGE(A1:A9) and the formula in cell A12 is =STDEV(A1:A9). The formula for calculating the variance s^2 in a spreadsheet is =VAR(A1:A9).

Most scientific calculators also have the ability to calculate the average $\bar{Y}$, the variance s^2, and the standard deviation s from a list of values. However, most calculators label the key for the mean as $\bar{x}$ rather than $\bar{Y}$.

Figure 2.4 compares the dot diagrams, sample means ($\bar{Y}$), and sample standard deviations ($s = \sqrt{s^2}$) for three sets of data. The center of the middle dot diagram is shifted further to the left than the dot diagram at the bottom of the figure, but the spread or variability is similar. This difference is best described by comparing the sample means (9.761 compared to 16.069).

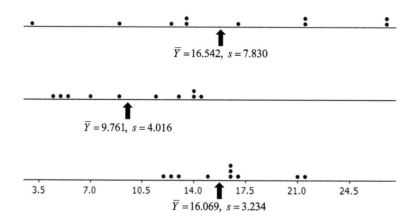

$\bar{Y} = 16.542, \ s = 7.830$

$\bar{Y} = 9.761, \ s = 4.016$

$\bar{Y} = 16.069, \ s = 3.234$

FIGURE 2.4: Comparison of Three Sets of Data

The variability or spread in the top dot diagram in the figure is greater than the dot diagram at the bottom of the figure, while the center is quite similar. This is best described by comparing the sample standard deviations (7.830 compared to 3.234).

2.2.3   Histograms

When there are many data points, computer software can be used to represent the variability of data as a histogram. A histogram is a bar chart where the height of the bars are the frequency counts of the number of data values that fall into various intervals. The intervals are a non-overlapping exhaustive set that covers the range of the data (i.e., minimum to maximum). As an example of a histogram, consider the pH measurements of weak acid solutions from a manufacturing process that are shown in Table 2.2.

TABLE 2.2
pH Measurements

4.2	4.0	4.2	4.4	4.6	5.0	3.6	3.8	4.2	4.0	3.8	3.8	3.6	3.6
4.0	4.2	4.0	3.9	3.9	4.0	4.1	4.0	3.4	4.2	4.0	4.1	4.4	3.0
3.5	3.9	3.9	3.5	3.5	3.5	4.2	4.3	4.0	4.2	3.4	3.3	3.5	3.2
4.3	3.8	4.0	4.2	4.2	3.4	3.2	3.4	3.4	3.0	3.1	3.4	3.4	3.2
3.8	3.3	3.6	3.9	3.5	3.8	3.6	4.0	4.9	4.9	4.9	5.0	3.5	3.2
4.0	3.8	3.8	3.4	4.0	3.9	4.4	4.4	3.9	4.4	3.5	3.5	3.9	3.9

A statistical software program divides the range of the data into intervals as shown in Table 2.3, and counts the number of data points that fall into each interval (the count column in Table 2.3). Next the program constructs the histogram shown in Figure 2.5. From this figure it can be seen that the average value is around 3.9, and that data values vary from around 3.0 to 5.0.

Histograms can also be created with spreadsheet programs like Excel© using the Data Analysis icon on the Data tab (in Excel 2010, check documentation for older versions of Excel) or with the open source program Open Office Calc©. To make a histogram using Calc© you must download and install the statistical data analysis extension.

The first step in making a histogram with the spreadsheet program is to divide the range of the data into intervals like those shown in Table 2.3.

TABLE 2.3

Frequency Counts

Interval			Count
3.0	–	3.2	7
3.2	–	3.4	10
3.4	–	3.6	14
3.6	–	3.8	8
3.8	–	4	21
4.0	–	4.2	11
4.2	–	4.4	7
4.4	–	4.6	1
4.6	–	4.8	0
4.8	–	5.0	5

Statistical analysis software, like Minitab© or the open source R, does this automatically. But with spreadsheet software this must be done manually. Deciding on the number of intervals or bars in the histogram is the first step. The number of intervals should be a function of the number of data points. If there is a lot of data, the range can be broken up into many intervals and there will be a sufficient number of data points in each interval. If there is little data, fewer intervals must be used. A rough guide to deciding the number of intervals (I) to use is expressed in Equation 2.3, where n is the number of data points. The value of I resulting from this equation should be rounded to the nearest integer.

$$I = [1.0 + 3.3 \log_{10}(n)]. \tag{2.3}$$

Making the number of decimal places in the interval boundaries one greater than the

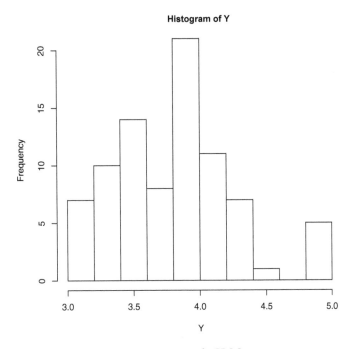

FIGURE 2.5: Histogram of pH Measurements

number of decimal places in the data will avoid having data values that fall exactly on one of the boundaries.

Figure 2.6 shows a symmetric histogram of data. Experimental error (that is the sum of random influences such as differences in experimental units, changes in lurking variables, and measurement errors) often follows this pattern.

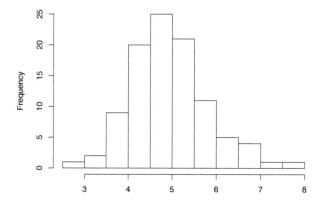

FIGURE 2.6: Symmetric Histogram

When data ranges over orders of magnitude (i.e., 0.1 to 1.0 to 10.0, etc.), a right skewed distribution like the histogram shown in Figure 2.7 may result. This histogram is called right skewed because the tail on the right is longer than the tail on the left.

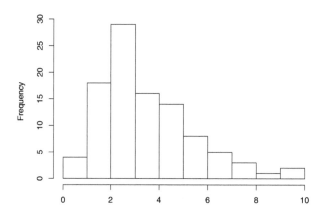

FIGURE 2.7: Right Skewed Histogram

2.2.4   Five-Number Summaries and Boxplots

Skewness in data can also be represented concisely in a five-number summary consisting of the minimum value, the 25th percentile, the median (or 50th percentile), the 75th percentile, and the maximum value. The median is the middle most number when an odd number of values are ordered from smallest to largest. When there are an even number of values, the median is the average of the two middle most numbers. The 25th percentile is the value that lies 25 percent of the way from the minimum to maximum, and 75th percentile is the value that lies 75 percent of the way from the minimum to maximum. If the difference between the 75th percentile and the median is larger than the difference between the median and 25th percentile, it indicates that the data are skewed to the right.

The five numbers can be easily calculated using a spreadsheet. Referring to Figure 2.3, for example, the minimum can be calculated with the formula `=MIN(A1:A9)`, the maximum can be calculated with the formula `MAX(A1:A9)`, and the median can be calculated with the formula `MEDIAN(A1:A9)`. The 25th percentile can be calculated with the formula `PERCENTILE(A1:A9,.25)` in Excel©, and by `PERCENTILE(A1:A9;.25)` in Open Office Calc©. The 75th percentile is similarly calculated by replacing the `.25` by `.75` in the formula. The five-number summary of the data in Table 2.1 is shown in Table 2.4.

TABLE 2.4
Five-Number Summary of
Data in Table 2.1

Minimum	0.5750
25th Percentile	0.5755
Median	0.5756
75th Percentile	0.5758
Maximum	0.5768

A convenient graphical way to represent the five-number summary is with a boxplot as shown in Figure 2.8, which is a boxplot of the data in Table 2.1. The box in a boxplot extends from the 25th percentile to the 75th percentile with a line at the median. In a simple boxplot, the whiskers extend from the minimum on the left to the 25th percentile, and from the 75th percentile on the right to the maximum. Simple boxplots drawn vertically (without the line at the median) can be created using a spreadsheet program by using the open-low-high-close stock chart with the five-number summary as the input. The median line will have to be added manually.

Figure 2.8 is a more sophisticated boxplot created automatically by a statistical software program. The whiskers in the more sophisticated boxplot extend from the quartiles to the largest (on the right) or smallest (on the left) values that are within 1.5 times the interquartile range from the nearest quartile. The interquartile range is the difference between the 75th and 25th percentiles. In Figure 2.8, the minimum value is more than 1.5 times the interquartile range below the 25th percentile, so it is shown by a small circle. The whisker on the lower side extends to the next to smallest value (which is within 1.5 times the interquartile range below the 25th percentile). On the right, the maximum value is more than 1.5 times the interquartile range from the 75th percentile, so it is also shown by a small circle. The whisker extends to the largest value within 1.5 times the interquartile range above the 75th percentile. Labeling individual values on the left or right makes it easy to identify "odd" values that don't appear to fit the pattern of the other data.

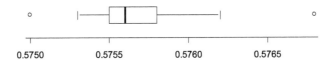

FIGURE 2.8: Boxplot of Bore Diameters

Boxplots are very useful for comparing two or more groups of data. For example the boxplots in Figure 2.9 compare data from a symmetric and skewed distribution similar to the histograms in Figures 2.6 and 2.7. When there are more than two large groups of data, the graphical comparison can be represented much more concisely with boxplots than with histograms.

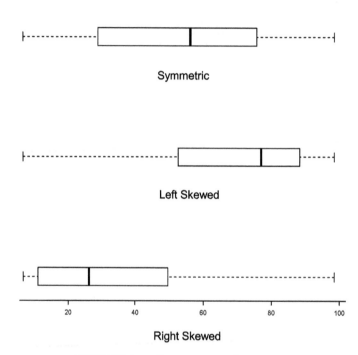

FIGURE 2.9: Comparison of Boxplots

2.3   Graphical and Numerical Summaries of the Relation Between Variables

A simple way to visualize the relationship between two quantitative variables (x and y) is to construct a scatterplot. In a scatterplot each pair of (x, y) coordinates are plotted on a Cartesian coordinate system. This can be done using the `XY Scatter` option on the chart wizard for a spreadsheet program.

Consider the following example. Small quantities of hydrogen can embrittle ultra high strength steel causing static failure. Johnson and Leone [1964b] present the data in Table 2.5 from a stress test where several samples of cadmium plated steel were stressed in a hydrogen environment at high pressure (measured in 1000 pounds per square inch units). The incubation time is the time required for sufficient hydrogen build-up to initiate a crack.

TABLE 2.5
Applied Stress in 1000 PSI and Log Incubation Time
in Minutes

Applied Stress (X).	Log incubation time (Y)
200	0.1139
185	0.0
185	0.0
175	0.3010
175	0.3010
160	0.2553
150	0.2788
135	0.3010
135	0.4150

The relationship between X = stress level and Y = Log incubation time is not immediately obvious from the table, but in the scatterplot, shown in Figure 2.10, it can be clearly seen that as the stress increases, the time to initiate a crack decreases along a somewhat linear trend.

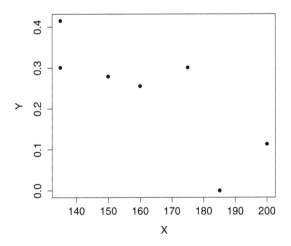

FIGURE 2.10: Scatterplot of Stress Test Data

A numerical measure of the strength of a linear relationship between two variables (X and Y) is the sample correlation coefficient r. This can be used in verbal or written descriptions of relationships that do not allow for graphics.

The formula for the correlation coefficient r is given in Equation 2.4. This is a unit-less measure that can take on values between -1 and 1. A correlation coefficient of $r = \pm 1.0$ would indicate that Y is a linear function of X (i.e., $Y = mX + b$ where m and b are constants), and a correlation coefficient of $r = 0.0$ would indicate that there is no linear relationship between X and Y.

$$r = \frac{\sum_{i=1}^{n}(X_i - \bar{X})(Y_i - \bar{Y})}{\sqrt{[\sum_{i=1}^{n}(X_i - \bar{X})^2] \times [\sum_{i=1}^{n}(Y_i - \bar{Y})^2]}}. \qquad (2.4)$$

Figure 2.11 illustrates five different values of the correlation coefficient to illustrate how it represents the strength of a linear relationship.

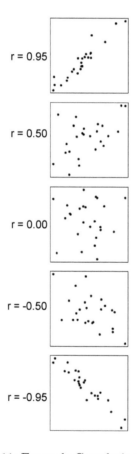

FIGURE 2.11: Example Correlation Coefficients

The correlation coefficient can be calculated in spreadsheet programs. If the X data in Table 2.5 were in the first 9 rows of column A in a spreadsheet, and the Y data were in the first 9 rows of column B in the spreadsheet, the formula =CORREL(A1:A9,B1:B9) in Excel (or =CORREL(A1:A9;B1:B9) in Calc) will calculate the correlation coefficient. For the data in Table 2.5, the correlation coefficient is $r = -0.791$, indicating the strength of the linear relationship is somewhere in between the two bottom graphs in Figure 2.11.

2.4 What to Expect from Theory

2.4.1 Probability Density Functions and Probability

To help us understand random experimental error, it is useful to think of simple physical devices that will generate data that vary randomly. For example the spinner shown in Figure 2.12 will generate random values between 0 and 360. We call this a random variable and denote it by the capital letter Y. If the spinner is fair and balanced, each time it is spun,

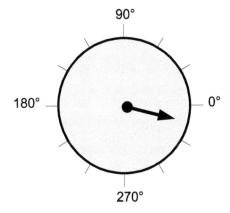

FIGURE 2.12: Spinner for Generating Random Data

a random number, which is equally likely to be anything between 0 and 360, will result. A mathematical way of representing this fact is with the density function shown in Figure 2.13. In this figure y represents the value of the random variable resulting from a spin, and $f(y)$ is the density. Since all values of y between 0 and 360 are equally likely, $f(y) = \frac{1}{360}$ is constant. Probability is defined to be a number between zero and one that indicates the

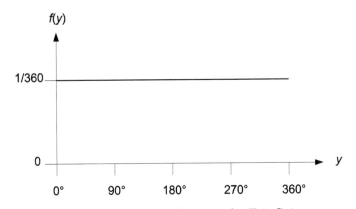

FIGURE 2.13: Density Function for Fair Spinner

certainty that a particular event will occur. Events with probabilities near zero are unlikely, while events with probability near 1 are near certain. The area under the density function in Figure 2.13 represents probability. The probability of observing a value y between 0 and 360 when the spinner is spun is 1 (or certain) since the area under $f(y) = \frac{1}{360}$ from 0 to 360 is the area of a rectangle ($360 \times \frac{1}{360}$). The probability of observing a number between 90 and 180 is the area of the rectangle shown in Figure 2.14, which is $\frac{1}{4}$.

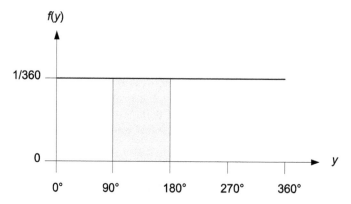

FIGURE 2.14: Probability of Being Between 90° and 180°

Mathematically this area can be represented as the integral

$$P[90 \leq y \leq 180] = \int_{90}^{180} f(y)dy = \int_{90}^{180} \frac{1}{360}dy = \frac{1}{4}. \tag{2.5}$$

The probability that the result, y, falls in any interval on the axis can be expressed as the area of the rectangle above the interval, or equivalently the integral of $f(y)$ over the interval.

2.4.2   Expected Values, Variance, and Standard Deviation

The expected value or mean, μ, of a random variable is defined as the center of gravity of its respective density function. It represents the center of the density function in the same way that the sample average $\bar{Y}$ represented the center of a dot diagram of data in Section 2.2.2. When data can be observed, but not the physical device that generated the data nor the density function, then the sample average $\bar{Y}$ can be used as an approximation or an estimate of μ. Likewise μ can be thought of as the long-run sample average of an infinite number of observations. When the density function is known, μ can be expressed as an integral shown in Equation 2.6.

$$\mu = \int_y yf(y)dy. \tag{2.6}$$

For the random variable resulting from flipping the spinner, the density function is $f(y) = \frac{1}{360}$ for $0 \leq y \leq 360$ (shown in Figure 2.13) and the mean is:

$$\mu = \int_0^{360} y\left(\frac{1}{360}\right)dy = \left(\frac{360^2}{2}\right) \times \left(\frac{1}{360}\right) = 180.$$

The variance, σ^2, of a random variable is defined as the expected squared distance from the mean as shown in Equation 2.7.

$$\sigma^2 = \int_y (y - \mu)^2 f(y) dy. \tag{2.7}$$

The standard deviation $\sigma = \sqrt{\sigma^2}$ is measured in the same units as the mean μ and is an indicator of the spread in a density function, in the same way that the sample standard deviation s represented the spread or variability in a dot diagram (as described in Section 2.2.2).

For the random variable resulting from flipping the spinner, the variance is

$$\sigma^2 = \int_0^{360} (y - 180)^2 \frac{1}{360} dy = 10,800,$$

and the standard deviation is $\sigma = \sqrt{10,800} = 103.92$. The mean and standard deviation can be pictured relative to the density function as shown in Figure 2.15.

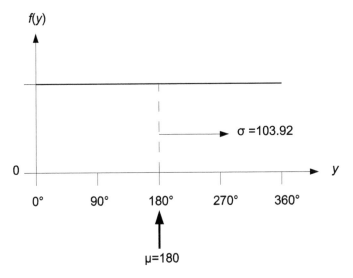

FIGURE 2.15: Mean and Standard Deviation of Spinner Random Variable

When random data can be observed, but not the device that generates the data nor the density function, then both the sample mean $\bar{Y}$ and the sample standard deviation s are used as estimates of μ and σ. For example, Figure 2.16 shows the spinner density function, combined with a dot diagram of the results of a sample of 9 spins. The sample mean $\bar{Y} = 181.8$, represented in the graph, is an estimate of $\mu = 180$, and the sample standard deviation $s = 126.1$ is an estimate of $\sigma = 103.92$. The accuracy of $\bar{y}$ and s as estimates will increase as the number of data points in the sample increases.

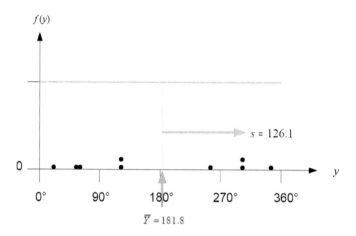

FIGURE 2.16: Mean and Standard Deviation of Spinner Random Variable

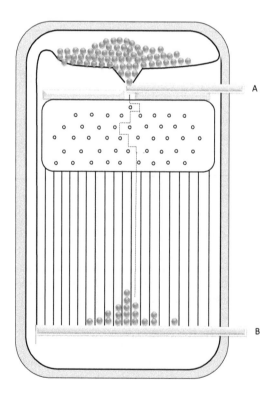

FIGURE 2.17: Quincunx

2.4.3 Normal Distribution

Although the spinner shown in Figure 2.12 can generate random data, that data would not be representative of the pattern usually seen in replicate experiments conducted at the same factor settings. To generate data simulating replicate experiments, another physical device, the quincunx shown in Figure 2.17, would be more appropriate. In the quincunx, a plastic bead is released through the funnel at the top of the device by pulling bar A to the right. The bead will hit the peg directly below the funnel and will be deflected randomly to the left or the right. Next, it will hit a peg in the second row of pegs and again be deflected randomly to either the left or right. This continues until the bead passes through all the rows of pegs and enters one of the channels below the maze of pegs.

Since it is likely that the plastic bead will bounce to the left about 50% of the time, and to the right 50% of the time, most of the beads that are dropped accumulate near the center as shown in the figure. Occasional beads bounce to the right the majority of the time, and others bounce to the left the majority of the time, ending up either in the extreme left or right. However, this is a less frequent occurrence, and the number of beads accumulating decreases as a function of the distance from the center.

A density function that would characterize random variability of the kind generated by the quincunx is the normal distribution. Its density function is given in Equation 2.8, and its shape is the "Bell Curve" shown in Figure 2.18.

$$f(y) = \frac{1}{\sqrt{2\pi}\sigma}e^{-\frac{(y-\mu)^2}{2\sigma^2}} \tag{2.8}$$

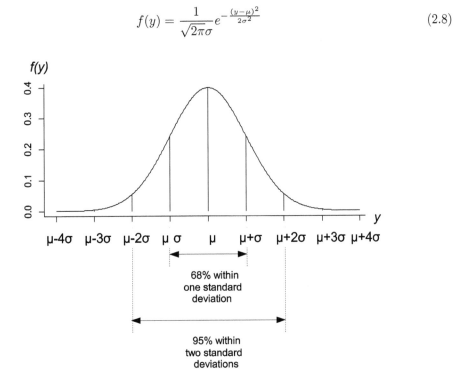

FIGURE 2.18: Normal Distribution

The mean μ of the normal density is at the center and the standard deviation σ measures the width of the distribution. The variability in replicate experiments conducted at the same factor settings can frequently be represented by a normal distribution.

2.4.4   Central Limit Theorem and Distribution of Sample Means

Figure 2.19 illustrates another important characteristic of random data. The open circles represent 15 sets of 4 observation that were each obtained from a device that generates random data. The histogram (shown vertically on the left edge of the graph) can be thought of as an estimate of the density function for this data. The solid black dots connected by lines represent the sample averages, $\bar{Y}_i$, of the $i = 1, \ldots, 15$ groups. The dot diagram on the right margin represents the distribution of these sample averages. The length of the gray arrow to the left of the histogram represents the sample standard deviation of all of the data points and the length of gray arrow to the right of the dot diagram represents the standard deviation of the 15 sample averages.

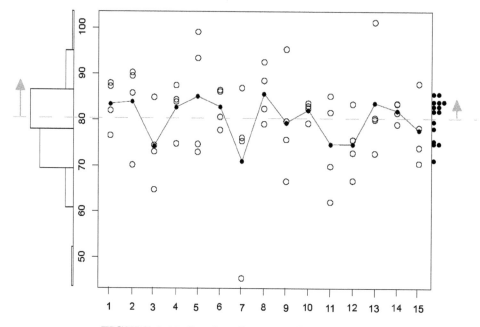

FIGURE 2.19: Random Data and Group Averages

Clearly there is less variability in sample averages than in individual random observations. This is true because high observations get averaged with low observations, pulling sample averages closer to the grand average, which is indicated by the dashed gray horizontal line in the figure. In practice this has led to the experimenters' belief that they have more faith in an average of several replicate experiments than the outcome of a single experiment. It can be expressed with the equation:

$$\sigma_{\bar{Y}} = \frac{\sigma_Y}{\sqrt{n}} \tag{2.9}$$

which says the standard deviation of an average of n observations is equal to the standard deviation of an individual observation divided by the square root of the number of observations n. Another way to express the same thing is:

$$\sigma_{\bar{Y}}^2 = \frac{\sigma_Y^2}{n} \tag{2.10}$$

which says the variance of an average is the variance of an individual divided by the number of observations.

Also, it is unlikely that any group of 4 observations will all be above the grand average or will all be below the grand average. Therefore, after many groups of data have been observed, the sample averages will tend to congregate around the grand average and extreme values will occur less frequently (similar to the distribution of beads in the quincunx). This remarkable fact can be expressed concisely as the *Central Limit Theorem*.

The Central Limit Theorem states that as the number of observations n obtained from a random generating device increases, the density function (or distribution) of sample averages will become closer to a normal distribution with the same mean μ as the individual data, but with a standard deviation equal to $\sigma/\sqrt{n}$, where σ is the standard deviation of the original data. This will be true regardless of the density or distribution of the original data.

Since the density function for a normal distribution (shown in Equation 2.8) is really a family of curves, indexed by their mean μ and standard deviation σ, it is useful to standardize normal random variables by subtracting the mean and dividing by the standard deviation. The standardized normal random variable then follows the *standard normal* distribution with mean 0 and standard deviation 1. For example, if a sample average, $\bar{Y}$, follows a normal distribution with mean μ and standard deviation $\sigma/\sqrt{n}$, then

$$Z = \frac{\bar{Y} - \mu}{\sigma/\sqrt{n}} \tag{2.11}$$

follows the standard normal distribution with mean equal to 0, and standard deviation equal to 1. The ratio in Equation 2.11 is often referred to as the signal-to-noise ratio, where $\bar{Y} - \mu$ is the signal that indicates how far a sample mean deviates from the expected, and $\sigma/\sqrt{n}$ is the noise or measure of variability in the sample mean. Cumulative areas under the standard normal density curve can be calculated using spreadsheet programs like Microsoft's Excel©, Oracle's Calc©, or LibreOffice Calc. For example to calculate the cumulative area under the curve to the left of $Z = -1.50$, enter the formula `=NORMSDIST(-1.5)` and you will find the area to be 0.066807. This value can also be looked up in Table A.1 in Appendix A. To do this locate the first digit to the left and right of the decimal place in Z in the left most column of the table (i.e., 1.5) then go across to the column that represents the second digit to the right of the decimal (in this case x.x0) and read the value in the cell of the table to be 0.0668. Note, since the curve is symmetric as indicated by the figure at the top of the table, this value indicates either the area under the standard normal density to the left of $Z = -1.50$, or to the right of $Z = 1.50$.

When the standard deviation σ is unknown and it is replaced by its estimate s in Equation 2.9, the standardized value no longer follows the standard normal distribution because the uncertainty in the denominator causes the signal-to-noise ratio to vary over a wider range. W.S. Gosset under the pseudonym "Student" (Student [1908]) derived the distribution of the standardized value shown in Equation 2.12, and it has become known as the student's t-distribution, which is very useful for interpreting small samples of research data. The t-distribution is again a family of curves indexed by the degrees of freedom that is defined to be $n - 1$, where n is the number of observations used in calculating $\bar{Y}$ and s.

$$t_{n-1} = \frac{\bar{Y} - \mu}{s/\sqrt{n}}. \tag{2.12}$$

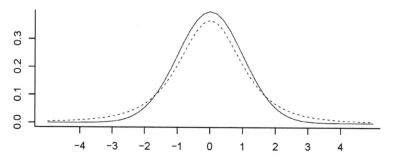

FIGURE 2.20: Standard Normal Compared to Student-t with 3 df

Figure 2.20 compares the standard normal distribution (the solid line) to the t-distribution with 3 degrees of freedom (dashed line). In this figure it can be seen that there is more area under the curve in the extreme left and right tails of the t-distribution than there is under the tails of the standard normal distribution. Therefore it is more likely to observe extremely low or high values of the signal-to-noise ratio, if the standard deviation (σ) is unknown and must be estimated from the data by (s). As the number of observation n and the degrees of freedom $n - 1$ increases, the sample standard deviation becomes a better estimator of σ and the t-distribution of the signal-to-noise ratio becomes closer to the standard normal distribution.

Tail areas under the t-distribution density can also be calculated in spreadsheet programs. The formula = `TDIST(1.5,3,1)` in Excel or = `TDIST(1.5;3;1)` in Calc calculates the area under the curve to the right of 1.5, which turns out to be 0.11529. The second argument (3) in the formula = `TDIST(1.5,3,1)` is the degrees of freedom, and the third argument is either a 1, or a 2, depending on whether you want the one-tail area (i.e., area under the curve to the right of 1.5) or the two-tail area, which is the sum of the area under the curve to the left of -1.5 and the area under the curve to the right of 1.5. Since the t-distribution density is symmetric, the area under the curve to the right of 1.5 is the same as the area under the curve to the left of -1.5. The area under the t-distribution density to the left of -1.5 (0.11529) is larger than the corresponding area under the standard normal density (0.066807) as would be expected from examination of Figure 2.18. If the degrees of freedom increased to 1000 (i.e., = `TDIST(1.5,1000,1)`), the result (0.066965) becomes very close to the corresponding area under the standard normal density.

If two independent sets of data are collected from the same random generating device, then the sample means $\bar{Y}_1$ and $\bar{Y}_2$ should be independent estimates of the mean μ produced by the device, and the two sample standard deviations s_1 and s_2 are independent estimates of the standard deviation of values produced by the device. However, due to random variability, it is very unlikely that the two sample means or that the two sample variances will be identical. How large a difference can be expected in these calculated quantities due to the random variability? For the sample means the t-distribution again provides an answer since the signal-to-noise ratio:

$$t_{n_1+n_2-2} = \frac{\bar{Y}_1 - \bar{Y}_2}{s_p\sqrt{\frac{1}{n_1} + \frac{1}{n_2}}} \tag{2.13}$$

can be shown to follow the t-distribution with $n_1 + n_2 - 2$ degrees of freedom, where n_1 is the number of experiments in the first independent set and n_2 is the number of experiments in the second independent set, and

$$s_p = \sqrt{\frac{(n_1 - 1)s_1^2 + (n_2 - 1)s_2^2}{n_1 + n_2 - 2}}$$

is the measure of noise formed by taking the square root of the weighted average of the sample variances s_1^2 and s_2^2 from the two independent samples (this weighted average is often referred to as the "pooled variance"). By referring to the t-distribution we can determine whether the difference in sample means of sets of data are within the range of what would be expected due to random variability.

When the data come from a normal distribution, the ratio:

$$F_{n_1-1,n_2-1} = \frac{s_1^2}{s_2^2} \tag{2.14}$$

can be shown to follow a probability density function known as the F-distribution (or variance ratio) with $n_1 - 1$ degrees of freedom for the numerator and $n_2 - 1$ degrees of freedom for the denominator. It can be used to determine how large a difference could be expected in sample standard deviations s_1 and s_2, or sample variances s_1^2 and s_2^2, due to experimental error. A random value following an F-distribution has an expected value, or mean, of 1.0. The F-distribution is again a family of density functions that are truncated at zero on the left and have long tails on the right (i.e., right skewed), as shown in Figure 2.21.

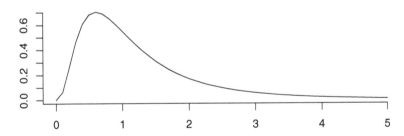

FIGURE 2.21: F-distribution with 8, and 8 degrees of freedom

Right tail areas under the F-distribution density function can again be evaluated using a spreadsheet program. The formula = FDIST(2,8,8) in Excel or = FDIST(2;8;8) in Calc evaluates the area under the F-distribution density to the right of 2, with 8 degrees of freedom in the numerator and 8 degrees of freedom in the denominator. The result of this formula is 0.1733, which means there is a greater than 17 percent chance of observing two sample variances that differ by a ratio of 2 or more from two independent sets of data from the same generating device (with $n = 9$ observations in each set).

2.5   Using Theory to Help Interpret Experimental Data

2.5.1   Comparing the Mean of Experimental Results to a Standard

Although it is very unlikely that an observed sample average, $\bar{Y}$, calculated from experimental data will exactly equal μ, determined from past experience or theory, a question remains. Is the difference within the range of what might be expected due to experimental error, or is there evidence that the current experimental conditions are different than past history or theory would suggest? The t-distribution and Equation 2.10 is very useful for answering this type of question.

Consider the following example. In a production process to produce dichlorobutadiene from chlorobutadiene, a small amount of the undesirable by-product trichlorobutene is always formed. From a long history of the process there is on average $\mu = 1.6$ percent of this by-product. After a change was made to the process, Peters et al. [1976] show the percent by-product for 5 batches of dichlorobutadiene. They were: 1.46, 1.62, 1.37, 1.71, and 1.52. Is there evidence that the process change has reduced the percent by-product from the historical average or is this within the range of expected batch-to-batch variation?

The sample mean $\bar{Y} = 1.536$ is smaller than the historical average, but the sample standard deviation is $s = 0.1331$. The way to answer this question statistically is to make a hypothesis test. In a hypothesis test, a null hypothesis is defined. For the example above the null hypothesis would be:

$$H_0 : \mu = \mu_0 = 1.6. \tag{2.15}$$

The null hypothesis can never be proven without collecting of an infinite amount of data. However, it is set up like a straw man (or the claim of innocence in a court of law) and must be proven false beyond a reasonable doubt by collecting data. If the data support rejecting the null hypothesis, then it is shown to support an alternative hypothesis. The natural alternative hypothesis for the example above is:

$$H_1 : \mu \neq \mu_0 = 1.6. \tag{2.16}$$

To determine the weight of the evidence against the null hypothesis, we can calculate the t-ratio, given in Equation 2.10, along with the probability of observing a t-value as extreme or more extreme than that determined from the data. In this case,

$$t_{5-1} = \frac{\bar{Y} - \mu}{s/\sqrt{5}} = \frac{1.536 - 1.600}{0.1331/\sqrt{5}} = -1.075.$$

The probability of obtaining a t-ratio greater than 1.075 in absolute value can be calculated as the area under the t-distribution density with 4 degrees of freedom. Using the spreadsheet formula (=TDIST(1.075,4,2)), this is 0.342896. Therefore, a sample average that differs from the historical average by more than the absolute value of $1.535 - 1.60$, or a t-value greater than 1.075 in absolute value, could be obtained more than one third of the time due to batch-to-batch variability, even if no change had been made to the process. This is hardly enough evidence against the null hypothesis to quell all reasonable doubt. Therefore, it would be said that there is not enough evidence to reject the null hypothesis.

In statistical terms the probability of obtaining a signal-to-noise ratio as extreme or more extreme than that calculated from the data is called the P-value. If the P-value is less than 0.05, then there is evidence that the difference in the mean μ and the hypothesized mean μ_0 is *statistically significant* (or, in nonstatistical terms, there is evidence beyond a reasonable doubt that the mean is different than the hypothesized value). By convention, when the P-value is less than 0.05, we can claim a statistically significant difference, and when the P-value is less than 0.01, we can claim a highly (statistically) significant difference.

An alternative to calculating the P-value is to compare the t-statistic calculated from the data to a critical limit. The 0.05 critical limit for a t-statistic with 4 degrees of freedom (labeled $t_{4;0.05}$) is the t-value that would result in a P-value of 0.05. The spreadsheet formula = TINV(0.05,4) in Excel or = TINV(0.05;4) in Calc returns this value. For the case with 4 degrees of freedom, $t_{4;0.05} = 2.7764$. This critical value is illustrated in Figure 2.22. The sum of the gray areas under the curve to the right of 2.776 and to the left of -2.776 is 0.05. In order to claim a statistically significant difference between the calculated mean $\bar{Y}$ and the hypothesized mean μ_0, the calculated t-statistic or signal-to-noise ratio (Equation 2.10) must fall either to the right of the upper critical limit (2.776) or to the left of the lower critical limit (-2.776).

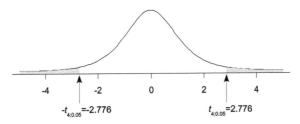

FIGURE 2.22: t-distribution with Critical Limits

The critical limits for the t-distribution can also be looked up in Appendix Table A.3. To do this, locate the row in the table that has the appropriate degrees of freedom in the left column. Next, move to the column whose label is equal to 1 minus the critical P-value. For example, if the critical P-value is 0.05 go to the column labeled 0.95, and if the critical P-value is 0.01 go to the column labeled 0.99. Finally read the critical t-value from the entry in that row and column. For example, the value 2.776 is read from the row with 4 degrees of freedom and the column labeled 0.95.

In this example the calculated t-statistic was -1.075, which is in between the critical limits and no significant difference can be claimed. The critical limit for a highly statistically significant difference $t_{4;0.01} = 4.60$, which means the calculated t-statistic in the example above would have to be greater than 4.60 in absolute value to claim a highly significant difference.

When testing a hypothesis with variable data, there is always a chance that the wrong conclusion will be reached. If the null hypothesis is rejected, when it is actually true, this mistake is referred to as a type I error. If the null hypothesis is false, but there is not enough data to reject null hypothesis, this is referred to as a type II error.

A type I error occurs when $\mu = \mu_0$ but experimental error causes the calculated t-statistic to be larger than the critical limit so that the null hypothesis is rejected. The chance of this error is very small (i.e., 0.05 if $t_{n-1;0.05}$ is the critical limit) and is called the significance level of the test or α.

When the mean μ is not equal to the hypothesized value μ_0, there is a chance, due to experimental error, that the calculated t-value will not exceed the critical limit and a type II error will be made. The probability of this happening is called β, and it will be a function of how much the mean, μ, differs from the hypothesized value μ_0. The power of the test is equal to $1 - \beta$, and it is the probability of making the correct decision based on experimental data when the null hypothesis is false. The power of the test can be increased by increasing n, the number of experimental runs. In Chapter 4 a formula will be presented that will allow an experimenter to determine how large the sample size must be in order to have a high (.90–.95) probability of detecting a practical difference in the means.

In general, the results of a hypothesis test of the mean of experimental data to a historical standard must be interpreted with extreme caution. The critical assumption is that the experimental data is a representative sample from the new process conditions and the historical standard is representative of past conditions. If a hypothesis test of the difference in the mean of experimental conditions to a historical standard is significant, it is like discovering a correlation (as described in Section 1.4), and it may be difficult to a assign a specific cause for this difference. Although a change was made to the process of producing dichlorobutadiene from chlorobutadiene, in the example above, biases could have been present due to the fact that other unplanned and unknown changes may have occurred in background or lurking variables. Instead of making critical assumptions about the representativeness of the experimental data, constancy of background variables, and the assumed value of the historical mean, it is better to change back and forth between two conditions in a random pattern to reveal the specific effect. The random changes make it very unlikely that the effect will be biased, and guarantee the validity of the test. Consider the example in the next section.

2.5.2   Comparing the Means of Two Experimental Conditions

Bowker and Lieberman [1955] present the data in Table 2.6 comparing the endurance limit of steel. It was hypothesized that polishing the steel would increase the endurance limit. A simple experiment was performed: 16 specimens of steel were selected. Eight were randomly assigned to be polished, leaving the other eight unpolished. After the selected samples were polished, all specimens were subjected to an endurance test in random order. In this simple experiment, the experimental units were specimens of steel, and the factor was polishing (yes or no).

By randomly selecting the samples that would be polished, any unforeseen differences in the steel specimens would be unlikely to bias the calculated effect of polishing. Likewise the random ordering of the endurance testing would prevent any learning curve or trend in test results from biasing the results in favor of the polished or unpolished specimens. In this way the use of randomization reduces the need to make critical assumptions about the nature of the test data. Do the results of the tests in Table 2.6 support the hypothesis?

TABLE 2.6
Endurance Limit (reverse bending) of
4% Carbon Steel (100s of cycles)

	Polished	Unpolished
	865	826
	919	824
	894	817
	840	795
	899	794
	787	698
	875	799
	831	834
Means:	863.75	798.375
Variances:	1837.928	1877.411

The two sample t-statistic (given in Equation 2.13) can be used to answer the question, and the random selection of samples to be polished and the random ordering of tests guarantee the approximate validity of the procedure. The null hypothesis would be the opposite of what the experimenter would like to prove, i.e., $\mu_P = \mu_U$. To test the hypothesis $H_0 : \mu_P = \mu_U$ versus the natural alternative $H_1 : \mu_P \neq \mu_U$. The calculated t-statistic is shown below.

$$t_{n_P+n_U-2} = \frac{\bar{Y}_P - \bar{Y}_U}{s_p\sqrt{\frac{1}{n_P} + \frac{1}{n_U}}} = \frac{863.75 - 798.375}{43.100\sqrt{\frac{1}{8} + \frac{1}{8}}} = 3.034. \qquad (2.17)$$

This statistic is again a signal-to-noise ratio where the numerator $\bar{Y}_P - \bar{Y}_U$ is a measure of the effect of polishing or the change in the response caused by switching from unpolished to polished steel samples. The denominator is a measure of the noise or standard deviation of the difference in the two means.

By including just one measure of the noise in the denominator of Equation 2.15, it is assumed that the variance of the observations is the same in each level of the factor. This assumption can be checked by comparing the ratio of the two variances to the F-distribution (as in Equation 2.14) or by making comparative dot diagrams or boxplots like Figures 2.4 and 2.9. In this example, the two variances shown in Table 2.6 are quite similar. s_P in the formula for the denominator in Equation 2.15 is the square root of the weighted average (weighted by the degrees of freedom in each group) of the two variances. It is shown in the equation below.

$$s_p = \sqrt{\frac{(n_P - 1)s_P^2 + (n_U - 1)s_B^2}{n_U + n_U - 2}} = \sqrt{\frac{(8 - 1)1837.928 + (8 - 1)1877.411}{10}} = 43.100.$$

To determine whether there is a statistically significant difference in the two means, the calculated t-ratio can be compared to the critical limit $t_{14,0.05} = 2.144$, which can be computed by the formula = `TINV(0.05,14)` in Excel or by looking it up in Table A.3. In this example, the calculated t-statistic is greater (in absolute value) than the critical value, implying that there is sufficient evidence to claim that polishing the steel increases the endurance limit.

The P-value, calculated with the formula = `TDIST(3.034,14,2)`, is 0.00894. This means that there is less than a 1% chance of seeing a difference in means as large as observed in this example just due to random differences in experimental units or steel specimens. This gives evidence beyond a reasonable doubt that polishing caused the difference in average endurance.

2.5.3 Comparing the Means of Several Experimental Conditions

Ott and Schilling [1990] describe a situation where a critical electrical characteristic of manufactured electronic units was too variable. To determine the cause of the variability attention was directed toward an important ceramic component of the assembly. It was possible that the electrical characteristic of the manufactured components was influenced by the ceramic component. The ceramic was purchased in sheets from a vendor, and if measured differences in the sheets were the cause of variability in electrical characteristics of the circuits, it may be possible to reduce variation in the assembled units by selectively rejecting some ceramic sheets. The sheets were normally cut into several strips that were each used in the assembly of one electronic unit. In order to test the hypothesis that the mean value of the electrical characteristic was caused by differences in sheets of ceramic obtained from the vendor, seven components were made from strips cut from each of six ceramic sheets. The specific sheets were selected that had measured differences, and seven strips were cut from each sheet. The 42 electronic units were assembled and tested in a random order. Table 2.7 shows the results of the electrical tests on the 42 units.

TABLE 2.7

Measurements on an Electronic Assembly

Ceramic Sheet	1	2	3	4	5	6
Electrical	16.5	15.7	17.3	16.9	15.5	13.5
Characteristic	17.2	17.6	15.8	15.8	16.6	14.5
	16.6	16.3	16.8	16.9	15.9	16.0
	15.0	14.6	17.2	16.8	16.5	15.9
	14.4	14.9	16.2	16.6	16.1	13.7
	16.5	15.2	16.9	16.0	16.2	15.2
	15.5	16.1	14.9	16.6	15.7	15.9

The random order of assembly and testing again prevents biases from unknown factors and guarantees approximate validity of a hypothesis test. The null hypothesis in this case would be $H_0 : \mu_1 = \mu_2 = \mu_3 = \mu_4 = \mu_5 = \mu_6$, where μ_i is the mean electrical characteristic from units made with the ith ceramic sheet. The alternative hypothesis would be that there are differences in the mean electrical characteristics among circuits made from at least two different ceramic sheets.

A symbolic way of representing the data for several conditions is through the formula:

$$Y_{ij} = \mu_i + \epsilon_{ij} = \mu + \tau_i + \epsilon_{ij} \tag{2.18}$$

where y_{ij} is the electrical characteristic for the jth assembly made from the ith ceramic sheet, and $\mu_i = \mu + \tau_i$. Then an equivalent way of writing the null hypothesis is $H_0 : \tau_1 = \tau_2 = \tau_3 = \tau_4 = \tau_5 = \tau_6$.

One way to test the null hypothesis would be to calculate the two-sample t-statistic (given in Equation 2.13 as illustrated in the last section) for all 15 possible pairs of ceramic sheets. However, this would not be a good statistical procedure. If a significance level of .05 was set for each of the 15 tests, there would be a $1 - (.95)^{15} = .56$ chance that at least one of the comparisons would result in a significant t-statistic when the means for all ceramic sheets were actually equal.

A better test statistic can be developed using the fact that the overall variability in the data will be much larger if the means for each sheet differ. Based on this fact, Fisher [Fisher, 1925] developed a procedure called *the analysis of variance* or ANOVA method for comparing the means simultaneously. This procedure partitions the total sum of squares (*ssTotal* or numerator in the calculation of a variance shown in Equation 2.2) into two components. The first component (called the *ssT*) represents the variability among group means (i.e., μ_i or $\mu + \tau_i$). The second component (called *ssE*) represents the variability within groups. This partition is illustrated in Equations 2.19 and 2.20.

$$ssTotal = ssT + ssE \tag{2.19}$$

$$\sum_{i=1}^{t}\sum_{j=1}^{r}(Y_{ij} - \bar{Y}_{..})^2 = \sum_{i=1}^{t}\sum_{j=1}^{r}(\bar{Y}_{i.} - \bar{Y}_{..})^2 + \sum_{i=1}^{t}\sum_{j=1}^{r}((Y_{ij} - \bar{Y}_{i.})^2, \tag{2.20}$$

where $\bar{Y}_{i.}$ is the sample average of the measured electrical characteristic for the assemblies made using the ith ceramic sheet, and $\bar{Y}_{..}$ is the average of all 42 measurements. The F-statistic (Equation 2.14) can then be used to test the hypothesis that the means are equal.

If the random experimental variability between the data made from the same ceramic sheet each follow a normal distribution with the same mean and standard deviation, then the ratio of the ssT (divided by its degrees of freedom) and ssE (divided by its degrees of freedom) follows the F distribution. The F-statistic can be used to test the null hypothesis. The appropriate degrees of freedom for ssT and ssE are shown in Table 2.8, where t is the number of groups of data being compared and n is the total number of data values (i.e., $t \times r$ where r is the number of values in each group).

Table 2.8 also shows the F-ratio calculated as the ratio of ssT divided by its appropriate degrees of freedom and ssE divided by its appropriate degrees of freedom. If the area under the F-distribution to the right of the calculated F-statistic (i.e., the P-value for the test) is less than 0.05, then there is sufficient evidence to reject the null hypothesis and conclude there are differences among the group means.

TABLE 2.8
Analysis of Variance Table

Source	df	Sum of Squares	Mean Squares	F-ratio
Treatment	$t-1$	ssT	$msT = ssT/(t-1)$	$F = msT/msE$
Error	$n-t$	ssE	$msE = ssE/(n-t)$	
Total	$n-1$	$ssTotal$	$msTotal$	

The Sums of Squares, Mean Squares, and F-ratio shown in Table 2.8 can be calculated automatically with spreadsheet programs if the data are entered in a format similar to Table 2.7. For example Figure 2.23 shows the data of Table 2.7 in a Spreadsheet.

With the data in this format the analysis of variance table can be computed with the Excel© Analysis ToolPak add-in. Once this Tool Pack is loaded, the `Anova:single` factor option can be selected from the Data Analysis menu on the Data tab. Select `A3:F9` for the input range, and Excel produces a table as shown in Figure 2.24. The same table can be produced with the LibreOffice Calc spreadsheet by clicking on the Data → Statistics → Analysis of Variance ANOVA ... and selecting the input range.

	A	B	C	D	E	F	G
1			Sheet				
2	1	2	3	4	5	6	
3	16.5	15.7	17.3	16.9	15.5	13.5	
4	17.2	17.6	15.8	15.8	16.6	14.5	
5	16.6	16.3	16.8	16.9	15.9	16	
6	15	14.6	17.2	16.8	16.5	15.9	
7	14.4	14.9	16.2	16.6	16.1	13.7	
8	16.5	15.2	16.9	16	16.2	15.2	
9	15.5	16.1	14.9	16.6	15.7	15.9	

FIGURE 2.23: Data in a Spreadsheet

Using Open Office Calc©, a table similar to the analysis of variance table in the lower half of Figure 2.24 can be produced by typing the array function

=ONEWAYANOVAUNSTACKED(A2:F2;A3:F9) in cell H2 and then simultaneously holding down the Cntl-Shift-Enter buttons to invoke the array function.

	H	I	J	K	L	M	N
Anova: Single Factor							
SUMMARY							
Groups		Count	Sum	Average	Variance		
Column 1		7	111.7	15.95714	1.01619		
Column 2		7	110.4	15.77143	1.032381		
Column 3		7	115.1	16.44286	0.749524		
Column 4		7	115.6	16.51429	0.194762		
Column 5		7	112.5	16.07143	0.162381		
Column 6		7	104.7	14.95714	1.139524		
ANOVA							
Source of Variation		SS	df	MS	F	P-value	F crit
Between Groups		11.15619	5	2.231238	3.117153	0.019388	2.477169
Within Groups		25.76857	36	0.715794			
Total		36.92476	41				

FIGURE 2.24: ANOVA from Spreadsheet Function

In Figure 2.24 the critical limit with the 0.05 level of significance is shown in column N in the ANOVA table. Although it is not produced by the Open Office© Calc array function, it can be calculated with the formula = FINV(0.05;5;36). For the data in this example the calculated F-ratio (3.11) falls to the right of the critical limit (2.48) indicating the mean electrical characteristics are significantly different for the six ceramic sheets at the 0.05 level of significance. The averages in Figure 2.24 (computed by Excel) show the mean electrical characteristic for ceramic sheet 6 is less than the other 5 sheets. This supports the supposition that the electrical characteristic of assembled components may be reduced by rejecting ceramic sheets such as number 6.

By combining the variation within groups to form ssE and msE in Table 2.8, the analysis of variance F-test assumes the variability is approximately equal for each of the groups being compared with respect to their means. This assumption can be checked graphically by making comparative dot diagrams or boxplots like Figures 2.4 or 2.9.

If the number of observations in several groups of data is not equal (as in Table 2.7), the analysis of variance can still be used to test hypotheses that the group means are equal. To produce the ANOVA table for this case using Excel© or LibreOffice Calc the data should be formatted similar to Figure 2.23, with an unequal number of values in each column. Using Open Office Calc©, on the other hand, the data should be put in a single column with an adjacent column containing indicators for the group membership. Then use the array function = ONEWAYANOVA rather than the array function = ONEWAYANOVAUNSTACKED. The data format used by Calc© is the same as more general purpose statistical programs like Minitab© or the aov function in the open source program R.

2.5.4 Comparing Observed Data to the Normal Distribution

To determine whether a normal distribution adequately represents experimental error, a histogram of replicate experiments conducted at the same factor levels could be compared to the bell-shaped curve that is represented in Figure 2.18. However, a large number of replicates would be required to have enough data to make a histogram. A way to compare a small set of observed values to a normal distribution is to make a normal probability plot which is scatterplot the ranked data values vs. the expected ranked values from a standard normal distribution (or normal scores). The $z_{(i)}$ = normal scores can easily be calculated in a spreadsheet as shown in Figure 2.25.

	A	B	C
1	0.5755	3	-0.589455798
2	0.5753	2	-0.967421566
3	0.5756	4	-0.282216147
4	0.5762	8	0.967421566
5	0.5756	4	-0.282216147
6	0.5758	6	0.282216147
7	0.5758	6	0.282216147
8	0.5768	9	1.593218818
9	0.575	1	-1.593218818

FIGURE 2.25: Making a Normal Plot in a Spreadsheet

In this figure, the data are shown in column A, and the ranks are shown in column B. The rank shown in cell B1 was calculated with the spreadsheet formula =RANK(A1,A$1:A$9,1) in Excel© (or =RANK(A1;A$1:A$9;1) in Calc©) and then copied down to cells B2 through B9. The expected ith largest value from n sampled from a standard normal distribution can be approximated by the number that has $(i - \frac{1}{2})/n$ area to the left of it under the standard normal density function, i.e.,

$$z_{(i)} = \Phi^{-1}\left(\frac{i - \frac{1}{2}}{n}\right),\tag{2.21}$$

where,

$$\Phi(z_{(i)}) = \int_{-\infty}^{z_{(i)}} \frac{1}{\sqrt{2\pi}} e^{-\frac{1}{2}y^2} dy = \left(\frac{i - \frac{1}{2}}{n}\right).\tag{2.22}$$

The spreadsheet formula for computing $z_{(3)}$ in cell C1 is =NORMSINV((B1-0.5)/9) and this formula is copied down to cells C2 through C9 to produce the expected ranked values from a standard normal distribution. A normal plot with the expected ranks on the x-axis and the data in column B on the y-axis is shown in Figure 2.26.

If the data represented in a normal plot appear to come from a normal distribution, then the plot should reveal a straight line of points with positive slope, as Figure 2.26 seems to show. If data come from a skewed distribution, then the points on a normal plot of this data will follow a curved trend as shown in the right two graphs in Figure 2.27.

In Chapters 3 and 4 of this book, normal plots will also be used to identify abnormal values, as illustrated in Figure 2.28. In this figure, the same data are represented in a boxplot on the left and a normal plot on the right. The reference line is added to the normal plot to accentuate the odd values in the left and right tails of the distribution.

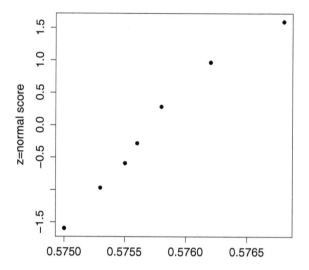

FIGURE 2.26: Normal Plot of Bore Diameters

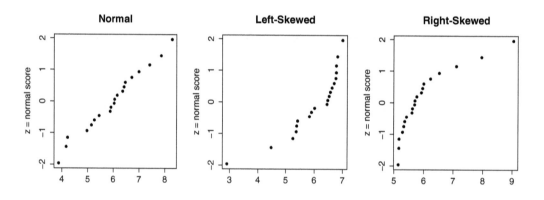

FIGURE 2.27: Normal Plots of Symmetric, Left-Skewed, and Right-Skewed Data

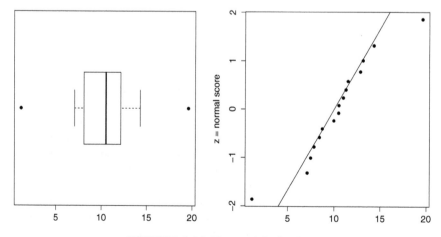

FIGURE 2.28: Data with Outlier

When the mean of the observed data is known to be zero, the subjectivity of placing a reference line and the process of identifying odd values can be improved by using a half-normal plot. The half-normal plot is a scatter plot of the absolute values of the observed data vs. the expected ranked absolute values from a standard normal distribution (or half-normal scores). The half-normal scores can be easily computed in a spreadsheet. For example, in Figure 2.29, the observed data with mean known to be zero are in column A. Cell B1 was calculated with the formula =ABS(A1), and this formula was copied down column B to get the absolute values of the data. Next, the rank of the absolute values was calculated by entering the formula =RANK(B1,B$1:B$15,1 into cell C1 and copying it down to cell C15. Finally, the half-normal scores are calculated with Equation 2.23:

$$|z|_{(i)} = \Phi^{-1}\left(\left(\frac{i - \frac{1}{2}}{n} + 1\right)\Big/ 2\right) \tag{2.23}$$

that is entered into cell D1 as =NORMSINV(((((C1-0.5)/15)+1)/2)) and copied down to D15.

	A	B	C	D
1	-2.1999	2.1999	10	0.902735
2	1.6891	1.6891	8	0.67449
3	-3.881	3.881	13	1.382994
4	-0.9537	0.9537	4	0.296738
5	-2.5797	2.5797	11	1.036433
6	2.8615	2.8615	12	1.191816
7	1.8333	1.8333	9	0.7835
8	-1.5997	1.5997	7	0.572968
9	0.3927	0.3927	2	0.125661
10	10.1	10.1	15	2.128045
11	-1.0261	1.0261	5	0.38532
12	-1.1527	1.1527	6	0.47704
13	0.2485	0.2485	1	0.041789
14	-0.6022	0.6022	3	0.210428
15	-9.1	9.1	14	1.644854

FIGURE 2.29: Making a Half-Normal Plot in a Spreadsheet

The half-normal plot is then constructed by plotting the absolute values of the data (in column B) vs the half-normal scores $|z|_i$ in column D. Figure 2.30 shows the half-normal plot of the data in Figure 2.29 along with a boxplot of the data in column A. The reference line is easier to draw on the half-normal plot, than the normal plot, because it starts at the origin (0,0) in the lower left of the half-normal plot and follows the points along a straight line. Odd values are also easier to identify because they always appear on the right as points well off the reference line (as illustrated with two odd values in Figure 2.30).

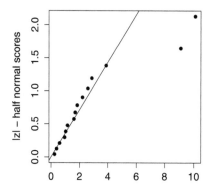 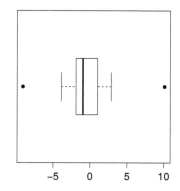

FIGURE 2.30: Half-Normal Plot and Boxplot of Data in Figure 2.29

2.6   Summary

In this chapter we discussed methods to summarize variable data using graphical and numerical techniques. We also discussed some concepts from probability theory that can help us decide whether data support or contradict preconceived beliefs in three different situations. The first situation was to determine whether a sample of data is representative of a process that has a known mean value. The second and more useful situation was comparing the means of two experimental conditions. Here the idea of randomization was introduced to prevent uncontrollable variation from biasing the conclusions. In the final situation the comparison of means was generalized to several experimental conditions. We also discussed graphical and numerical methods for checking the assumptions required by the methods we recommend.

2.6.1   Important Equations

$$\bar{Y} = \frac{\sum_{i=1}^{n} Y_i}{n}$$

$$s^2 = \frac{\sum_{i=1}^{n}(Y_i - \bar{Y})^2}{n-1}$$

$$s^2 = \frac{\sum_{i=1}^{n}(Y_i - \bar{Y})^2}{n-1}$$

$$r = \frac{\sum_{i=1}^{n}(X_i - \bar{X})(Y_i - \bar{Y})}{\sqrt{[\sum_{i=1}^{n}(x_i - \bar{X})^2] \times [\sum_{i=1}^{n}(Y_i - \bar{Y})^2]}}$$

$$Z = \frac{\bar{Y} - \mu}{\sigma/\sqrt{n}}$$

$$t_{n-1} = \frac{\bar{Y} - \mu}{s/\sqrt{n}}$$

$$t_{n_1+n_2-2} = \frac{\bar{Y}_1 - \bar{Y}_2}{s_p\sqrt{\frac{1}{n_1} + \frac{1}{n_2}}}$$

$$s_p = \sqrt{\frac{(n_1-1)s_1^2 + (n_2-1)s_2^2}{n_1+n_2-2}}$$

$$Y_{ij} = \mu_i + \epsilon_{ij} = \mu + \tau_i + \epsilon_{ij}$$

$$ssTotal = ssT + ssE$$

$$\sum_{i=1}^{t} \sum_{j=1}^{r} (Y_{ij} - \bar{Y}_{..})^2 = \sum_{i=1}^{t} \sum_{j=1}^{r} (\bar{Y}_{i.} - \bar{Y}_{..})^2 + \sum_{i=1}^{t} \sum_{j=1}^{r} ((Y_{ij} - \bar{Y}_{i.})^2$$

2.6.2 Important Terms and Concepts

The following is a list of the important terms and concepts covered in this chapter.

dot plot or dot diagram
sample mean and sample variance and standard deviation
histogram
boxplot
scatter plot and correlation coefficient
normal distribution and Central Limit Theorem
hypothesis test of $H_0 : \mu = \mu_0$
one sample t-test
two sample t-test
pooled variance and standard deviation
analysis of variance
normal probability plot

2.7 Exercise

1. Bennett and Franklin [1954] present the following data where the 10 determination of the percentage of chlorine were made in two different batches of a polymer.

Determination	Batch 1	Batch 2
1	58.59	55.71
2	58.45	56.65
3	59.64	56.72
4	58.64	57.56
5	58.00	58.27
6	57.03	56.58
7	57.33	57.08
8	57.80	57.13
9	58.04	57.92
10	58.41	56.21

   (a) Compute the sample mean and sample variance of the ten determinations for each batch.

(b) Make a normal probability plot of the determinations for each batch separately and comment on whether the data appears to follow a normal distribution.

(c) Calculate the pooled standard deviation s_p.

(d) Calculate the two-sample t-statistic for testing, whether the percentage of chlorine is the same in the two batches of polymer.

(e) Calculate the critical $t_{18,0.05}$

(f) Is there a significant difference in the two batches with respect to the percentage of chlorine?

2. Guttman et al. [1965] present the data below of the determinations of the ratio of iodine to silver in four different silver preparations.

Preparation A	Preparation B	Preparation C	Preparation D
1.176422	1.176411	1.176429	1.176449
1.176425	1.176441	1.176420	1.176450
		1.176437	

(a) Calculate the ANOVA table to compare the iodine to silver ratios between the four silver preparations.

(b) How is the F-statistic calculated?

(c) What is the critical value of the F-distribution to which the calculated statistic should be compared?

(d) Is there a significant difference in the four preparations?

Chapter 3

Basic Two-Level Factorial Experiments

3.1 Introduction

The experimental environments to be discussed in this book are shown again in Figure 3.1. An experimental investigation generally begins with a preliminary exploration design (if appropriate) or a screening design in order to get the number of factors that we need to examine down to a manageable level. Then we move on to an optimization strategy. So it would seem logical to begin our discussion of experimental designs with variance components or screening. However, we will not do that.

		Present⇓		Goal⇓	
0%			**Knowledge**		**100%**
Objective:	Preliminary Exploration	Screening Factors	Effect Estimation	Optimization	Mechanistic Modeling
No. of Factors		5 - 20	3 - 6	2 - 4	1 - 5
Model:	Variance Components	Linear	Linear + Interactions	Linear + Interactions + Quadratics	Mechanistic Model
Purpose:	Identify Sources of Variability	Identify Important Factors	Estimate Factor Effects + Interactions	Fit Empirical Model Interpolate	Estimate Parameters of Theory Extrapolate
Designs:	Nested Designs	Fractional Factorials	Two-level Factorials	RSM	Special Purpose

FIGURE 3.1: Objective of Constrained (or Crude) Optimization

Instead we will choose as our first topic for discussion the two-level factorial design usually used as the second or third step in an investigation (for crude optimization). The reason for this apparent jumbling of the order of our discussion is that the two-level factorial design is the building block for almost all of the experimental strategies discussed in this book (as was mentioned in Chapter 1). The typical screening design is formed by taking a piece, or fraction, of the full two-level factorial design. Moreover, the typical comprehensive, unconstrained optimization design is built up by adding a group of experiments to the two-level factorial design. Therefore, it makes good, pedagogical sense to begin our discussion with factorial designs. Once the two-level factorial design is fully understood, it is much easier to go back and discuss the fractional factorial designs used for screening (Chapter 7) and the composite designs used for unconstrained optimization (Chapter 9). The variance components designs are discussed in Chapter 5.

3.2   Two-Level Factorial Design Geometry

By definition, a factorial design consists of all combinations of the factor levels of two or more factors. For example, let us say that you are an engineer working for a soft drink bottler, and you are interested in obtaining more uniform fill heights in the bottles produced by the manufacturing process. The filling machine does not always fill each bottle to the target height, and you decided to study the effect that two of the variables have on the deviation from target. You wanted to look at four different fill pressures (24, 26, 28, and 30 psig) and three different line speeds (200, 225, and 250 bottles/min). The full 4 x 3 factorial design to do the job is shown in Table 3.1, and it can be seen to be all possible combinations of the factor levels.

TABLE 3.1: Full 4×3 Factorial Design to Check the Accuracy of a
Bottle Filling Machine

Run Number	Fill Pressure	Line Speed	Deviation From Target
1	24	200	—
2	26	200	—
3	28	200	—
4	30	200	—
5	24	225	—
6	26	225	—
7	28	225	—
8	30	225	—
9	24	250	—
10	26	250	—
11	28	250	—
12	30	250	—

Since frugality is a major virtue in any practical situation, we would like to run the fewest number of levels of the factors while still learning about their impacts. Since we must change a factor to discover what effect it has on our response, the smallest number of levels of each factor is two. Therefore, our cornerstone design will be the two-level factorial design in which all possible combinations of two levels for each of the factors are run as experiments. Figure 3.2 shows this design geometry for two factors, and Figure 3.3 shows a two-level factorial design for three factors.

The two levels of each factor are denoted symbolically by a "+" and a "−" to indicate a high and a low level of each particular factor. When the factor is quantitative, it is obvious what this means. For example, if one of the factors is line pressure, and it is being studied

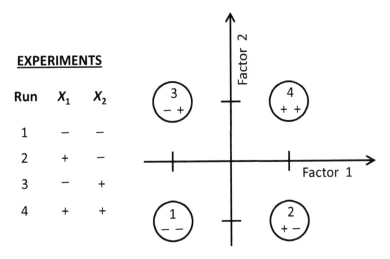

FIGURE 3.2: Two-Level Factorial Design for Two Factors (2^2 Factorial Design)

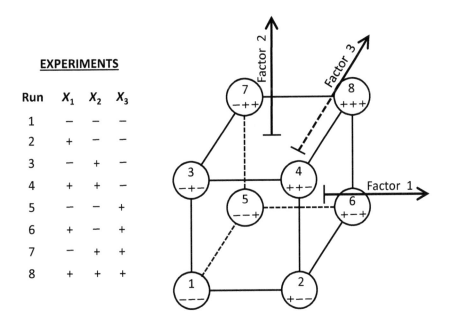

FIGURE 3.3: Two-Level Factorial Design for Three Factors (2^3 Factorial Design)

at 24 psi and 28 psi, the "+" would indicate 28 psi, while "−" would indicate 24 psi. If the factor is discrete such as a bottling machine (or operator, raw material lot, etc.), one of the bottling machines (or operators, lots, etc.) is arbitrarily called "+" and the other is called "−".

Plotting the experimental design points as in Figures 3.2 and 3.3 is useful and should be done whenever possible. This is because one of the basic goals of a good experimental design is to cover the region to be studied as thoroughly as possible with the number of runs to be made. Plotting the experimental design allows us to assess just how well we are covering the region. The factorial design, which looks at all the extreme points in the experimental region, is doing this job well. The only things that are missing are some points in the interior of the region. However, in practice, some replicated points in the center of the region are usually run. This topic will be discussed in more detail in Chapter 4, and for the time being we will assume only the factorial (corner) points are to be run.

3.3   Main Effect Estimation

Once we have run the experiments and measured a response (or several responses) at each set of conditions, we must analyze the data to determine the impacts of each of the factors under study. To see how this is done, let us look at the 2^2 factorial design shown in Figure 3.4. From this figure we can see that one way of thinking about the design is to view it as two one-at-a-time experiments in X_1: one at a low level of X_2 (Runs 1 and 2), and the other at a high level of X_2 (Runs 3 and 4). We can calculate the effect of X_1 for both of these pairs of runs. Since we want to summarize the effect of X_1 by one number, called the **main effect** of X_1, the most natural thing to do is to take the average of these two effects.

Effect of X_1 (at low X_2) $= Y_2 - Y_1$
Effect of X_1 (at high X_2) $= Y_4 - Y_3$
Average Effect of $X_1 = [(Y_2 - Y_1) + (Y_4 - Y_3)]/2$

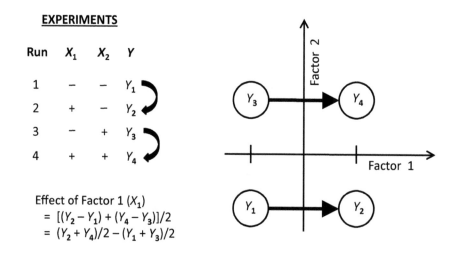

FIGURE 3.4: Estimating the Main Effect of Factor 1 for the 2^2 Factorial Design

An equivalent way of performing the calculation of this main effect is to average the runs at a high level of X_1 and subtract the average of the runs at a low level of X_1. For this reason, the main effect is often called the **average main effect.**

The power (efficiency) of two-level factorial designs compared to one-at-a-time designs is due to the fact that these same four experiments can be used to calculate the effect of X_2. However, in this case we are comparing Y_3 to Y_1, and Y_4 to Y_2 (see Figure 3.5). The calculation of the effect of X_2 is identical in principle to that for X_1.

Effect of X_2 (at low X_1) $= Y_3 - Y_1$
Effect of X_2 (at high X_1) $= Y_4 - Y_2$

Average Effect of $X_2 = [(Y_3 - Y_1) + (Y_4 - Y_2)] / 2$

Likewise, this expression can be written as the difference between the average of the Y's at the high X_2 minus the average of the Y's at the low X_2.

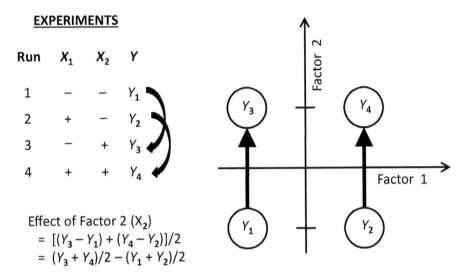

EXPERIMENTS

Run	X_1	X_2	Y
1	−	−	Y_1
2	+	−	Y_2
3	−	+	Y_3
4	+	+	Y_4

Effect of Factor 2 (X_2)
$= [(Y_3 - Y_1) + (Y_4 - Y_2)]/2$
$= (Y_3 + Y_4)/2 - (Y_1 + Y_2)/2$

FIGURE 3.5: Estimating the Main Effect of Factor 2 for the 2^2 Factorial Design

These concepts can be readily extended to factorial designs with more than two factors. Figure 3.6 shows the estimation of the main effect of X_1 for a 2^3 factorial design. In this case, we have four pairwise comparisons that give us estimates of the effect of X_1. The main effect of X_1 is simply the average of these four comparisons. Or, equivalently, the effect of X_1 is seen to be the average of the four Y's at the high X_1 minus the average of the four Y's at the low X_1. These same eight experiments can also be used to calculate the effects of X_2 and X_3 in a similar fashion (by subtracting the average of the Y's at the low level of the factor from the average of the Y's at the high level of the factor).

At this juncture, it may be useful to point out two more advantages of factorial designs over one-at-a-time designs (in addition to being space-filling). The simple 2^2 design shown in Figure 3.2 is used for illustration. The simplest one-at-a-time design for calculating the effects of two factors would have three points: a base point (−,−), a point at which only X_1 was changed (+,−), and a point at which only X_2 was changed (−,+). These three points allow the estimation of the effect of X_1 at low X_2 and the effect of X_2 at low X_1.

By the addition of only one more point (+,+), thus making a factorial design, we get: (1) greater accuracy, and (2) a broader basis for our conclusions. The accuracy comes about because of "hidden replication." Although no single pair of conditions is replicated to increase the accuracy of the estimate of the main effect, we measured the effect of each factor at two levels of the other factor. The average of the two estimates is then more precise than either single estimate. The one-at-a-time design would need to have two runs at each of the three conditions (a total of six runs) to achieve the same precision as the factorial design with four runs. This increase in efficiency becomes even more pronounced as the number of factors increases. We are also more confident in our conclusions about the effects of X_1 and X_2 on Y because we measured the effect of each factor at both levels of the other factor (or at all possible combinations of the other factors if there are more than two factors). Thus, our conclusions have a broader basis.

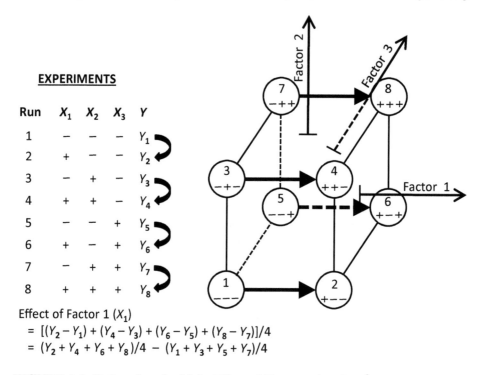

EXPERIMENTS

Run	X_1	X_2	X_3	Y
1	−	−	−	Y_1
2	+	−	−	Y_2
3	−	+	−	Y_3
4	+	+	−	Y_4
5	−	−	+	Y_5
6	+	−	+	Y_6
7	−	+	+	Y_7
8	+	+	+	Y_8

Effect of Factor 1 (X_1)

$$= [(Y_2 - Y_1) + (Y_4 - Y_3) + (Y_6 - Y_5) + (Y_8 - Y_7)]/4$$
$$= (Y_2 + Y_4 + Y_6 + Y_8)/4 - (Y_1 + Y_3 + Y_5 + Y_7)/4$$

FIGURE 3.6: Estimating the Main Effect of Factor 1 for the 2^3 Factorial Design

---

3.4   Interactions

Perhaps the most important advantage of the factorial design is that it allows for estimation of interactions between the factors. Interaction means that the effect of one factor depends on the settings of one or more of the other factors. The simplest type of interaction is called a two-way or two-factor interaction. This interaction is shown graphically for a number of situations in Figure 3.7. If X_1 and X_2 have a two-way interaction, this means that the slope of the linear plot of Y versus X_1 depends on the value of X_2. If there is no interaction, the slope of the line does not depend on X_2, although the position of the line might shift.

The magnitude of the interaction for a 2^2 factorial design is defined as one-half of the difference between the effect of X_1 at "high" X_2 and the effect of X_1 at "low" X_2 (see Figure 3.8). This will be shown later to be equal to the difference between the slopes of the two lines. Equivalently, the X_1X_2 interaction can be seen to be half the difference between the effect of X_2 at "high" X_1 and the effect of X_2 at "low" X_1. It is important that these two ways of calculating the X_1X_2 interaction (half the change in the effect of X_1 or half the change in the effect of X_2) come out the same, otherwise, it would matter which variable we chose to be Factor 1 and which variable became Factor 2. The X_1X_2-interaction definition is easily extended to larger factorials. It is still half the difference between the (average) effect of X_1 at "high" X_2 and the (average) effect of X_1 at "low" X_2.

For higher order factorials (three or more factors) we can extend the definition of the 2-factor interaction to higher order interactions, although these are harder to visualize and are, fortunately, quite rare. So, the $X_1X_2X_3$-interaction can be computed as half the difference

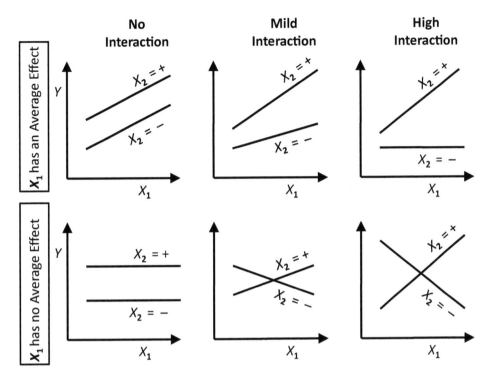

FIGURE 3.7: Different Degrees of Interaction Between Factor 1 and Factor 2

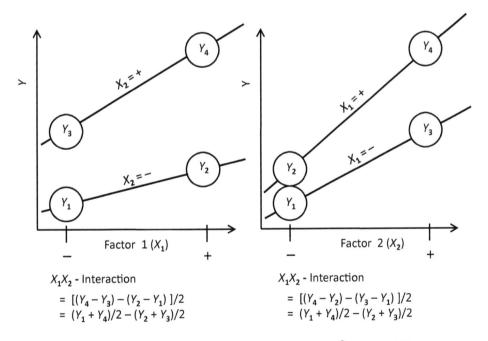

$X_1 X_2$ - Interaction

$= [(Y_4 - Y_3) - (Y_2 - Y_1)]/2$
$= (Y_1 + Y_4)/2 - (Y_2 + Y_3)/2$

$X_1 X_2$ - Interaction

$= [(Y_4 - Y_2) - (Y_3 - Y_1)]/2$
$= (Y_1 + Y_4)/2 - (Y_2 + Y_3)/2$

FIGURE 3.8: Calculation of the Interaction for a 2^2 Factorial Design

between the (average) $X_1 X_2$-interaction at "high" X_3 and the (average) $X_1 X_2$-interaction at "low" X_3. Equivalently, it could be computed as half the change in the $X_1 X_3$-interaction (as X_2 changes), or it could be computed as half the change in the X_2X_3-interaction (as X_1 changes). Note: we would get the same result regardless of which way we chose to compute the interaction.

3.5   General 2^k Factorial Designs

A factorial pattern of experiments can easily be written down for any number of factors, k. The pattern of experiments is shown in Figure 3.9 for up to five factors. The table shown in the figure is reproduced in Table B.2-1, which is in Appendix B.2 and on the website for this book (also in a tab labeled Appendix B.2). The total number of experiments is 2^k and represents all possible combinations of the k factors. The "+" and "−" signs follow an easily remembered pattern which can be readily written down. The first column alternates minuses and pluses, the second column alternates pairs of minuses and pluses, the third column alternates groups of four minuses and four pluses, etc. Remember that the "+" stands for one setting of a factor (the high setting if the factor is quantitative) and the "−" stands for another setting of the factor (the low setting if the factor is quantitative). Each of the rows corresponds to a set of experimental conditions which will be run. Any number of responses can be measured in each experiment. At the end of the experimental program, each response is analyzed separately.

Factorial designs for 2 to 5 levels, respectively, are given in Tables B.2-2 through B.2-5 in Appendix B.2. Spreadsheet .xls files for the same designs are on the website for this book. In addition to the columns of + and − signs for the factors, as shown in Figures 3.2 and 3.3, the files for Tables B.2-2 through B.2-5 contain additional columns for interactions. The interaction columns are used to define how the interaction effects are calculated. In Figure 3.8 it is shown that the X_1X_2-interaction in a 2^2 design could be calculated as:

$$X_1X_2\text{-interaction} = (Y_1 + Y_4)/2 - (Y_2 + Y_3)/2.$$

It is very important to notice that the interaction can be calculated as the difference between the average of half the responses and the average of the other half of the responses, even though the interaction is defined as something more complicated than that. Remember that this is exactly the way the effects of the factors are calculated. For the factors, we average the responses at the high level (+) and average the responses at the low level (−), and then we take the difference to get the effect. We would like to do the same thing to calculate the interactions if we only knew which responses were at the "high level of the interaction" and which responses were at the "low level of the interaction." That is exactly what the interaction columns in the tables give us. For example, it can be seen in Table 3.2 (copied from Table B.2-2 in the appendix) that the + and − signs in the X_1X_2 column define exactly how that interaction was calculated by having "+" signs for Runs 1 and 4 and "−" signs for Runs 2 and 3. Therefore, use of the table eliminates the need for going back to the definition (or, equivalently, resorting to a figure like Figure 3.8) to determine which observations need to be averaged and subtracted to calculate interactions. Using the table, the responses (the Y_i's) are simply added up with the appropriate sign and averaged to calculate the interactions (and effects). Therefore, Tables B.2-2 through B.2-5 are called computation tables. Their use is illustrated in Sections 3.7 and 3.9.

	Run	X_1	X_2	X_3	X_4	X_5
	1	−	−	−	−	−
k = 1	2	+	−			
	3	−	+	−		
k = 2	4	+	+	−		
	5	−	−	+	−	
	6	+	−	+	−	
k = 3	7	−	+	+	−	
	8	+	+	+	−	
	9	−	−	−	+	−
	10	+	−	−	+	−
	11	−	+	−	+	−
	12	+	+	−	+	−
	13	−	−	+	+	−
	14	+	−	+	+	−
k = 4	15	−	+	+	+	−
	16	+	+	+	+	−
	17	−	−	−	−	+
	18	+	−	−	−	+
	19	−	+	−	−	+
	20	+	+	−	−	+
	21	−	−	+	−	+
	22	+	−	+	−	+
	23	−	+	+	−	+
	24	+	+	+	−	+
	25	−	−	−	+	+
	26	+	−	−	+	+
	27	−	+	−	+	+
	28	+	+	−	+	+
	29	−	−	+	+	+
	30	+	−	+	+	+
k = 5	31	−	+	+	+	+
	32	+	+	+	+	+

FIGURE 3.9: The Factorial Pattern of Experiments

TABLE 3.2
Computation Table for 2^2 Design

		Computation Table		
		Design Table		
Run	Mean	X_1	X_2	$X_1 X_2$
1	+	−	−	+
2	+	+	−	−
3	+	−	+	−
4	+	+	+	+

These extra interaction columns were created by multiplying together the columns of the individual factors involved in the interaction. Specifically, to create any particular interaction column, each row is examined separately, and the + and − signs for the factors in the interaction are multiplied together to get the + or − sign for the interaction. This procedure is continued for each row (element) of the interaction. For example, in Table B.2-2 on the row for Run 1, the "−" under the X_1 column is multiplied by the "−" under the X_2 column to form the "+" in the $X_1 X_2$ column. As already stated, this multiplication is repeated for each row in the table.

3.6   Randomization

The order of the experiments shown in Figures 3.9 and Table 3.2 is called the Yates order or standard order. This order is convenient for listing experimental results or analyzing them, but the experiments should not be run in this order since the risk of bias would be very great. For example, if you were running a 2^4 factorial design (16 runs) and the response drifted down during the course of the experimental program (say, due to a catalyst aging) or shifted in the middle of the program (e.g., due to a raw material change) this effect would be mistakenly interpreted as an effect of X_4. This is shown in Figure 3.10A. If none of the four factors under study have any real impact, but the first eight runs were made at the low level of X_4 (the "−" level) and the next eight runs were made at the high level of X_4 (the "+" level), we would calculate the main effect of X_4 as the difference between the average response of the last eight runs, $\overline{Y}_+$, and the average response of the first eight runs, $\overline{Y}_-$. This difference can be seen to be large simply because of process drift. Consequently, even though X_4 has no effect at all, we would think it was very important.

To minimize this type of risk, the experiments should be run in random order. Any method of generating a random run order can be used, like writing the run numbers on pieces of paper, shuffling the pieces of paper, and using the shuffled run order. A list of random run orders for experiments can also be created in a spreadsheet that contains the list of experimental runs. The simplest procedure to use in a spreadsheet is to generate a column of n random numbers (using the RAND function for that purpose), and then running the RANK function on that column. That will give a random listing of the numbers from 1 to n.

The impact of randomization is shown in Figure 3.10B. Now when we calculate the main effect of X_4 (or any of the other insignificant factors) the "effect" will usually be small, even when there is drift in the process (or other sources of bias are present). The importance of running experiments in random order cannot be stressed too much. Randomization is usually expensive, but it is essential to achieving valid results. This point is so important, it is worth repeating. Randomizing may seem like an esoteric statistical nicety, but it is not. And we clearly recognize that in practice it is usually not easy to accomplish (i.e. it can be time consuming and therefore expensive). But system drifts or shifts are so common that not randomizing is simply begging for a non-randomized factor to be identified as important when it many not be. Therefore, if a particular factor is extremely difficult (expensive) to change, the it should be changed no fewer than twice during the course of the experimental program. For example, if X_4 in a 2^4 experiment design is extremely difficult to change, you might concede to run four runs at low X_4, eight runs at high X_4, and another four runs at low X_4. The other factors should be totally randomized. When a restricted randomization plan like this is used, the proper way to analyze the data will change. The proper analysis for experiments with restricted randomization plans (so called split-plot designs) will be discussed in Section 4.6

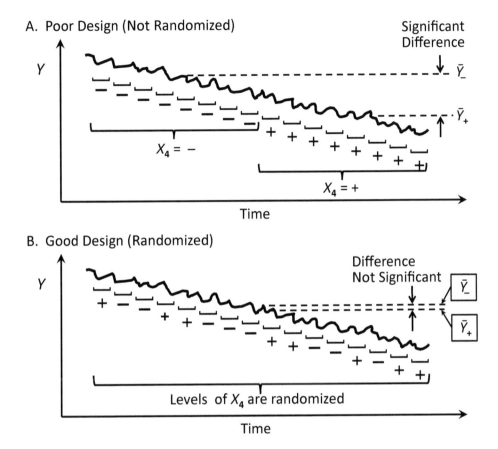

FIGURE 3.10: Example of Response Data from a Drifting Process

3.7 Example of a 2^3 Factorial Experiment

3.7.1 Background and Design

Coal is the main fuel in the production of electricity in the United States, with over two billion tons of coal burned each year. Unfortunately, coal is not pure carbon, but rather has some mineral matter that is left behind as ash after burning. Since most power plants pulverize the coal and inject the powdered coal into the burners, the ash goes up the stack with the other gases after combustion, and thus it is called *fly ash*. It is collected by one of several filtration technologies, and then it must be disposed of. Landfilling is one possibility, but it is much better to make some use of the material. One use that is becoming common is to include the fly ash as a component in concrete. An experiment was set up to study the compressive strength of such concrete made from fly ash from two different sources. The experiment factors chosen for study were:

(Factor 1):X_1 Fly Ash Source. The composition of the ash in different coals can be quite different. Hence, it is conceivable that the strength of the concrete could vary greatly, depending on the source of the ash. Therefore, it was important to investigate this as one of the factors. Two different sources of ash were used in the study, denoted by A and B.

(Factor 2):X_2 Water/Cement Ratio. It is important to add the appropriate amount of water, recognizing that what is appropriate may depend on the particular ash used (and its composition).

(Factor 3):X_3 Set-up Temperature. Since concrete is generally poured outdoors, it is important to find out if the strength of the concrete depends on the temperature at which it hardens. The factors and levels for this experiment are shown in Table 3.3.

TABLE 3.3

Factors and Levels

Factors and Response	Levels	
	(−)	(+)
X_1 = Fly Ash Source	A	B
X_2 = Water/Cement Ratio	0.35	0.55
X_3 = Set-up Temperature, °C	10	30
Y = 14 Day Compressive strength, mPa		

A data collection worksheet is shown in Figure 3.11. The first eight rows of the X_1, X_2 and X_3 columns were copied from the Appendix Table B.2-3 spreadsheet. Two replicate measurements were made at each condition; therefore a copy of the factor settings in rows 5–12 was made in rows 13 to 20. The actual factor settings were created in columns F–H and a random run order was created in column J.

	A	B	C	D	E	F	G	H	I	J	K
1			Coded Values				Actual Values			Run	Response
2	Run	X_1	X_2	X_3		Fly	Water/	Set-Up		Order	Y [=] mPa
3						Ash	Cement	Temp.			
5	1	−	−	−		A	0.35	10		5	43.4
6	2	+	−	−		B	0.35	10		14	50.1
7	3	−	+	−		A	0.55	10		15	27.3
8	4	+	+	−		B	0.55	10		1	47.0
9	5	−	−	+		A	0.35	30		8	39.4
10	6	+	−	+		B	0.35	30		12	43.8
11	7	−	+	+		A	0.55	30		10	25.0
12	8	+	+	+		B	0.55	30		9	40.7
13	9	−	−	−		A	0.35	10		4	42.8
14	10	+	−	−		B	0.35	10		7	49.5
15	11	−	+	−		A	0.55	10		2	29.1
16	12	+	+	−		B	0.55	10		3	48.4
17	13	−	−	+		A	0.35	30		11	38.4
18	14	+	−	+		B	0.35	30		16	45.0
19	15	−	+	+		A	0.55	30		6	24.2
20	16	+	+	+		B	0.55	30		13	42.9

FIGURE 3.11: A 2^3 Factorial Design to Study the Use of Fly Ash in Concrete

To create the column with random run orders, the formula =RAND() was entered into the row opposite the first experimental run in column L (not shown). This formula was copied down to the row in column L that is opposite run 16 in the table. This creates a column of random numbers between zero and one. Next, the ranks of the random numbers were computed in column M using the RANK formula as described in Section 3.6. Finally, the values created by the rank formula in column M were copied into column J and the column label Run Order was typed into cells J1 andJ2. The experiments were completed in the random run order with Run 4 completed first. It used fly ash source B, a water/cement ratio of 0.55, and a set-up temperature of 10 degrees centigrade. The resulting compressive strength of 47.0 mPa was recorded in its proper place in column K. Run 11 was completed second, and eventually all 16 data points were obtained and listed in column K of Figure 3.11.

The data were summarized with some calculated statistics in the spreadsheet shown in Figure 3.12. Specifically, in addition to the individual compressive strengths at each of the eight experimental conditions, the average strength and the variances of the replicate runs at each condition were calculated. The average in cell H4 was calculated using the formula AVERAGE(E4:F4) and this formula was copied into cells H5 through H11. The variance in cell I4 was calculated with the formula =VAR(E4:F4), and this formula was copied into cells I5 to I11. The variances will be important later when we calculate the statistical significance of our effects and interactions. For now, however, this information is extra and will not be used.

	A	B	C	D	E	F	G	H	I	J
1		Factor Levels				Response,		Avg		
2	X_1	X_2	X_3			Y [=] mPa		Y	Variance	DF
3										
4	−	−	−		43.4	42.8		43.1	0.18	1
5	+	−	−		50.1	49.5		49.8	0.18	1
6	−	+	−		27.3	29.1		28.2	1.62	1
7	+	+	−		47.0	48.4		47.7	0.98	1
8	−	−	+		39.4	38.4		38.9	0.50	1
9	+	−	+		43.8	45.0		44.4	0.72	1
10	−	+	+		25.0	24.2		24.6	0.32	1
11	+	+	+		40.7	42.9		41.8	2.42	1

FIGURE 3.12: Summary of Average Compressive Strengths of Concrete and Variability

3.7.2 Calculation of Effects and Interactions

The data from any 2^3 experiment can be conveniently displayed on a cube which represents the experimental region as in Figure 3.3. It always helps to get a visual appreciation of the data before tackling the numerical analysis. For the fly ash in concrete experiment, this is done in Figure 3.13. In this figure, called a cube plot, the direction of the arrows indicates the higher setting of each factor. We can learn quite a bit from this figure alone. If 35 kPa or higher is a good compressive strength, we can see that six out of the eight conditions meet this criterion. The strengths at the high level of X_1 (fly ash source B) are all good, while those at the low level of X_1 (fly ash source A) depended on the water/cement ratio. The set-up temperature, X_3, seemed to matter, but not enough to produce poor strengths at

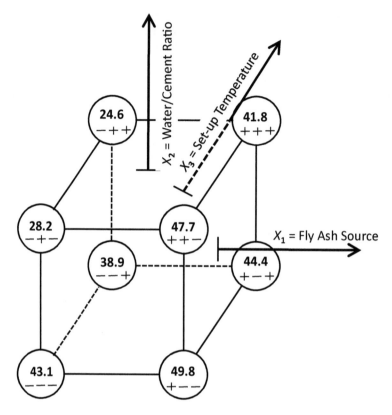

FIGURE 3.13: Graphical Display of Data from a 2^3 Factorial Design for Fly Ash in Concrete

one temperature and good strengths at another. All of these temporary conclusions should be quantified and tested for statistical significance.

We begin the numerical analysis with the calculation of effects and interactions. The calculation of the average main effects and interactions as defined in Sections 3.3 and 3.4 is accomplished via the use of the computation table B.2-3 in Appendix B, which was described in Section 3.5. The $+1$'s and -1's in the table indicate what sign is associated with responses for each run in the calculation of the various effects. Table B.2-3 was copied into the spreadsheet shown in Figure 3.14, with the average compressive strength added to the right. Then, to calculate any of the effects, the Y's are simply added up (with the appropriate sign) and averaged by dividing by the number of $+$'s for the given effect.

Let us start at the beginning and compute the main (average) effect of X_1

$$\text{Effect of } X_1 = \text{Average Response when } X_1 \text{ is high } (+)$$
$$- \text{Average Response when } X_1 \text{ is low } (-)$$
$$= (Y_2 + Y_4 + Y_6 + Y_8)/4 - (Y_1 + Y_3 + Y_5 + Y_7)/4.$$

If we merge the two groups of responses into one large group, we get

$$\text{Effect of } X_1 = (-Y_1 + Y_2 - Y_3 + Y_4 - Y_5 + Y_6 - Y_7 + Y_8)/4,$$

which is what the computation table indicates. For our example,

	A	B	C	D	E	F	G	H	I	J	K
1		Run	Mean	X_1	X_2	X_3	X_1X_2	X_1X_3	X_2X_3	$X_1X_2X_3$	Avg Y
2											
3		1	1	-1	-1	-1	1	1	1	-1	43.1
4		2	1	1	-1	-1	-1	-1	1	1	49.8
5		3	1	-1	1	-1	-1	1	-1	1	28.2
6		4	1	1	1	-1	1	-1	-1	-1	47.7
7		5	1	-1	-1	1	1	-1	-1	1	38.9
8		6	1	1	-1	1	-1	1	-1	-1	44.4
9		7	1	-1	1	1	-1	-1	1	-1	24.6
10		8	1	1	1	1	1	1	1	1	41.8
11											
12	SUMPRODUCT		318.5	48.9	-33.9	-19.1	24.5	-3.5	0.1	-1.1	
13		Effect	39.81	12.23	-8.48	-4.78	6.13	-0.88	0.02	-0.28	

FIGURE 3.14: Worksheet to Compute Effects and Interactions from 2^3 Factorial Design for Fly Ash in Concrete

$$\text{Effect of } X_1 = (-43.1 + 49.8 - 28.2 + 47.7 - 38.9 + 44.4 - 24.6 + 41.8)/4$$
$$= 48.9/4 = 12.225.$$

The numerator of this effect is calculated in cell D12 of the spreadsheet with the formula =SUMPRODUCT(D3:D10,$K3:$K10), and the actual effect is calculated in cell D13 as D12/4. Notice that the high and low levels in this spreadsheet are indicated by 1 and −1 rather than just + and − so that the multiplication performed by the SUMPRODUCT function makes sense.

Similarly, we can simply use the computation column for the rest of the effects and interactions. For example,

$$X_1X_3\text{-Interaction} = (+Y_1 - Y_2 + Y_3 - Y_4 - Y_5 + Y_6 - Y_7 + Y_8)/4$$
$$= (+43.1 - 49.8 + 28.2 - 47.7 - 38.9 + 44.4 - 24.6 + 41.8)/4 = -3.5/4 = -0.88.$$

The calculation for the rest of the effects is made by simply copying the formula in cell D12 across to cell J12 and D13 across to J13. If the formula in cell D12 is copied to cell C12, the result is the sum of all the averages in column K, and dividing cell C12 by 8 and storing the result in cell C13 results in the overall average of the response data.

3.8 Significance of Effects and Interactions

3.8.1 Statistical Significance of Results

Experimentally measuring the effects the factors have on the response(s) and the interactions of the factors is only part of our job. We must keep in mind that our measurements are subject to experimental error. Even when a particular factor has no real impact on a response, it would be very unusual to measure an effect of exactly zero. Therefore, we have to be careful that we don't draw unwarranted conclusions from our data. As discussed earlier, if a measured effect or interaction can be reasonably explained by experimental variation, we will assume it is zero. Only if the effect or interaction is so large in comparison to experimental variation that it cannot reasonably be said to be zero will we claim that the factor has a statistically significant effect on the response. Often, in this text we might leave out the term "statistically" and just say there is a significant effect of a factor, but that should not be confused with practical significance. We are always referring to statistical significance. Consequently, how do we determine the significance of our results? We actually already know how. Since each of the average effects (both main effects and interactions) are differences of two averages (i.e., $\overline{Y}_+ - \overline{Y}_-$) the statistical significance of these effects can be judged by computing a signal-to-noise t-ratio exactly as in Section 2.5.2. The numerator of the t-ratio is the effect or interaction (minus the hypothesized value which is presumed to be zero). Recall that the denominator of the signal-to-noise ratio is the standard deviation of the difference in the numerator:

$$s_E = \sqrt{s^2(\frac{1}{n_+} + \frac{1}{n_-})} \tag{3.1}$$

where s_E = the standard deviation of an effect or interaction (the same),
  s = the estimate of the standard deviation of individual responses
  n_+ = the number of observations in the average response,
    at the high level of the effect, $\overline{Y}_+$, and
  n_- = the number of observations in the average response
    at the low level of the effect, $\overline{Y}_-$.

Factorial designs put half of the points at the high level of any factor (or interaction) and the other half of the points at the low level of that factor. So, $n_+ = n_- = n_F/2$, where n_F is the total number of factorial points (including replication). Taking that relationship into account, the equation for s_E simplifies to:

$$s_E = \sqrt{s^2(\frac{1}{n_F/2} + \frac{1}{n_F/2})} = \sqrt{s^2(\frac{4}{n_F})} = 2s/\sqrt{n_F}. \tag{3.2}$$

The signal-to-noise t-ratios, denoted by t_E, are given by the formula:

$$t_E = \text{Effect}/s_E = \text{Effect}/(2s/\sqrt{n_F}). \tag{3.3}$$

The statistical significance of each effect and interaction is judged by comparing its t-ratio, t_E, to the critical limit, denoted by $t_{\nu_p,\alpha}$, where ν_p is the degrees of freedom associated with s and $1 - \alpha$ is the confidence level. The critical limit $t_{\nu_p,\alpha}$ can be calculated with the spreadsheet formula `TINV()` as explained in Section 2.5.1 or looked up in Table A.3.

3.8.2 Pooled Variance

We would now like to proceed with our signal-to noise ratios, but a critical item, the standard deviation of our measurements, s, must be known. If we are fortunate, we may have a prior estimate of s. But usually we do not. And even if we do, it is a good idea to verify that it is valid for our current set of experiments. Where do we look for a current estimate of s? We can get an estimate from conditions that were run more than once. As discussed in Section 2.2.2, we estimate the variance from repeat measurements by the formula:

$$s^2 = \frac{1}{n-1}\sum_{i=1}^{n}(Y_i - \overline{Y})^2. \tag{3.4}$$

If we have more than one set of conditions that were replicated, we can calculate the variance from each group of replicates. If we assume that the variances for each group are the same (an assumption that should be checked), we can average the estimates of the variance to get a **pooled variance**, denoted by s_p^2. If the groups have the same number of observations, we can simply average the estimates of the variances. (Note: it is NOT correct to average the standard deviations, s, but rather the variances, s^2.) In general, the groups may not have the same number of observations, and we need to take into account the higher accuracy of the estimates from larger groups. We do this by weighting the various estimates by the degrees of freedom for each particular estimate. If we use $\nu_j = n_j - 1$ to denote the degrees of freedom for a particular variance estimate, s_j^2, the general formula when we have m groups of data, and therefore m estimates of the variance, is:

$$s_p^2 = \sum_{j=1}^{m} \nu_j s_j^2 / \sum_{j=1}^{m} \nu_j, \tag{3.5}$$

and the degrees of freedom for s_p^2 (denoted by ν_p), is just the sum of the degrees of freedom of all of the estimates being pooled:

$$\nu_p = \sum_{j=1}^{m} \nu_j. \tag{3.6}$$

The square root of this pooled variance, a.k.a. the pooled standard deviation, s_p, is used for all of the t-tests which determine the statistical significance of the calculated effects and interactions. The pooled degrees of freedom, ν_p, is the degrees of freedom used in these tests. The formula for s_P here is a generalization of the formula for s_P in Section 2.5.2 for two sets of replicates.

In a replicated factorial experimental design, we have 2^k different sets of replicates – one set at each experimental condition. For example, in the fly ash experiment we have 2^3 = 8 pairs of replicates. From each set (pair) of replicates we can compute a variance, and these variances were given in Figure 3.12. If we assume that the variability of replicates is the same at each experimental condition, we can come up with one pooled variance using Equation 3.5. The value of the pooled variance for the fly ash example is:

$$s_p^2 = [(1)(0.18) + (1)(0.18) + (1)(1.62) + \ldots + (1)(2.42)]/[1+1+1+ \ldots +1], \text{ or}$$

$$s_p^2 = [6.920]/[8] = 0.865,$$

and the pooled standard deviation that we need for the t-tests is:

$$s_p = \sqrt{0.865} = 0.930 \text{ with } \nu_p = 8 \text{ degrees of freedom.}$$

As already stated above, this pooled value is calculated assuming that the variability is the same at all experimental conditions. This assumption should generally be checked. A simple plot of the variance or standard deviation versus the average at each experimental condition is useful for this purpose. This is often called the plot of cell standard deviations versus cell means, where cell refers to an experimental condition or run. For example, Figure 3.15 shows this plot for the fly ash data. In this plot there is no apparent pattern or trend. Therefore, our assumption that each sample standard deviation is estimating the same σ appears to be reasonable.

If there had been a pattern in this plot, it would indicate that the variability tends to change as the average response changes. That would contradict our assumption of equal variability at all experimental conditions. In that situation, transformations of the data to achieve a constant variance may be required before a valid determination of the significance of effects can be made. But for the fly ash experiment, we feel somewhat confident that the value of s_p we calculated represents the common variance of the data at all conditions.

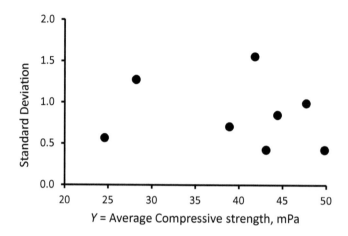

FIGURE 3.15: Plot of Cell Standard Deviation Versus Cell Mean for Fly Ash in Concrete Example

3.8.3   Statistical Significance of Results for Fly Ash

We will now use our value of the pooled standard deviation to estimate the standard deviation of the effects, s_E . Then we can compute the signal-to-noise t-ratios to assess the statistical significance of the effects and interactions (given in Figure 3.14). Using Equation 3.2, and remembering that n_F is the number of factorial points including replication ($n_F = 16$), we find:

$$s_E = 2s_p/\sqrt{n_F} = 2(0.930)/\sqrt{16} = 0.465.$$

With this value of s_E in hand, the signal-to-noise t-ratio, t_E, was calculated for each effect and interaction using Equation 3.3. These t-ratios are given immediately below the effects in Figure 3.16. This figure is a repeat of Figure 3.14, with the signal-to-noise ratios, t_E's, added below each main effect and interaction. The statistical significance of the effects and interactions is judged by comparing the t_E's to the critical t-value, $t_{8,.05}$, which can be obtained using the spreadsheet formula =T.INV(p,ν) as explained earlier, or by looking the value up in Table A.3 (in Appendix A). In this case the $t_{8,.05}$-value is looked up with 8 degrees of freedom, since that is the degrees of freedom in s_p, our estimate of σ, which then becomes our degrees of freedom for s_E. At the 95% confidence level, $t_{8,.05} = 2.306$. It can be seen that all of the main effects were significant, as well as the $X_1 X_2$ interaction.

3.8.4   Interpretation of Results for Fly Ash Example

The results of the statistical analysis indicate that all three factors (fly ash source, water/cement ratio, and set-up temperature) are important in determining the 14-day com-

	Run	Mean	X_1	X_2	X_3	X_1X_2	X_1X_3	X_2X_3	$X_1X_2X_3$	Avg Y
1										
3	1	1	-1	-1	-1	1	1	1	-1	43.1
4	2	1	1	-1	-1	-1	-1	1	1	49.8
5	3	1	-1	1	-1	-1	1	-1	1	28.2
6	4	1	1	1	-1	1	-1	-1	-1	47.7
7	5	1	-1	-1	1	1	-1	-1	1	38.9
8	6	1	1	-1	1	-1	1	-1	-1	44.4
9	7	1	-1	1	1	-1	-1	1	-1	24.6
10	8	1	1	1	1	1	1	1	1	41.8
12	SUMPRODUCT	318.5	48.9	-33.9	-19.1	24.5	-3.5	0.1	-1.1	
13	Effect	39.81	12.23	-8.48	-4.78	6.13	-0.88	0.02	-0.28	
15	t_E		26.29	-18.22	-10.27	13.17	-1.88	0.05	-0.59	

FIGURE 3.16: Significance of Effects and Interactions for Fly Ash in Concrete Example

pressive strength of concrete made with fly ash as one of the components. The effects are summarized in Table 3.4.

TABLE 3.4
Average Strength of Concrete to Show Effects

Factor	Level (−)	$\bar{Y}_-$	Level (+)	$\bar{Y}_+$	Effect $(\bar{Y}_+ - \bar{Y}_-)$
X_1 = Fly Ash Source	A	33.70	B	45.92	12.22
X_2 = Water/Cement Ratio	0.35	44.05	0.55	35.58	−8.48
X_3 = Set-up Temperature, °C	10	42.20	30	37.42	−4.78

It can be seen that concrete using fly ash source B had a compressive strength that was, on average, 12.22 mPa greater than concrete made using fly ash source A. Higher water/cement ratios dropped the compressive strengths by 8.48 mPa, again, on average. And the warmer set-up temperature gave a lower compressive strength by 4.78 mPa. But, remember that these effects are all describing what happens on average. If there are no significant interactions, then these effects tell the whole story. But since there is a large interaction between fly ash source and water/cement ratio, we need to say more. Specifically, if we stopped now with our description of this system, we would conclude that fly ash source A was unsatisfactory for use in concrete. But let us keep going and examine the interaction between fly ash source and water/cement ratio. The average compressive strengths for the two fly ash sources at the two water/cement ratios are given in Table 3.5.

It can be seen from this table that fly ash source B is good at both water/cement ratios (as expected), but fly ash source A is also more than adequate (having a compressive strength well above the 35 mPa required) at the lower water/cement ratio of 0.35. Thus, it would be very wrong to conclude that fly ash source A would not be useful in making concrete based on the average result alone. The significant interaction tells us that we need to examine the results more carefully. This interaction is shown graphically in Figure 3.17, which is often a more easily understood representation of the results.

In general, results of factorial experiments should be interpreted as in this example. In-

TABLE 3.5
Average Strength of Concrete to Show the $X_1 X_2$ Interaction

		X_2 = Water/Cement		Effect
		0.35	0.55	$(\bar{Y}_+ - \bar{Y}_-)$
X_1 = Fly Ash Source	A	41.00	28.40	−14.60
	B	47.10	44.75	−2.35
$X_1 X_2$ Interaction = $(1/2)[(-2.35)-(-14.60)]$ =				6.13

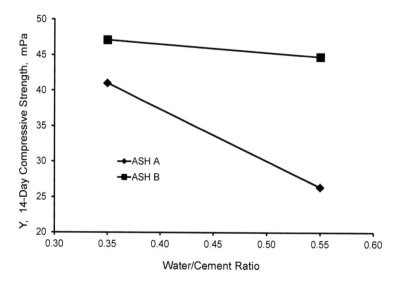

FIGURE 3.17: Interaction Plot of Fly Ash Source and Water/Cement Ratio

significant effects can be ignored. Significant main effects can be interpreted as the average difference in the response or dependent variable as the factor is changed from low to high level. And any factors having significant interactions must be interpreted by referring to a table of averages as given in Table 3.5 or a graph like Figure 3.17, because the effects alone are an oversimplification of how the system under study behaves. Ignoring the interactions and pretending that the (average) effect describes the system is like the old joke about the statistician with his feet in the freezer and his head in the oven: on average he was comfortable!

3.8.5   Example of Computer Analysis of Data

Many computer programs have been written which can automate the tedious analysis of data as shown in Figure 3.16. As an example, Figure 3.18 shows a page of output produced by the program MINITAB^{TM}(Version 17). This figure shows the calculated effects (along with some other information) that are the same as those previously calculated in Figure 3.16. The column labeled "Coef" gives the coefficients when using the variables in an equation. These coefficients (which will be discussed fully later in this chapter) are simply half of the effects. The next column, "SE Coef," gives the standard error (standard deviation) of the coefficients, which is half the standard deviation of the effects, $s_E = 0.465$. The column labeled "T" gives the t-values. The column labeled "P" represents the probabilities of obtaining an effect of greater magnitude than those observed by chance if the real effect is zero. This column saves us the effort of looking up a critical $t_{\nu,\alpha}$ value; any P's less than 0.05 are statistically significant at the 95% confidence level. Therefore, all main effects and the $X_1 X_2$ interaction are significant, as we determined by hand. The "Constant" is the grand mean of all the data.

Factorial Fit: Compressive Strength versus Fly Ash Source, Water/Cement, and Temperature

Estimated Effects and Coefficients for Compressive Strength (coded units)

Term	Effect	Coef	SE Coef	T	P
Constant		39.813	0.2325	171.23	0.000
Fly Ash Source	12.225	6.112	0.2325	26.29	0.000
Water/Cement	-8.475	-4.238	0.2325	-18.22	0.000
Temperature	-4.775	-2.387	0.2325	-10.27	0.000
Fly Ash Source*Water/Cement	6.125	3.063	0.2325	13.17	0.000
Fly Ash Source*Temperature	-0.875	-0.437	0.2325	-1.88	0.097
Water/Cement*Temperature	0.025	0.012	0.2325	0.05	0.958
Fly Ash Source*Water/Cement* Temperature	-0.275	-0.137	0.2325	-0.59	0.571

S = 0.930054   R-Sq = 99.39%   R-Sq(adj) = 98.86%

Analysis of Variance for Compressive Strength (coded units)

Source	DF	Seq SS	Adj SS	Adj MS	F	P
Main Effects	3	976.31	976.307	325.436	376.23	0.000
2-Way Interactions	3	153.13	153.128	51.043	59.01	0.000
3-Way Interactions	1	0.30	0.302	0.302	0.35	0.571
Residual Error	8	6.92	6.920	0.865		
Pure Error	8	6.92	6.920	0.865		
Total	15	1136.66				

FIGURE 3.18: Sample Output from MINITAB^{TM} for Fly Ash Example

Figure 3.18 also shows an Analysis of Variance table that has more sources of variation than the one shown in Section 2.5.3. For now, suffice it to say the additional sources in the table allow checking of the significance of groups of effects or interactions. An F-ratio is used to calculate significance rather than the t-value. The other important item of information given in that part of the table is the "Pure Error" adjusted mean square ("Adj MS") which is our estimate of the variance, $s_p^2 = 0.865$, with 8 degrees of freedom. Thus $s_p = \sqrt{0.865} = 0.930$, all of which agrees with the calculations in Section 3.8.2.

In addition to tabular output, computer programs often produce graphs that make it easier to interpret the data and results of the statistical analysis. For example, if desired, MINITAB^{TM} produces the cube plot, graphs of the main effects of each of the three factors, and interaction plots. This software, and other products like it, makes the analysis of experimental data much more convenient.

3.9   Example of an Unreplicated 2^4 Design (Stack Gas Treatment)

In this section we present another example of a 2^k experiment. This example will again illustrate the interpretation of results from a factorial experiment and introduce additional techniques for analysis.

3.9.1   Background and Design

This example is taken (a bit loosely) from a study done at the University of North Dakota Energy and Environmental Research Center [Weber and Erjavec, 1987]. The overall research goal was to control SO_2 emissions during coal combustion in power plants via the injection of dry sodium carbonate into the flue gas upstream of the baghouse (used to collect particulate emissions). The method worked well in the laboratory, but when implemented on a plant scale some NO_2 occasionally was formed resulting in a brown plume from the stack, which was totally unacceptable. To investigate the conditions under which NO_2 would be produced, a factorial experimental design was run in the lab.

The response measured was the amount of NO_2 formed (ppm), and the four factors studied, along with the levels of each, are given in Table 3.6. The factors were SO_2 concentration (X_1), temperature (X_2), oxygen concentration (X_3), and the amount of moisture (X_4). A full factorial design was planned. The design, in both coded and uncoded X's, is given in Table 3.7.

There was an important "wrinkle" in this study that needed to be recognized. It was known that the variability in the measured response, NO_2 concentration, was Normally distributed, but it did not have a constant variance. Rather, the standard deviation of the error in the measurements was a constant percentage, namely 25%. Since our analysis assumes that all the data have the same precision (constant variance), the response had to be transformed to make that assumption valid. The log transformation accomplished this goal; constant percentages become constant amounts on a log scale. Therefore, Y = log (NO_2) became the actual response to be analyzed. The standard deviation was known to be 25% of the measurement (as stated already), so on the logarithmic scale the standard deviation is the difference of the logs, i.e., $\sigma_Y = \log(1.25) - \log(1.0) = 0.10 - 0.00 = 0.10$.

TABLE 3.6

Stack Gas Treatment Example of a 2^4 Factorial Design

Factors and Response	Levels	
	$-$	$+$
$X_1 = SO_2$ concentration, ppm	0	3000
$X_2 = $ Temperature, °F	150	350
$X_3 = O_2$ concentration, %	0	6
$X_4 = $ moisture, %	0	20

Response: Concentration of NO_2 formed, ppb
$Y = \log(NO_2$

Variability: NO_2 determination has a $\sigma = 25\%$ (known)
$\sigma_Y = \log(1.25) = 0.10$

TABLE 3.7
Stack Gas Treatment Example of a 2^4 Factorial Design and Data

Run	Coded Values X_1	X_2	X_3	X_4	Actual Values SO_2	Temp	O_2	Moist	Run Order	NO_2 ppb	Y log(NO_2)
1	−	−	−	−	0	150	0	0	3	130	2.11
2	+	−	−	−	3000	150	0	0	13	150	2.18
3	−	+	−	−	0	350	0	0	16	210	2.32
4	+	+	−	−	3000	350	0	0	4	200	2.30
5	−	−	+	−	0	150	6	0	15	110	2.04
6	+	−	+	−	3000	150	6	0	1	18,000	4.26
7	−	+	+	−	0	350	6	0	8	110	2.04
8	+	+	+	−	3000	350	6	0	5	150,000	5.18
9	−	−	−	+	0	150	0	20	7	130	2.11
10	+	−	−	+	3000	150	0	20	2	150	2.18
11	−	+	−	+	0	350	0	20	10	250	2.40
12	+	+	−	+	3000	350	0	20	6	140	2.15
13	−	−	+	+	0	150	6	20	14	140	2.15
14	+	−	+	+	3000	150	6	20	9	22,000	4.34
15	−	+	+	+	0	350	6	20	12	150	2.18
16	+	+	+	+	3000	350	6	20	11	170,000	5.23

Once the design and response were settled, the next step was to run the experiments. Of course, they could not be run in the order listed in Table 3.7, or a large risk of biases invalidating the results would have been incurred. The tool that was absolutely necessary to use was randomization. In this study the factorial points were run in completely random order; a random number generator on a calculator was used to obtain the sequence. The 16 data points were then collected (i.e., each experiment was run in the appropriate order and the NO_2 concentration was measured). The results are also given in Table 3.7.

3.9.2 Graphical Analysis

Before mechanically cranking out a numerical analysis of a design, it is a good idea to get as much insight into the situation as possible. The first step in doing this is to make simple plots of the data. Figure 3.19 is a line graph of the data in the order the experiments were conducted. In this figure we see no patterns or trends. If there were an increasing or decreasing trend, or a consistent cycling pattern in the plot, it would lead us to suspect that changes in some lurking variable might have added variability to our experimental results. In that case, we might make additional plots of the data versus any other recorded variables that were known to have changed during the course of the experiment. If we could find a graphical relationship between the response and an uncontrolled variable, we might be able to adjust for this variable and reduce the standard error of the effects using regression analysis, which is discussed in Chapter 8. Recall that we can never adjust for biases in effects caused by changes in lurking variables if we failed to randomize our experiments. Furthermore, regression analysis will only allow us to correct the effect of lurking variables on the standard error of the response if that standard error is estimated from the data.

Simple boxplots of the data separated by levels of each factor can give us a general feeling about the importance of the factor effects. Figure 3.20 shows Boxplots separated by levels of each of the factors. In this figure, it can be seen that SO_2 and O_2 seem to have the largest effects on Y=log(NO_2), while moisture seems to have a negligible effect.

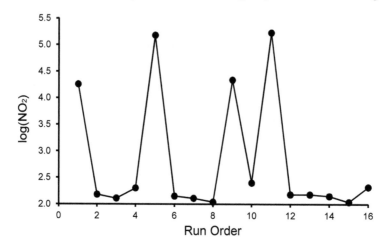

FIGURE 3.19: Plot of Data vs. Random Run Order from Stack Gas NO_2 Experiments

The wider spread in the response at the high levels of SO_2 and O_2 may indicate that these factors interact with others. If a Boxplot program is not available, a simple scatter plot of the response on the y-axis and factor levels on the x-axis will give essentially the same information as Figure 3.20.

A good way to begin visualizing potential interactions is to plot the data on the cube which represents the experimental region. For our gas treatment experiment, this is shown in Figure 3.21(a) and 3.21(b). From a comparison of the cubes for low moisture and for high moisture, it can be judged that moisture had very little impact, if any. Therefore, the two moisture levels were averaged and the averages shown in Figure 3.21(c).

From this figure, it seems like all three remaining variables had positive effects. It also seems that a $X_1X_2X_3$-interaction exists, since the interaction of X_1X_2 (i.e., temperature and SO_2) is negligible at low $X_3 \equiv O_2$ (the front of the cube), but appears quite large at high $X_3 \equiv O_2$ (at the back of the cube). By the interaction of temperature and SO_2 we mean the effect of temperature is different at high SO_2 than at low SO_2. The negligible interaction, or difference in temperature effects, can be seen at the front of the cube as the $Log(NO_2)$ increases from 2.11 to 2.36 as temperature increases (with low SO_2) and again increases from 2.18 to 2.22 as temperature increases with high SO_2. On the other hand at the back of the cube (i.e., high O_2), the $Log(NO_2)$ increases from 2.10 to 2.11 as temperature increases (with low SO_2), but increases from 4.30 to 5.20 (a much larger increase) as temperature increases with high SO_2.

3.9.3   Numerical Analysis

In order to determine the exact average effects and interaction, and even more importantly, whether or not they are statistically significant, a numerical analysis is necessary. The calculation of effects and interactions is accomplished via the use of the computation table for a 2^4 factorial design listed in Appendix B.2-4. The table was copied into a spreadsheet and the results are shown in Figure 3.22. The response, Y = log (NO2), was appended to the right for handy reference. Then, to calculate each effect, the Y's are simply added up (with the appropriate sign) and averaged by dividing by the number of observations at the high level (or low level) of each factor (or interaction).

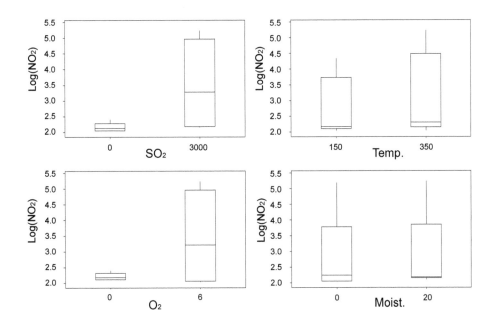

FIGURE 3.20: Boxplots of Data vs. Factor Levels from Stack Gas NO_2 Experiments

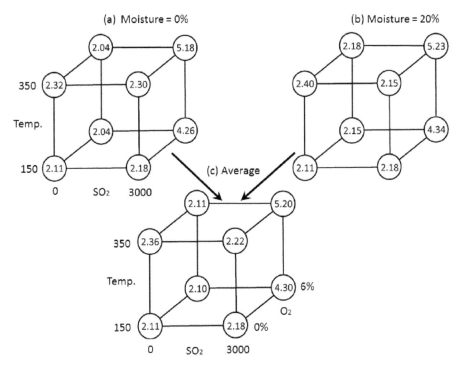

FIGURE 3.21: Cube Plots of Data from Stack Gas NO_2 Experiments

The calculation of effects and interactions was shown in the example in Section 3.7. As a review, let us calculate one of the interactions.

Run	Mean	X_1	X_2	X_3	X_4	X_1X_2	X_1X_3	X_1X_4	X_2X_3	X_2X_4	X_3X_4	$X_1X_2X_3$	$X_1X_2X_4$	$X_1X_3X_4$	$X_2X_3X_4$	$X_1X_2X_3X_4$	$Y=\log(NO_2)$
1	1	-1	-1	-1	-1	1	1	1	1	1	1	-1	-1	-1	-1	1	2.11
2	1	1	-1	-1	-1	-1	-1	-1	1	1	1	1	1	1	-1	-1	2.18
3	1	-1	1	-1	-1	-1	1	1	-1	-1	1	1	1	-1	1	-1	2.32
4	1	1	1	-1	-1	1	-1	-1	-1	-1	1	-1	-1	1	1	1	2.30
5	1	-1	-1	1	-1	1	-1	1	-1	1	-1	1	-1	1	1	-1	2.04
6	1	1	-1	1	-1	-1	1	-1	-1	1	-1	-1	1	-1	1	1	4.26
7	1	-1	1	1	-1	-1	-1	1	1	-1	-1	-1	1	1	-1	1	2.04
8	1	1	1	1	-1	1	1	-1	1	-1	-1	1	-1	-1	-1	-1	5.18
9	1	-1	-1	-1	1	1	1	-1	1	-1	-1	-1	1	1	1	-1	2.11
10	1	1	-1	-1	1	-1	-1	1	1	-1	-1	1	-1	-1	1	1	2.18
11	1	-1	1	-1	1	-1	1	-1	-1	1	-1	1	-1	1	-1	1	2.40
12	1	1	1	-1	1	1	-1	1	-1	1	-1	-1	1	-1	-1	-1	2.15
13	1	-1	-1	1	1	1	-1	-1	-1	-1	1	1	1	-1	-1	1	2.15
14	1	1	-1	1	1	-1	1	1	-1	-1	1	-1	-1	1	-1	-1	4.34
15	1	-1	1	1	1	-1	-1	-1	1	1	1	-1	-1	-1	1	-1	2.18
16	1	1	1	1	1	1	1	1	1	1	1	1	1	1	1	1	5.23
SUMPRODUCT	45.17	10.47	2.43	9.67	0.31	1.37	10.73	-0.35	1.25	-0.07	0.45	2.19	-0.29	0.11	0.07	0.17	
Effects	2.82	1.31	0.30	1.21	0.04	0.17	1.34	-0.04	0.16	-0.01	0.06	0.27	-0.04	0.01	0.01	0.02	
t(E)		26.18	6.08	24.18	0.78	3.43	26.83	-0.87	3.13	-0.17	1.13	5.48	-0.73	0.28	0.18	0.43	

FIGURE 3.22: Calculation of Effects and Interactions for Stack Gas NO_2 Example

$$X_1X_2 = [Y_1 - Y_2 - Y_3 + Y_4 + \ldots - Y_{15} + Y_{16}] \,/\, 8$$

or $X_1X_2 = [\,2.11 - 2.18 - 2.32 + 2.30 + \ldots - 2.18 + 5.23\,] \,/\, 8 = 0.17$

The simultaneous calculation of all the effects was accomplished using the SUMPRODUCT function in the spreadsheet. The results of all the computations are shown in Figure 3.22. The next question that should come to mind is: Are any of these calculated effects real, or could they just be due to the random variability of the data? To help us with this decision we divide each of the effects by their standard error as described in Section 3.8. In other words, we calculate the signal-to-noise ratios, denoted by t_E (also given in Figure 3.22).

There is a difference between the calculation of the standard error of effects for the gas treatment experiment and the fly ash experiment described in Section 3.8.2. Recall that in the fly ash experiment, a measure of variability called the pooled standard deviation, s_p, was calculated from the replicates in the experiment, and then the standard error of the effects or interactions was computed using the formula $s_E = 2\,s_p \,/\, \sqrt{n_F}$. In the gas treatment example, there were no replicates and thus no estimate of the variability could be made from the data. However, there was a known constant percentage error that allowed us to determine that the standard deviation of $Y = \log(NO_2)$ was $= \log(1.25) = 0.10$. Therefore, the standard error of an effect or interaction was computed below as:

$$\sigma_E = 2\,\sigma/\sqrt{n_F} = 2\,(0.10)/\sqrt{16} = 0.050,$$

and the signal-to-noise ratios in Figure 3.22 were computed as:

$$t_E = \text{Effect (or Interaction)}/ \sigma_E.$$

These were compared to the $t_{\nu,\alpha}$ values in Table A.3. The degrees of freedom (ν) associated with the critical statistic $t_{\nu,\alpha}$ are always the degrees of freedom associated with the estimate of variability. Recall that in the fly ash experiment we used 8 degrees of freedom, which were the degrees of freedom for s_p. In the gas treatment experiment, the measure of variability $\sigma = \log(1.25) = 0.10$ was known, so the degrees of freedom are essentially infinite. The last row of Table A.3, with degrees of freedom $= \infty$, is equivalent to the Standard Normal Table A.1, and our signal-to-noise ratios, t_E, are actually the same as the z-statistics shown in Equation 2.11. The critical value with 95% confidence is $t_{\infty,.05} = 1.96$, and it can be seen in Figure 3.22 that the significant effects are X_1, X_2, X_3, X_1X_2, X_1X_3, X_2X_3, and $X_1X_2X_3$. The moisture content of the stack gas (X_4) is not important, nor does it have any significant interactions with other factors.

3.9.4 Interpretation of Results

The results of the experiment show that there is not enough evidence to indicate that the moisture content of the stack gas had any effect on NO_2 production. This result was a big surprise, and it meant that wet scrubbing could be used if desired. All three of the other variables ($X_1 \equiv SO_2$ concentration, $X_2 \equiv$ temperature, and $X_3 \equiv O_2$ concentration) do affect the amount of NO_2 formed. However, the highest order interaction that is significant, namely $X_1X_2X_3$, must take precedence in interpreting the effects. The three-way interaction indicates there is a three-way dependence in the way X_1, X_2, and X_3 affect NO_2 formation, and neither the average main effects nor even the two-way interactions should be interpreted alone. In most circumstances it will be extremely rare to discover significant interactions of order higher than two. But in unusual cases like this one where they are significant, the simple cube plots like Figure 3.21(c) can help to make the interpretation straightforward. This figure shows that NO_2 levels (or to be exact, $\log(NO_2)$) increase only when both SO_2 and O_2 concentrations increase, and NO_2 increases most when all three variables (SO_2 and O_2 concentrations and temperature) simultaneously increase. In other words, $\log(NO_2)$ is essentially constant except when there is high level of SO_2 and a high level of O_2, and in that case increasing temperature also increases NO_2 production. This would most likely be the condition causing a brown plume of NO_2 to be emitted from the stack. Notice that this result would never have been discovered using the classical one-at-a-time approach to experimentation, wherein each variable would have been separately varied between low and high conditions while the other variables were held constant at their low level.

In presenting the results, the fact that interactions are significant makes life more difficult. It means that more is going on than can adequately be described by the average effect of each variable. Therefore, our final results should be summarized and displayed graphically as was done in Figure 3.17 for the fly ash experiment and Figure 3.21(c) for the gas treatment experiment, or through use of a prediction equation that will be described in the next chapter.

3.10   Judging Significance of Effects When There Are No Replicates

In the gas treatment experiment presented in Section 3.9, there were no replicate experiments, but there was a known estimate of σ that was used as a yardstick to judge the significance of effects and interactions. This was a special and unusual case, however. Typically there is no known estimate of σ, and in experiments without replicate runs the signal-to-noise ratios cannot be computed to judge the significance of effects. We might seem to be at an impasse, but there are still two options available to us if enough runs were made (at least 16).

The first option is to examine the interactions of third order or higher. These are extremely unlikely to be real, and hence they can be used to estimate s_E directly. If we have q of these interactions, then the formula for estimating s_E is:

$$s_E = \sqrt{\sum_{i=1}^{q} I_i^2/q}. \tag{3.7}$$

Note that we did not lose a degree of freedom for estimating the mean of our data, because we assume that the mean is zero. Once we have s_E, we can compute the signal-to-noise ratios (i.e., t-statistics), and then judge the significance of the effects and two-factor interactions. The only thing that you need to be careful of is the assumption that all the interactions of third order or higher are not significant. If one of them is large, it would inflate the estimate of s_E, and increase the chance of missing an important effect. If one of them does look large, as is the case in the gas treatment example, use the approach described next.

An alternative that makes no assumptions as to which interactions are real, and, in fact, gets away from computing signal-to-noise ratios entirely is to use graphical methods to judge the significance of effects. The approach is as follows. If none of the effects are actually significant, all of the observations would be random replicates. The factorial main effects and interactions would be differences of averages of random data, and by the Central Limit Theorem discussed in Chapter 2, they would be approximately Normally distributed with a mean of zero. Therefore, a Normal probability plot of the effects would appear as a straight line of points, like that described for Normal data in Section 2.5.4. If any effects are significant, they would stick out from the line of points as an outlier in a sample of Normal data. This method works quite well as long as we have a large amount of data ($n \geq 16$) and the majority of effects are unimportant and fall along a straight line.

To illustrate the idea, we will use the effects from the gas treatment experiment and assume no known estimate of σ was available. Table 3.8 shows the ranked effects (from Figure 3.22) along with their expected ranked values or calculated z-scores. Figure 3.23 is a Normal plot of the effects (z-scores or Normal scores on the y-axis and ordered effects on the x-axis. From the graph, the effects that really stick out like outliers (namely X_1, X_2, X_3, X_1X_2, X_1X_3, X_2X_3, $X_1X_2X_3$.) are labeled. They are the same effects that were found to be significant using the signal-to-noise t-ratios in Figure 3.22. Drawing a straight line through the first eight pairs $(E_{(i)}, z_i)$ in Figure 3.23 helps in the determination.

The reciprocal of the slope, the straight line drawn on the Normal plot, is a direct estimate of the standard error of an effect (see Chapter 2), $s_E = 0.058$. This is actually very close to the $\sigma_E = 0.05$ that was determined from the known standard deviation. Thus, when no replicates are present in an experiment, we may still be able to get a graphical estimate of σ_E. (Note: Since we have already determined which effects and interactions are

TABLE 3.8
Worksheet for Calculation of z-Scores of Effects for Gas Treatment Example

Order Number (i)	Ordered Effect E_i	Effect Label	Percentile $p_i=(i-1/2)/15$	Z-score $z_i = \Phi^{-1}(p_i)$
1	−0.044	X_1X_4	0.033	−1.83
2	−0.036	$X_1X_2X_4$	0.100	−1.28
3	−0.009	X_2X_4	0.167	−0.97
4	0.009	$X_2X_3X_4$	0.233	−0.73
5	0.014	$X_1X_3X_4$	0.300	−0.52
6	0.021	$X_1X_2X_3X_4$	0.367	−0.34
7	0.039	X_4	0.433	−0.17
8	0.056	X_3X_4	0.500	0.00
9	0.156	X_2X_3	0.567	0.17
10	0.171	X_1X_2	0.633	0.34
11	0.274	$X_1X_2X_3$	0.700	0.52
12	0.304	X_2	0.767	0.73
13	1.209	X_3	0.833	0.97
14	1.309	X_1	0.900	1.28
15	1.341	X_1X_3	0.967	1.83

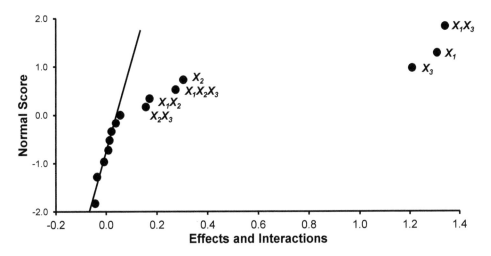

FIGURE 3.23: Normal Probability Plot of Effects for Gas Treatment Example

significant from the graph, we are obtaining s_E for the sake of curiosity only. We are not going to use it for anything. We could, however, if we were interested, use it to estimate σ from the relationship: $\sigma = (1/2)\sigma_E\sqrt{n_F}$.

An even simpler graphical approach to determining the significant effects is to make a bar chart of the effects, ordered by their absolute value, as shown below in Figure 3.24. This style plot was suggested by Lenth [1989]. The dotted ME lines on the plot represent the margin of error. The margin of error is computed as $\pm K \times PSE$ where PSE, the pseudo-standard error, is given by:

$$PSE = 1.5 \times \text{median}(|E_i| : |E_i| < 2.5s_0) = 0.05625, \qquad (3.8)$$

where E_i is the ith effect in Table 3.9, and

$$s_0 = 1.5 \times \text{median}(|E_i|) = 1.5 \times (0.056) = .084. \qquad (3.9)$$

TABLE 3.9
Effects Ordered by
Their Absolute Value

Factor(s)	Effect
X_2X_4	-0.009
$X_2X_3X_4$	0.009
$X_1X_3X_4$	0.014
$X_1X_2X_3X_4$	0.021
$X_1X_2X_4$	-0.036
X_4	0.039
X_1X_4	-0.044
X_3X_4	0.056
X_2X_3	0.156
X_1X_2	0.171
$X_1X_2X_3$	0.274
X_2	0.304
X_3	1.209
X_1	1.309
X_1X_3	1.341

The multipliers K, needed to get the margin of error (ME) limits, are given in Table A.6 in Appendix A, where they are indexed by m, the number of effects calculated. They are smoothed values from the table given by Lenth. The table also gives multipliers for the simultaneous margin of error (SME). Comparing each effect to the margin of error can increase the chance that an effect is determined to be significant purely due to random error. Comparing the effects to the simultaneous margin of error reduces the probability of a type I error. For $m = 15$ effects, as in the example above, $K = 2.157$ for ME, and $K = 4.207$ for SME.

In Figure 3.24 we can separate the vital few important effects from the trivial many unimportant effects. The effects whose bars protrude beyond the margin of error lines would be considered significant, and again we can see that the most important effects are: X_1X_3, X_1, X_3, X_2, $X_1X_2X_3$, X_1X_2, and X_2X_3.

Figure 3.25 shows the same effects compared to the simultaneous margin of error limits. In this plot we see that X_1X_2 and X_2X_3 do not protrude beyond the simultaneous margin of error lines, and therefore could only be considered marginally significant using both ME and SME to judge.

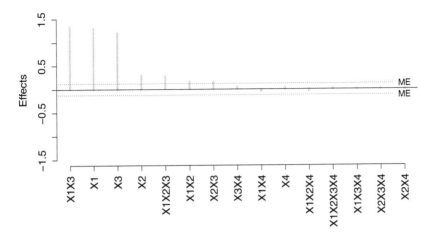

FIGURE 3.24: Lenth Plot of Effects for Gas Treatment Example with ME Limits

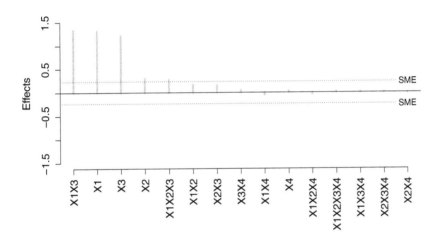

FIGURE 3.25: Lenth Plot of Effects for Gas Treatment Example with SME Limits

3.11 Summary

3.11.1 Analysis of 2^k Experiments

Once data are collected, they must be analyzed. We showed two examples of analysis and interpretation of results. A simple flow diagram is given in Figure 3.26 to help in remembering the various methods of analysis (both graphical and numerical), as well as when and where to use each. In the figure, we see the three steps that are usually followed in the analysis of a 2^k experiment. First is the exploratory stage where we examine the data graphically to see roughly what effects and interactions may be important. In the exploratory step we also check the data to see whether the assumption of equal variance at all experimental conditions appears reasonable. The second step in the analysis is the data summary step. Here we calculate the effects and interactions and variability summaries. We test the significance of the effects using either the signal-to-noise ratios if there are replications in the data, or the Normal plot of effects (or Pareto Diagram of absolute effects) if there are no replicates. The final step of the analysis is the interpretation and presentation of results. In this step, verbal descriptions of the effects and interactions are made. These verbal descriptions are usually improved if simple graphical displays like main effect plots and interaction plots are made and described.

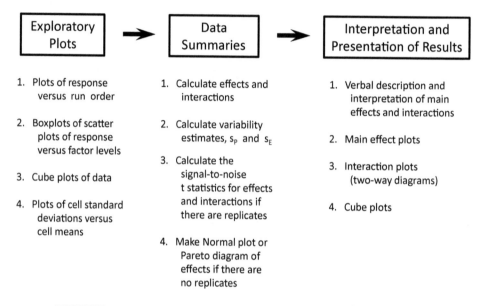

FIGURE 3.26: Flow Diagram for the Analysis of a 2^k Experiment

3.11.2   Important Equations

Effect or Interaction $= \bar{Y}_+ - \bar{Y}_-$

$s_E = 2s_p/\sqrt{n_F}$

$t_E = \text{Effect}/s_E$

$s_p^2 = \sum_{j=1}^{m} \nu_j s_j^2 / \sum_{j=1}^{m} \nu_j$

3.11.3   Important Terms and Concepts

The following is a list of the important terms and concepts covered in this chapter.

two-level factorial design
main effect
interaction
computation table
randomization
cell means

cell standard deviations
pooled variance and pooled standard deviation
standard deviation of an effect
signal-to-noise ratio
main effect plots
two-way plot (for interactions) or interaction plot
cube plot
significance of effects when there is no replication

3.12 Exercise

1. The cutting of small, square, metal plates (used to make square nuts) is done using a shearing process. The variables in the process include shearing pressure, shearing blade width, and the temperature of the metal to be cut. The response is the crispness of the cut, rated on a scale of 1 to 10. The measured (reported) crispness is an average crispness of 10 sheared plates. A process engineer decided to study the system to find the settings of the variables that would maximize crispness. The data are listed in the Table 3.10 below.

TABLE 3.10
Data to Study Effect of System Settings on Shearing of Metal Plates

Run Order	X_1 Pressure $- = 200$ $+ = 400$	X_2 Width $- = 0.2"$ $+ = 1.0"$	X_3 Temperature $- = 300$ °F $+ = 500$ °F	Crispness
7	−	−	−	3.2
1	+	−	−	8.1
5	−	+	−	3.8
10	+	+	−	8.9
2	−	−	+	3.9
3	+	−	+	6.7
6	−	+	+	8.0
8	+	+	+	9.4
4	−	−	−	4.1
14	+	−	−	9.1
12	−	+	−	4.6
15	+	+	−	9.1
11	−	−	+	3.6
16	+	−	+	7.4
13	−	+	+	8.4
9	+	+	+	9.9

(a) Plot the response versus run and comment on what you see.

(b) Calculate the mean and variance of the replicate responses for each run, and calculate the pooled standard deviation.

(c) Make a plot of cell standard deviations versus cell means to see if the equal variance assumption is justified.

(d) Calculate the main effects, two-way interaction effects, and the three-way interaction effects.

(e) Make boxplots of the response at each level of the three factors, and comment on what you see.

(f) Calculate the standard error of an effect, s_E, and determine which effects and interactions are significant at the 95% confidence level.

(g) Graph any significant interactions and write a sentence or two interpreting each significant effect or interaction.

(h) What conditions would maximize crispness?

Chapter 4

Additional Tools for Two-Level Factorials

4.1 Introduction

The last chapter dealt with the basics of two-level factorial designs, including design geometry, main effect and interaction estimation, significance testing, and checking for significance if there are no repeat runs. On the surface, that might seem to be all you need to know in order to charge ahead and run one of these designs. But, in reality, there are a number of issues that must still be addressed in almost any practical application of two-level factorial designs. Some of these issues are: (a) Will the number of runs from the full 2^k design give the accuracy needed? (b) How do you obtain a linear prediction equation from the effects and interactions that were calculated, and how do you check for the adequacy of that model? (c) What do you do with known background or nuisance variables that might influence the results but are not really of interest, like batch of raw material? (d) What can be done when the order of experimentation is difficult or impossible to randomize? These very important questions will be dealt with in this chapter.

4.2 Number of Replicates Needed for Desired Precision

Although factorial designs give estimates of the effects of factors which are more precise than any other experimentation strategy, it may still be necessary to replicate the design to obtain the precision that is desired.

Remember our basic assumption when we examine the results of our experiments: If a measured effect could reasonably have occurred as a result of the variability in our measurements, then we say the variable had no effect at all (or, more precisely, no statistically significant effect). We don't want this to happen if the effect is important. In other words, we don't want to end up ignoring any factor if its effect is *economically* significant. So, to make sure that the risk of that oversight happening is small, we use one of the tools in our statistics arsenal to increase the precision of our results—replication.

To facilitate our discussion, let us use the symbol, δ, to denote the size of the smallest effect that we do not want to overlook. (Of course, any effects that are bigger are easier to "see.") And if s_P, as usual, is the estimate of the standard deviation of a single observation, then the number of factorial runs needed to "do the job" is given by

$$n_F^* = (8s_P/\delta)^2. \tag{4.1}$$

If the number of factorial points, 2^k, is less than the number of runs needed for the desired precision, n^*_F, then the whole factorial design should be replicated enough times (denote the number of replicates by r) to get n_F at least as large as n^*_F ($n_F = r \times 2^k$).

This will guarantee that the risk of overlooking an effect of size δ is less than 10% (usually closer to 5%), and will allow for testing effects at the 95% confidence or significance level.

Equation 4.1 is given without justification for the time being. So it is important to examine it for reasonableness. It should first be noted that as s_P increases, it requires more experiments for a given accuracy. That makes sense. It should next be noted that it requires more and more experiments to pick out smaller and smaller δ's; this is also as one would expect. So, at least the direction of the impacts of s_P and δ are correct.

When δ is about the same size as s_P, 64 experiments are required. Generally speaking, precision greater than that is impractical (unless experiments are quite inexpensive). One should also be aware that it does not matter whether the experiments come from a large factorial or a smaller, replicated factorial; the precision will be the same. So if 32 experiments are required for adequate precision, and the number of factors to be studied is k = 4, then the 16 run (= 2^4) full factorial must be repeated. Alternatively, another factor can be added to the study, and a 32 run (= 2^5) full factorial could be run. The precision will be the same for the two designs: it depends only on the total number of factorial runs carried out, not on how many factors are studied.

Equation 4.1 is extremely important, and it should always be used before beginning an experimental program of any reasonable size. In the form above, it gives the "price tag," n_F, associated with any particular precision, δ, desired. But in practice it is often used (very appropriately) "backwards," as given in Equation 4.2:

$$\delta = 8 s_P / \sqrt{n_F}. \tag{4.2}$$

In this form, the equation gives the precision you can "buy" for a given budget, n_F. If the precision is not good enough, a decision must then be made to either increase the budget (n_F), or to scrap the whole experimental program (or to look for a way to decrease σ, which is not a statistical problem). Scrapping the program is not a desirable thing to do, but it is better than spending the money (running the experiments) and not being able to draw any conclusions from the data, because the variability was too great.

Often students ask, "Where does δ come from?" The answer is that it is decided on a case-by-case basis from the economics of the situation. Although you would always love to have infinite precision (i.e., $\delta = 0$) so that any factor with a non-zero effect on the response would be found to be statistically significant, this is not practical. So δ is then ultimately decided by the sponsor of the experiments (whoever is paying the bills). They must decide how much accuracy is affordable by the use of Equation 4.2.

Before leaving this topic, let us discuss briefly how Equation 4.1 was derived. In order to understand it, we must remember what we are going to do with the effects that we calculate—namely, we are always going to test them for statistical significance. There is some value, denoted by $E^* = t_{\nu,\alpha} \times s_E$, which we feel is the largest an experimentally determined effect can be just due to chance (when the true effect is zero). That critical value, E^*, depends, of course, on the variability in our measured effects, s_E. Therefore, any effects that we measure to be greater than E^*, we will call real. In other words, we will say that the factors are **significant**. (We should really say that the factors are **statistically significant**.) We must keep in mind that we may be wrong. The true effect of the factor may be zero, but the random errors happen to give us an apparent effect. The chance of being wrong is known (in statistical jargon) as the α-risk, which was discussed in more detail in Chapter 2. The amount of α-risk we take is totally under our control, since we decide how big E^* is. The bigger an effect has to be before we are willing to call it real (i.e., the larger the value of E^* we select), the smaller our α-risk. This is shown in Figure 4.1 (A).

The critical thing to keep in mind is that we can control E^* (for a given α-risk) by the

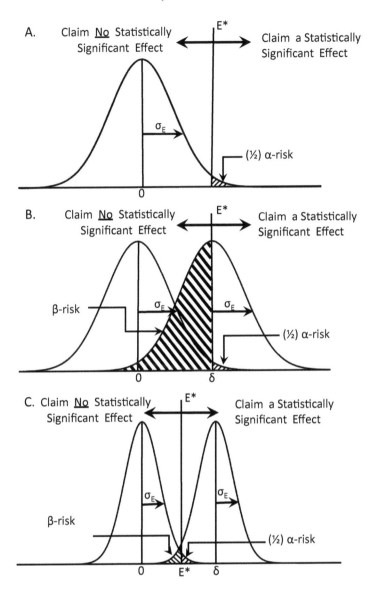

FIGURE 4.1: Replication Needed for Adequate Accuracy: (A) Control α-risk by selection of E*, (B) E* at δ gives β-risk of 50%, (C) E* well within δ gives acceptable β-risk

selection of the number of experiments we run (which, in turn, determines s_E). How small should E* be relative to δ? The first response most people would make is that E* and δ should be the same. This seems to make a lot of sense since δ is the size of an effect we do not want to overlook, and when we measure an effect of size E* we will say it is real. The problem is that we "never" measure the true effect of a factor. It is always clouded by experimental error. So sometimes what we measure will be less than δ and sometimes it will be more, as shown in Figure 4.1(B). The chance of it being less than E* (in this case, less than δ) is the risk we run of calling the factor unimportant. We should say "the effect is not statistically different from zero." That risk is known, in statistical jargon, as β-risk (which was also discussed in more detail in Chapter 2). To repeat, β-risk is the risk of missing an important effect (like a doctor failing to detect that a person is sick, and telling them they are fine). When E* = δ, the β-risk is 50%, which is usually considered much too high. In order to get the β-risk down to an acceptable level, E* must be a good bit smaller than δ; just how much smaller depends on exactly what level of β-risk is acceptable (see Figure 4.1(C)). Again, we can make E* as small as we want because we are able to make the standard deviation of an effect, s_E, as small as we want by taking enough data; the more data, the smaller s_E will be.

If α-risk is selected to be 5%, and we have a Normal Distribution for the measured effects, then E* = 1.96 σ_E. If β-risk is also set to 5%, then δ = E* + 1.64 σ_E. (It is different from the α-risk distance because only one side of the distribution is relevant.) Combining the two relationships gives us:

$$\delta = 1.96\sigma_E + 1.64\sigma_E = 3.60\sigma_E.$$

Earlier in the previous chapter (Section 3.8.1) we developed the relationship between the standard deviation of an effect and the number of factorial points (Equation 3.2), which is repeated here for convenience:

$$\sigma_E = 2\sigma/\sqrt{n_F}.$$

As previously mentioned, σ is the experimental variability of a single data point. By inserting this relationship into the previous equation for δ, we can express δ as a function of n_F.

$$\delta = 3.60\sigma_E = 3.60(2\sigma/\sqrt{n_F}) = 7.2\sigma/\sqrt{n_F}.$$

Rearranging this equation gives us the result we were looking for,

$$n_F = (7.2\sigma/\delta)^2,$$

which is valid for a Normal Distribution. Since we do not usually have a Normal Distribution (because we do not usually know σ), but rather we typically have a t-Distribution, the constant, 7.2, must increase a bit. To be precise about how much it should be increased would require taking into account how well we will know σ as estimated by s_P (i.e., how many "degrees of freedom" we will have in our estimate, s_P). A value of 8.0 rather than 7.2 is recommended by Wheeler [1974] as being reasonable in most instances. Therefore,

$$n_F = (8\ s_P/\delta)^2$$

for a t-Distribution. This is Equation 4.1.

To illustrate how this equation is used, we will present two examples. In the first example an experiment is being planned to study the purification process for a crystalline product. In the second example we will revisit the stack gas treatment problem from Section 3.9.

Example 1: In the purification process for a crystalline product, the product is first dissolved in a solvent. Next a second liquid is added to the solution and stirred in. Finally the purified crystalline product is precipitated out by steam distillation. In the plans for experimentation, it was decided to vary the temperature of the dissolution solvent, the type of liquid added at the second stage, the stirring time, and the steam distillation time. Two levels were chosen for each of the four factors, and a 2^4 factorial experiment was planned. Based on the dollar value of the purified product, it was determined that it would be important to detect any main effect or interaction that was 0.05 or more. Therefore, in this example the size of an important effect $\delta = 0.05$. From past history of the purification process, the standard deviation in final crystalline purity is expected to be $s_P = 0.02$. This was an estimate based on run-to-run variability with the factor levels constant. The number of factorial points needed was calculated, using Equation 4.1, to be:

$$n_F^* = [8\ (0.02)/(0.05)]^2 = 10.24.$$

Since 10.24 is less than $2^4 = 16$, the number of runs in one replicate of the design gave more than adequate precision, and therefore $r = 1$. In this case, with one replicate of the factorial experiment giving us more precision than required, we might ask the reverse question: "How small an effect can we reasonably expect to detect with our budget of $n_F = 16$ experiments?" Using Equation 4.2, we calculate:

$$\delta = 8\ s_P/\sqrt{n_F} = 8(0.02)/\sqrt{16} = 0.04,$$

which is even better than required, as expected.

Example 2: Recall that in the stack gas treatment experiment from Section 3.9, the variability in measured response (NO_2) was expressed as a constant percentage rather than a constant. Therefore the log transformation was used before the data were analyzed. In order to determine how much data are needed for this situation, we must agree on what precision is desired on the percent scale, then use the log transformation before applying Equation 4.1 or 4.2.

In this example, it was decided that any factors that affected the NO_2 concentration by 50% or more were important to detect. Therefore, $\delta = \log(1.5) = 0.18$. The standard deviation, on the percentage scale, was known to be 25%, so $\sigma = \log(1.25) = 0.10$. Thus, the number of factorial points needed is $n_F = [7.2\ (0.10)/(0.18)]^2 = 16$. How fortunate! In this extremely unusual example, the number of runs needed for precision was exactly one replicate of our 2^4 design! Notice that Equation 4.1 used to calculate n_F was modified (that is, a constant of 7.2 was used instead of 8) because we knew the value of σ, which is fairly unusual. If we were not confident that we knew σ, then we ran fewer runs than necessary. If we decided not to trust our "known" value of σ and use the t-distribution in our data analysis instead of the Normal distribution, the δ to be expected with 16 runs is:

$$\delta = 8\ s_P/\sqrt{n_F} = 8(0.10)/\sqrt{16} = 0.20.$$

This corresponds to an error in NO_2 concentration of $\log^{-1}(0.20) = 10^{0.20} = 1.59$, or an increase in NO_2 concentration of 59%. This is not greatly different from the desired detectable increase of 50%, so we are satisfied.

4.3   Results in Equation Form

In order to use data from a factorial experiment to predict the response at combinations of factor settings that have not been run in the experiment, it is necessary to develop a mathematical model (also called a prediction equation) for the response. The mathematical model which is implicit in the 2^k factorial design is:

$$\hat{Y} = b_0 \qquad\qquad\qquad\qquad \leftarrow \text{ constant term} \qquad (4.3)$$
$$+ b_1 X_1 + b_2 X_2 + \ldots + b_k X_k \qquad\qquad \leftarrow \text{ linear terms}$$
$$+ b_{12} X_1 X_2 + b_{13} X_1 X_3 + \ldots + b_{k-1,k} X_{k-1} X_k \quad \leftarrow \text{ cross-product terms}$$
$$+ b_{123} X_1 X_2 X_3 + \ldots \qquad\qquad \leftarrow \text{ higher order terms terms}$$

where $\hat{Y}$ denotes the predicted value of Y, and X_i is the coded value for factor i. For qualitative factors (like operator or machine) X_i can only take on the values of -1 or $+1$. For quantitative factors, X_i can take on any value, but it should be kept in mind that extrapolating outside of the range of -1 to $+1$ is very risky. The coding of X_i for quantitative factors is:

$$X_i = \left[\frac{\text{factor value} - \text{center}}{\text{high factor value} - \text{center}} \right] \text{ and} \qquad (4.4)$$
$$center = \left[\frac{\text{high factor value} + \text{low factor value}}{2} \right].$$

The unknown constants in the equation are readily obtainable from the factorial effects that we have already calculated.

$$b_0 = \bar{Y} \qquad\qquad\qquad\qquad\qquad\qquad (4.5)$$
$$b_i = [Effect\ of\ Factor\ X_i]/2$$
$$b_{ij} = [Interaction\ of\ Factors\ X_i X_j]/2$$
$$b_{ijk} = [Interaction\ of\ Factors\ X_i X_j X_k]/2$$

The coefficients are half the value of the effects because coefficients represent the slope or change in Y for a one-unit change in X (e.g., from $X = 0$ to $X = 1$), whereas the effects represent the change in Y for a two-unit change in X (from -1 to $+1$).

Normally when writing the prediction equation for a factorial design, *we only include terms in the model for effects and interactions that were found to be statistically significant.* By doing this we average or smooth away noise in predictions. If we include all possible terms in the model, the prediction equation will simply give us back the average of the data at each of the factorial points with no smoothing. This is not helpful.

To see an example of this consider the fly ash experiment presented in Section 3.7. The prediction equation for this experiment can be written as:

$$\hat{Y} = 39.81 + \left[\frac{12.23}{2} \right] X_1 + \left[\frac{-8.48}{2} \right] X_2 + \left[\frac{-4.78}{2} \right] X_3 + \left[\frac{6.13}{2} \right] X_1 X_2$$
$$\hat{Y} = 39.81 + 6.11 X_1 - 4.24 X_2 - 2.39 X_3 + 3.06 X_1 X_2$$

where 39.81 is the mean of the data, and the rest of the coefficients are half of the statistically significant main effects and interactions. As an aside, if it turns out that there is a significant interaction, all lower order terms involving those factors are also included in the model (by most analysts) whether they are significant or not. This is not an issue for the fly ash example, but it does come up occasionally.

It could be even more meaningful to write the equation in terms of the actual factors, rather than as the coded X's. For the fly ash example, even fly ash source, which is a qualitative factor, can be written in this way as the percentage of fly ash source B. When fly ash source B is less than 100%, fly ash source A makes up the rest of the mixture. Of course, the percentages put into the equation must lie between 0% and 100%. The factors are:

$X_1 = (\%\ \text{Source B} - 50\%)/50\%$
$X_2 = (\text{Water/Cement Ratio} - 0.45)/0.10$, and
$X_3 = (\text{Set-up Temperature} - 20)/10$.

The equation in uncoded factors becomes:

$$\hat{Y} = 39.81 + 6.11\left[\frac{\%B - 50\%}{50\%}\right] - 4.24\left[\frac{R - 0.45}{0.10}\right] - 2.39\left[\frac{T - 20}{10}\right] + 3.06\left[\frac{\%B - 50\%}{50\%}\right]\left[\frac{T - 20}{10}\right]$$

In this way predictions of the compressive strength at any combination of fly ash source, water/cement ratio (R), and set-up temperature (T) can be made by substituting the values directly into the equation. For example the predicted strength at 100%B, R = 0.55, and T = 30 would be:

$$\hat{Y} = 39.81 + 6.11\left[\frac{100\% - 50\%}{50\%}\right] - 4.24\left[\frac{0.55 - 0.45}{0.10}\right] - 2.39\left[\frac{30 - 20}{10}\right]$$
$$+ 3.06\left[\frac{100\% - 50\%}{50\%}\right]\left[\frac{30 - 20}{10}\right],$$

or $\hat{Y} = 39.81 + 6.11(1) - 4.24(1) - 2.39(1) + 3.06(1)(1) = 42.35$.

Note that the data at those conditions had an average strength of 41.8 mPa, but the equation gives a better prediction of future runs at those conditions because of the smoothing involved in developing the coefficients in the equation.

Another way of writing the prediction equation would be to multiply through the brackets and collect the constants and multipliers. However, in this form the constants in the model lose their direct interpretation as the grand average and half effects for the experiment. Also, it becomes less obvious if the equation is being used to predict the response inside the region of experimentation (with X values between -1 and $+1$, which is fine) or outside the region of experimentation (X values greater than 1 in absolute value, which is very tenuous). For this reason, the first form of the model is generally preferred.

The prediction equation for the fly ash experiment is shown graphically in Figure 4.2. Here we can see the actual data from the experiment (represented as black dots) and the predictions (represented as the wire-frame surface). This 3-D graph gives us a good feeling for how well the prediction model actually predicts the data. A prediction, made at a certain combination of fly ash source, water/cement ratio, and set-up temperature, will represent our estimate of the long run average compressive strength at those conditions. We can expect individual strength readings at the same conditions to vary from our prediction, just as the experimental data points in Figure 4.2 differ from the wire-frame prediction surface. Since the pooled standard deviation, s_p, is a measure of the variability of individual response values from their long run average, we can use it to get a rough idea how far a single strength reading will vary from a predicted value. Assuming a Normal Distribution of individual responses at the same conditions, we could roughly say that we are 95% sure that a single measurement should be within $\hat{Y} \pm 2s_p$. More exact formulas for confidence intervals on predictions will be given in Chapter 8.

To verify the validity of a mathematical model and predictions derived from it, we should always check the assumptions upon which the model is based. The assumptions made for the prediction equation derived from a factorial experiment are: (1) linearity of the underlying relationship, (2) equal variance of response throughout the experimental region, and (3) a Normal Distribution of the responses around the true relationship. A rough check of all of these assumptions can be made by looking at the relative position of the data points to the

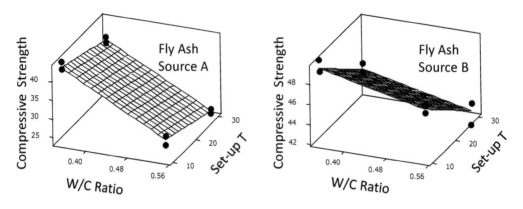

FIGURE 4.2: 3-D Graph of Prediction Equation for Fly Ash Experiment

prediction surface in a picture like Figure 4.2. But, this is a simple situation with only two quantitative factors and one discrete factor. In more complicated situations, it is impossible to make a figure like Figure 4.2. Instead, various plots of residuals must be made. The residuals are the differences between the actual observed response data and their predicted values from the equation. Chapter 8 will show examples of calculating the residuals and plotting them to check the validity of a prediction equation. Another (more accurate) way of checking the linearity of the underlying relationship will be described in the next section.

Another common way of displaying the prediction equation is a contour plot. Contour plots are similar to a topographic map. In a contour plot, two factors are shown on the x and y-axes and different contour lines represent the level of the predicted response. If there are more than two factors in the prediction equation, the most important factors are put on the axes and multiple plots are made at different levels of the less important factors. Figure 4.3 is an example of contour plots for the fly ash prediction equation.

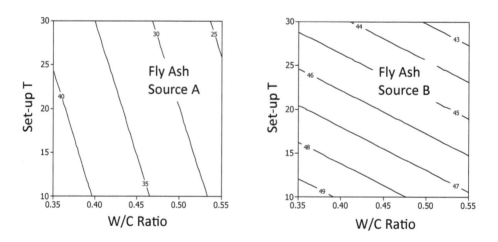

FIGURE 4.3: Contour Plots of Prediction Equation for Fly Ash Experiment

4.4 Testing for Curvature

Replicate experiments at a center point (as shown in Figure 4.4 for one, two, and three factors) are a valuable adjunct to a two-level factorial design for two reasons: (1) they provide additional replicates from which an estimate of error can be calculated, and (2) they provide a check of the adequacy of the linear prediction model. Both of these are extremely important benefits. Since there is so much to be gained from the center points, many factorial designs used in practice do incorporate center points in the total experimental design. Let's talk about both benefits now in a bit more detail.

The center points provide us with a "pure" estimate of error, since they are replicated points. Without center points (or replication at all the design points), the only way an estimate of error could be obtained would be by assuming some higher order interactions are zero. If the interactions are not zero, then our estimate of error is inflated. This, in turn, inflates our estimate of s_E, which causes our calculated t-values to be too small, which could result in some of our significant effects or interactions being labeled insignificant. Thus, without the estimate of error we get from the replicated design points, we could easily overlook some important effects.

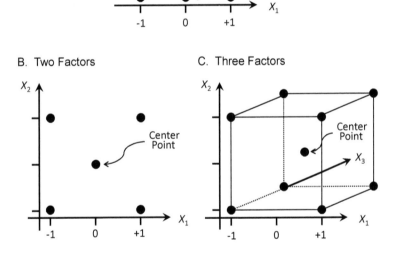

FIGURE 4.4: Geometry of Center Points in Two-Level Factorial Designs

The analysis of data from a factorial design also assumes that a linear model adequately describes the changes in the response resulting from changes in the independent variables. With only two levels of each factor, there is no way to check the validity of this assumption. You cannot, after all, fit anything more complicated than a straight line through two points! If there are nonlinearities (curvature), the factorial design will yield poor predictions except near the actual design points (corners of the cube). Yet, you would like to make predictions about the response throughout the experimental region. How can you tell if you are justified in doing that? Center points come to our rescue. Center points allow us to check for curvature by seeing if there is a big difference between the actual response at the center point and the value that we expect (or predict) at those middle conditions if a linear model is adequate. This is shown graphically in Figure 4.5 for one factor. It should be noted that we are demonstrating curvature with one factor because it is the easiest situation to visualize, although it is not of great importance in practice. However, the concept generalizes readily to more factors. For example, curvature is also shown in Figure 4.6 for two factors.

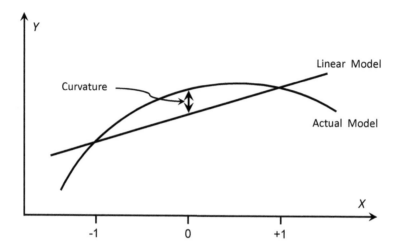

FIGURE 4.5: Curvature in Two-Level Factorial Design with One Factor

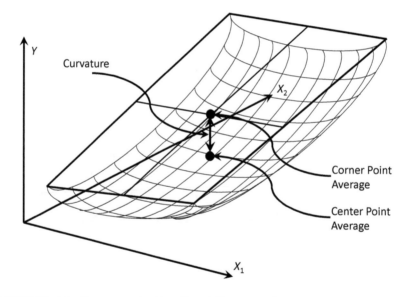

FIGURE 4.6: Curvature in Two-Level Factorial Design with Two Factors

To simplify our discussion, let us use the term, *curvature*, to denote specifically the difference between the actual response in the middle and the linear prediction, and let us give it the symbol, C. Thus:

C = Actual response at the center
    − Predicted (via linear model) response at the center

The actual response at the center is estimated by averaging the center points. The prediction at the center using a straight line model through the factorial points is obtained very easily by just averaging the factorial (corner) points. That this procedure works is obvious in the case of only one factor, but it is also true for any number of factors. So the computation of C becomes:

$$C = \overline{Y}_{Center} - \overline{Y}_{Factorial}. \tag{4.6}$$

When we calculate C, which we will call curvature for simplicity, it will almost never be exactly zero (due to random error), even when a linear model is perfect. So before we can say that we do or do not have curvature, we must test for the statistical significance of the calculated curvature. In order to do that, we must know how much variability there is in the curvature estimate, which is quantified by its standard deviation, s_C. This s_C is easily determined via the same procedure as we used for s_E in Section 3.8. Since C is the difference between two averages, its standard deviation is given by:

$$s_C = \sqrt{s_P^2 \left(\frac{1}{n_C} + \frac{1}{n_F} \right)} \tag{4.7}$$

where
n_C = the number of observations in the average response at the center, $\overline{Y}_C$, and
n_F = the number of observations in the average response at the factorial points, $\overline{Y}_F$.
It must be remembered that n_F includes replication; it is not just the number of corner conditions.

To test the statistical significance of the measured curvature, C, we could calculate the t-statistic (signal-to-noise ratio),

$$t_C = C/s_C \tag{4.8}$$

and see if it is larger than could reasonably be explained by chance alone (as compared to the $t_{\nu,\alpha}$ values in Table A.3 of Appendix A). To illustrate the test for curvature, consider the following example.

Example 3: In a chemical process one reactant is held in a vessel for cooling. When it reaches the desired temperature, a second reactant is quickly charged in an aqueous/organic mixture. Experiments were performed on this process to optimize the yield. Two factors were varied in the experiments. The first was the temperature in the vessel, and the second was the aqueous to organic ratio of the charge. The response was the yield in percent. A 2^2 factorial experiment was performed with two replicates at each of the four experimental conditions, and since both factors were quantitative, four additional replicates were made at a center point. The coded factor levels were $X_1 = \left[\frac{\text{Temperature} - 113}{9} \right]$ and $X_2 = \left[\frac{\text{Ratio} - 0.55}{0.45} \right]$ and a worksheet with the results are shown in Figure 4.7.

	A	B	C	D	E	F	G	H
	Run	Temp X_1	Aq/Org X_2	Observed Yields (Y)		Means $\bar{y}$	Variances σ^2	df
2	1	-1	-1	68.5	68.3	68.400	0.020	1
3	2	+1	-1	72.0	72.0	72.000	0.000	1
4	3	-1	+1	73.0	70.0	71.500	4.500	1
5	4	+1	+1	72.4	73.0	72.700	0.180	1
6	5	0	0	74.0	75.0	74.125	0.396	3
7				74.0	73.5			

FIGURE 4.7: Yields from Chemical Experiment and Summary Statistics

Figure 4.8 shows the average response at each of the five experimental conditions. Since the average response at the center point (74.12) appears to be higher than any of the factorial points, a linear prediction equation probably would not be appropriate for this data.

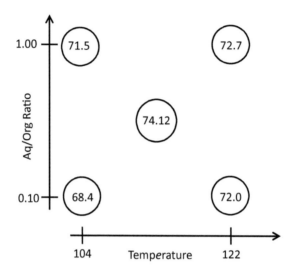

FIGURE 4.8: Average Yields for Chemical Reaction Experiment

To perform a formal test of the statistical significance of the amount of curvature we start by calculating the pooled variance. In this case we pool (or average) the variances from each of the four replicated factorial points and the replicated center point using Equation 3.5 (given in Section 3.8.2). Notice that s_p^2 is a weighted average of the five estimates of the variance, and the degrees of freedom are different at the center point than at the factorial points. The degrees of freedom for this pooled variance are the sum of the degrees of freedom for the individual variances, or 7.

$$s_P^2 = [1{\times}0.020 + 1{\times}0.000 + 1{\times}4.500 + 1{\times}0.180 + 3{\times}0.396]/[1{+}1{+}1{+}1{+}3]$$
$$= 0.841,$$

$$s_P = \sqrt{0.841} = 0.917.$$

Curvature, as defined by Equation 4.3, is:

$$C = \text{Center Point Average} - \text{Factorial Average}$$

$$C = 74.12 - [68.4 + 72.0 + 71.5 + 72.7]/4 = 74.12 - 71.15 = 2.97$$

The standard deviation or standard error of curvature, C, is:

$$s_C = s_P \sqrt{1/n_C + 1/n_F} = 0.917\sqrt{1/4 + 1/8} = 0.562,$$

and the signal-to-noise t-ratio is:

$$t_C = C/s_C = 2.97 \,/\, 0.562 = 5.30.$$

When this calculated value is compared to the critical value of the Student's t-statistic from Table A.3 with 95% confidence for 7 degrees of freedom, which is $t_{7,.05} = 2.365$, we find that there is significant curvature. When significant curvature is found, as in this example, the linear prediction equations developed in Section 4.3 will not be valid for making predictions within the experimental region. To come up with a valid interpolation equation in this case, more experiments are needed as well as techniques for fitting a quadratic model (which will be described in detail in Chapters 9 and 10).

In summary, it is worth reviewing what we can conclude after we test for significant curvature. If no significant curvature is found, our experimental work is done assuming the factors are constrained to lie within the region of experimentation! Hurrah! We can use our linear model to describe our system and arrive at an optimum. This optimum will be somewhere along the boundary of our experimental region. If the factors are not constrained, the linear model gives us a direction to go in seeking an optimum. A good strategy is to follow the path of steepest ascent (or descent if we are seeking a minimum) starting at the conditions that are predicted to give the best response. On the other hand, if significant curvature is found, our experimental work is not done. Bummer! The linear model is NOT adequate to describe our system. And we have no way of telling (without further experiments) which factor or factors is/are responsible for the curvature. To sort out what is going on, more experiments are required. Exactly which experiments need to be run next is the subject of Chapter 9. We will hold off further discussion of those experimental designs and quadratic model fitting until then.

4.5   Blocking Factorial Experiments

In cases where a known background variable can change and affect the results of an experiment, we should take that variable into account to avoid biasing the effects of the factors we are studying. We can take known background variables into account by grouping the experiments (in a factorial design) into sets of runs which have common levels of this variable. This is called blocking. For example, Table 4.1 compares a blocked and unblocked method for running two replicates of a simple 2^2 design for studying the effects of temperature and time on the yield of a chemical reaction.

In the unblocked design on the left, the order of experiments, or runs, is completely randomized and run in two days (runs 1–4 on Day 1 and runs 5–8 on Day 2). In the blocked design on the right, one complete replication of the design is run on each day or block, and the order is randomized within each block. The presumption is that uncontrolled conditions on a given day are more alike (consistent) than conditions from one day to the next.

There are three purposes for blocking:

1. First of all, we want to prevent the effects of known background variables from biasing or confusing the treatment effects of interest.

2. The second reason is to broaden the basis for conclusions.

3. And the third is to increase the precision of the treatment effects.

These concepts will be illustrated with the simple chemical reaction experiment described in Table 4.1.

TABLE 4.1

Unblocked and Blocked Factorial Designs to Study the Effects of Time and
Temperature on the Yield of a Chemical Reaction

Temp X_1	Time X_2	Run Order	Day	Yield	Temp X_1	Time X_2	Block Day	Run Order	Yield
		Unblocked Design					Blocked Design		
$-$	$-$	5	2	Y_1	$-$	$-$	1	1	Y_1
$-$	$-$	7	2	Y_2	$+$	$-$	1	3	Y_2
$+$	$-$	4	1	Y_3	$-$	$+$	1	2	Y_3
$+$	$-$	1	1	Y_4	$+$	$+$	1	4	Y_4
$-$	$+$	6	2	Y_5	$-$	$-$	2	2	Y_5
$-$	$+$	3	1	Y_6	$+$	$-$	2	3	Y_6
$+$	$+$	2	1	Y_7	$-$	$+$	2	4	Y_7
$+$	$+$	8	2	Y_8	$+$	$+$	2	1	Y_8

First, blocking prevents bias caused by known background variables. Let us say that eight runs are needed in total, but only four can be completed in one day. If we use the random run orders listed in the unblocked design, it would cause three of the four experiments at the high level of Factor 1 (temperature) to be run on Day 1, and three of the four for low temperature to be run on Day 2. If there was any change in a background or lurking variable between Day 1 and Day 2 of experimentation, the effect for temperature could be seriously biased. This bias can be completely prevented if the experiments are blocked as in the right side of Table 4.1. The block can be treated just like another factor in the experiment. The systematic randomization in the blocked design makes only the block factor confused or biased by any lurking variables that change from Day 1 to Day 2. The average effects of X_1 and X_2 are completely unconfused with blocks and any lurking variables that change from Day 1 to Day 2.

The second reason for blocking is to broaden the base for conclusions. If only the first block of experiments (Day 1) was completed, and all runs used the same raw materials (i.e., reagents from one batch), then the conclusions regarding the effects of time and temperature are only valid with respect to that particular day or batch of raw materials. However, if the second block of experiments was also completed, and used a different batch of raw materials, then the conclusions can be broadened. If the two batches of raw materials (represented by blocks or days in Table 4.1) are randomly selected from several batches available, then the conclusions regarding the effects of time and temperature are more likely to hold true for all the batches of materials that were available.

The final purpose of blocking is to increase precision. The purposes of blocking are best achieved by grouping homogeneous sets of experimental units together in blocks, and letting the characteristic of the units vary widely between blocks. By doing this, treatment effects are compared to the variability of the homogeneous experimental units within the blocks and the precision is increased. In the example in Table 4.1 the effects of time, temperature, and their interaction would be compared to the variability of batches within a day (where the same batch of raw materials were used).

The blocked experiment on the right side of Table 4.1 looks like a 2^3 experiment with the levels of the third factor (block) run in sequence rather than in a randomized order. For this reason many beginning users of experimental designs make an unknowing mistake in conducting their experiments. If they think their list of factorial experiments would be difficult or impossible to run in a completely random order, because it is difficult to change back and forth between high and low levels of one factor, then they include the hard to vary factor as a block (like the example shown in Table 4.1). However, when a factor is included as a block with no randomized sequence of changes between high and low levels, its effect will be biased by other background variables that change during the course of experiments.

If we are really not interested in a block variable (like the Day in Table 4.1) the bias will not bother us. The purpose of blocking is to give us greater precision, by removing from s_E the variability caused by the blocking variable, and to give our conclusions a broader base by repeating experiments at different levels of the blocking variable. But the purpose of blocking is not to estimate the effect of the blocking variable! If it is difficult to run experiments in completely random order, because levels of one or more factors are hard to change, a split-plot type experiment (to be explained in the next section, Section 4.6) should be used, not a blocked experiment.

Originally blocked designs were used in agricultural experimentation where the experimental units were plots of ground. Grouping similar and adjacent plots of ground together in blocks could prevent (or block out) the effects of soil fertility, etc., from biasing treatment effects or inflating the estimate of s_E. The effects of the different soil plots were not of interest in agricultural experiments, but by choosing blocks or sets of plots from widely different areas and soil types, conclusions from agricultural experiments could be more generally applied.

4.5.1 Calculating the Error of an Effect (s_E) in Blocked 2^k Experiments

Calculation of effects in blocked factorials is no different than calculating effects in the randomized factorials described earlier in this chapter. In a blocked design like the right side of Table 4.1, the main effects and interactions of X_1 and X_2 are calculated exactly as was shown in Sections 3.3 and 3.4. But, the addition of the block factor (Day) has changed this from a 2^2 experiment with replicates to a 2^3 design without replicates. Without replicates in a blocked design, the standard error of the effects cannot be computed by Equation 3.2 (in Section 3.8). We can still judge the significance of effects using graphical methods (see Section 3.10), or a more exact method using t-statistics, that will be explained next.

To employ t-statistics, we still need an estimate of the standard error of an effect, s_E. We take advantage of the premise that, in a blocked design, there should be no interactions of the block effects with the other factors in the design (otherwise we could not generalize a conclusion about a consistent effect over all blocks). Therefore, the estimates of the block interaction terms are a direct indication of the variability that can manifest itself in the effects. In other words, the interactions of the block factors should all be zero, but the actual measured interactions will exhibit scatter around zero with a standard deviation equal to the standard deviation of the other effects and interactions, s_E. Therefore, s_E can be estimated by calculating the standard deviation of the block interactions around zero (i.e., the square root of the average of the squared block interaction terms). To be more specific consider the following: The usual formulas for estimating the variance of a random variable, Y, are:

If we know μ, the mean of the distribution,

$$s^2 = \sum_{i=1}^{n}(Y_i - \mu)^2/n. \tag{4.9}$$

If we do not know μ, and need to estimate it by $\overline{Y}$,

$$s^2 = \sum_{i=1}^{n}(Y_i - \overline{Y})^2/(n-1). \tag{4.10}$$

In this case, we are not dealing with individual observations, but rather with effects (and interactions) directly. So we will not be calculating s^2 with our formula, but rather we will be calculating s_E^2. Also note that this is an example of the fairly unusual case in which we happen to know the mean of the distribution, μ, which is zero. So, we will use Equation 4.9, with blocking variable interactions (which we will denote by BI's) instead of the Y's.

$$s_E^2 = \sum_{i=1}^{m}(BI_i - 0)^2/m \tag{4.11}$$

or

$$s_E = \sqrt{\sum_{i=1}^{m}BI_i^2/m} \tag{4.12}$$

where BI_i are the block interactions, and m is the number of block interactions. Therefore, m is also the degrees of freedom our estimate of s_E . Once we have s_E, and its degrees of freedom, the t-statistics, E/s_E, can be calculated as usual, and used to test the significance of our measured effects.

4.5.2   Two Block Example

Figure 4.9 shows pictorially the apparatus for a laboratory experiment conducted to study two factors that may effect the half-life of a chemical pesticide in soil. The experimental unit was a container of soil treated with the pesticide. Each container was stored in a growth chamber at constant temperature and moisture. The sign on each container in the figure denotes the moisture level. The moisture level of each container was monitored over time, and water was added periodically to maintain a nearly constant soil moisture level. The half-life of the pesticide was determined for each container by sampling the soil weekly and analyzing for the concentration of the pesticide. Typical, exponential decay curves resulted, as shown in Figure 4.10, that allowed estimation of the half-life.

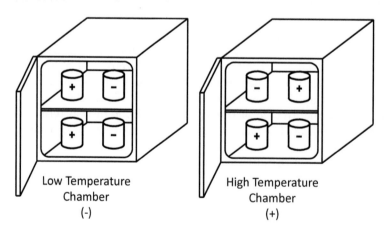

FIGURE 4.9: Growth Chambers for Pesticide Degradation Experiment

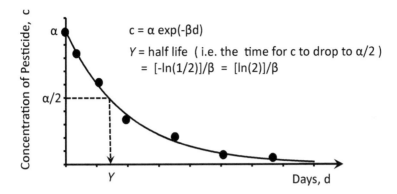

FIGURE 4.10: Typical Degradation Curve Used to Determine Half-Life of Pesticide

The temperature level was varied by having two identical growth chambers maintained at different temperatures. Half of the soil containers in each chamber were maintained at a high moisture level, and half were maintained at a low moisture level. Positions of the high and low moisture containers were randomly assigned to shelf positions in the chambers in order to avoid bias caused by any positional effects. The entire set of four experiments was duplicated simultaneously using two different soil types [1 = sandy-loam (SL), 2 = silt-clay (SC)], so that the conclusions could be extended to more than one soil type. This created a blocked design.

The experiments and results are listed in Table 4.2. The four containers of soil from each soil type were randomly assigned to receive one of the temperature and moisture combinations, and the container numbers are shown in the eighth column of the table. Whenever the response (half-life for this experiment) ranges over an order of magnitude or more (in this case 31.5 – 886) it is common to perform the analysis on the logarithm. (Incidentally, it makes no difference whether base 10 or base e is used for the logarithms.) The transformation is often necessary to ensure the validity of the two major assumptions which are invoked in the analysis of the data from a factorial design: 1) a linear model and 2) a constant variance. See Section 3.9 for another example of the use of logarithms. The half-life data are listed in the ninth column of the table, and the logarithms are listed in the tenth column.

TABLE 4.2
Blocked Factorial Design to Study Pesticide Degradation in Soil

Run No.	X_1	X_2	Block	Temp	% Moist	Soil Type	Soil Con- tainer	Half Life HL	Y ln(HL)	Effects
1	−	−	1	50°F	15%	SL	2	886	6.79	Mean=5.02
2	+	−	1	80°F	15%	SL	3	188	5.24	1 =−1.17
3	−	+	1	50°F	25%	SL	1	229	5.43	2 =−0.63
4	+	+	1	80°F	25%	SL	4	130	4.87	12 =0.08
5	-	-	2	50°F	15%	SC	3	168	5.12	B=−1.13
6	+	-	2	80°F	15%	SC	4	65.2	4.18	1B =−0.11
7	-	+	2	50°F	25%	SC	1	156	5.05	2B =0.23
8	+	+	2	80°F	25%	SC	2	31.5	3.45	12B =−0.41

A graphical summary of the response, $Y = \ln(\text{half-life})$, is shown in Figure 4.11. It can be seen that increasing temperature and increasing moisture tend to reduce the half-life (i.e. they have negative effects). To confirm what can be seen graphically, the effects were calculated and are given in the last column of Table 4.2.

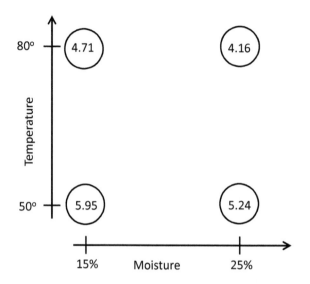

FIGURE 4.11: Graphical Analysis of Pesticide Half-Life Results

To judge the significance of the effects we calculate the variance of the effects, s_E^2, from the block interactions (X_1B, X_2B, and X_1X_2B) which are assumed to be zero. Using Equation 4.11, s_E^2 is estimated as the average of the block interactions squared. The standard deviation of an effect, s_E, is then the square root of the result, as shown below:

$$s_E = \sqrt{\sum_{i=1}^{m} BI_i^2/m} = \sqrt{[(-0.106)^2 + (0.232)^2 + (-0.410)^2]/3} = \sqrt{0.233/3},$$

$$s_E = 0.279.$$

The degrees of freedom for this estimate are 3 (the number of block interactions). The signal-to-noise t-ratios are calculated for the main effects X_1 and X_2 and their interactions, by dividing the effects by $s_E = 0.279$, as:

$$t_{X_1} = -1.116/0.279 = -4.18,$$
$$t_{X_2} = -0.630/0.279 = -2.26,$$
$$t_{X_1X_2} = 0.081/0.279 = 0.29.$$

The critical $t_{3,.05}$ from Table A.3 is 3.182 with 95% confidence level and 2.353 with 90% confidence. Therefore, there is 95% confidence that temperature had a significant effect, but less than 90% confidence that moisture had an effect.

4.5.3   Designs for Blocked Factorials

Factor effects and interactions for 2^k factorials run in blocks can always be computed using the worksheet method described earlier in this chapter. However, the block effects and block interaction effects cannot be computed by these methods unless the number of blocks in the design is a power of two (i.e., $2^1 = 2$, $2^2 = 4$, etc.). For the more general blocked factorial designs, where the number of blocks is arbitrary, the analysis and significance tests must be completed using the Analysis of Variance (ANOVA) that will be described in Chapter 5. However, there are still many useful blocked designs where the number of blocks is a power of two. Table 4.3 shows how the computation tables in Appendix B.2 can be used to create and analyze various blocked designs with this property.

TABLE 4.3
Blocked Full Factorial Designs

Fac.	Des.	No. Blks.	App. Table	Terms to Define Blocks	Terms for Calculating Block Effects	Terms Used to Calculate Error (Factor Subscripts Given in Table)
2	2^2	2	B.2-3	X_3	X_3	13, 23, 123
2	2^2	4	B.2-4	X_3, X_4	X_3, X_4, X_3X_4	13, 14, 134, 23, 24, 234 123, 124, 1234
2	2^2	8	B.2-5	X_3, X_4, X_5	$X_3, X_4, X_5, X_3X_4, X_3X_5, X_4X_5, X_3X_4X_5$	13, 14, 15, 134, 135, 145, 1345, 23, 24, 25, 234, 235, 245, 2345, 123, 124, 125, 1234, 1235, 1245, 12345
3	2^3	2	B.2-4	X_4	X_4	14, 24, 34, 124, 134, 234, 1234
3	2^3	4	B.2-5	X_4, X_5	X_4, X_5, X_4X_5	14, 15, 145, 24, 25, 245, 34, 35, 345, 124, 125, 1245, 134, 135, 1345, 234, 235, 2345, 1234, 1235, 12345
4	2^4	2	B.2-5	X_5	X_5	15, 25, 35, 45, 125, 135, 145, 235, 245, 345, 1235, 1245, 1345, 2345, 12345

To demonstrate the use of this table, consider extending the pesticide half-life experiment presented in the last section to four blocks. According to Table 4.3 a 2^2 in four blocks can be created through use of Table B.2-4, by using the X_3 and X_4 terms to define the blocks. A $(-,-)$ combination for X_3 and X_4 would be Block 1, a $(-,+)$ combination would be Block 2, a $(+,-)$ would be Block 3, and a $(+,+)$ would be Block 4. Table 4.4 shows the results of the extended experiment. The analysis would proceed by calculating the effects with use of Table B.2-4 and the standard error of an effect by the square root of the average of the squared effects of the block interactions: $X_1X_3, X_1X_4, X_1X_3X_4, X_2X_3, X_2X_4, X_2X_3X_4, X_1X_2X_3, X_1X_2X_4$, and $X_1X_2X_3X_4$. The details will be left for an exercise.

4.6 Split-Plot Designs

In many multifactorial industrial experiments, the experimental unit is different for one factor than it is for another. One of the most common causes for this difference is the inability to completely randomize the order of experiments. In some experimental situations, the levels of one or more factors may be easy to change or manipulate throughout the

TABLE 4.4

Example of a 2^2 Design with Four Blocks (pesticide degradation)

Run No.	Temp X_1	Moisture X_2	Block	X_3	X_4	Random Soil Container	Half-Life
1	−	−	1	−	−	2	886
2	+	−	1	−	−	3	188
3	−	+	1	−	−	1	229
4	+	+	1	−	−	4	130
5	−	−	2	+	−	3	168
6	+	−	2	+	−	4	65.2
7	−	+	2	+	−	1	156
8	+	+	2	+	−	2	31.5
9	−	−	3	−	+	1	254
10	+	−	3	−	+	3	65.8
11	−	+	3	−	+	4	188
12	+	+	3	−	+	2	54.7
13	−	−	4	+	+	2	241
14	+	−	4	+	+	4	75.5
15	−	+	4	+	+	3	166
16	+	+	4	+	+	1	45.2

course of experimentation, while other factor levels may be more difficult or expensive to manipulate. This fact restricts the normal practice of randomization, and in these cases a block of experiments at the low level of the difficult to vary factor are run first, followed by a block of experiments at the high level (or vice versa). When this happens, the effect of the difficult to vary factor could be seriously biased by any background variables that change during the course of experimentation. One way to reduce the chance of bias is to repeat the two blocks of experiments, while possibly reversing the order of the levels of the difficult to vary factor. When a set of experiments is performed in this way, the experimental unit for the difficult to vary factor is a block of runs, but the experimental unit for the easy to vary factors is an individual run.

Whenever the experimental unit is different for one factor than it is for another, we call the experimental design a ***split-plot experiment***. The name comes from its agricultural origin, where some factors (such as row spacing) were difficult to vary (from narrow to wide) within small plots of ground, but other factors such as seed variety or fertilizer rate could be easily varied within a small plot. The levels of a factor that are difficult to change are called ***whole-plot factors*** because their levels were assigned at random to larger blocks (or plots in the agricultural setting). On the other hand, factors whose levels were easy to change were randomized to smaller experimental units, within the larger blocks, and were called the ***split-plot factors***.

In a split-plot experiment, we must calculate a different standard error of an effect for the whole-plot factors than we will calculate for the split-plot factors. Unless the list of experiments is repeated, so that there are at least two blocks of experiments for each level of the whole-plot factor, it will not be possible to calculate the standard error of the whole-plot effect. A frequent mistake in the conduct of experiments is failure to recognize a split-plot situation, and failure to replicate the list of experiments. In this case the whole-plot factor effect is biased by changes in background variables, and no significance test can be performed on it since there is no estimate of its standard error.

In experiments with restricted randomization, the standard error for the whole-plot effects often can be quite different than the standard error for the split-plot factors. Errors can easily be made in judging the significance of effects if the two separate standard errors are not properly calculated and used in making the signal-to-noise t-ratios.

Split-plot experiments occur frequently when experiments are conducted in processes that are comprised of many steps. In these situations changing the levels of some factors involve changes at one processing step, while changes in the levels of other factors may involve changes at other processing steps. This usually results in different experimental unit sizes for different factors. The following example illustrates an example in a multi-step process. In this example, we will illustrate the proper method of replicating the whole-plot and calculating the correct standard errors for the whole and split-plot factors.

4.6.1   Split-Plot Example

In the integrated circuit industry, individual circuits, called die, are formed on a silicon wafer. Contact window forming is one of the more critical processing steps. A window is a hole of about 3 μm diameter etched through an oxide layer of about 2 μm. The purpose of the windows is to permit interconnections to the micro-circuit. For this reason the windows are called contact windows. The contact windows are formed by a photo-lithography process, as diagrammed in Figure 4.12. In this process the photo-resist is applied to spinning oxide coated silicon wafers in order to leave a uniform layer. Next, lots of wafers are baked to dry the photo-resist. The third step is to expose the photo-resist coated wafers to ultraviolet light through a mask. The light passes through the mask in the window areas and causes the photo-resist in those areas to become soluble in an appropriate solvent. After that, the wafers are dipped in a developer which removes the photo-resist in the window areas and exposes the oxide. In a high vacuum chamber, a plasma is established that etches the oxide faster than the remaining photo-resist layers. At the end of the etching process, windows are left through the oxide to the underlying silicon.

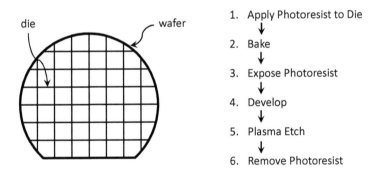

FIGURE 4.12: Contact Window Formation

A critical characteristic of contact windows is uniformity of size, and in the manufacturing process it is important to produce window sizes very near the target dimension. Windows that are not open, or are too small, result in loss of contact of the device. Windows that are too large will result in shorted device features.

In order to study the effects of the photo-resist baking time, exposure time, and developer time on the uniformity of contact windows, an experiment was set up. Test mask patterns were made in the upper left corner of each wafer. In the test patterns, soluble lines of photo-resist were imprinted that were later developed and etched. The width of the lines could be measured, and the uniformity of these measurements (as indicated by the variance, s^2) was the response or dependent variable of interest.

In conducting the experiment, the photo-resist exposure and development steps were applied individually to each wafer, and it was easy to change the exposure or development time between individual wafers. Thus the experimental unit for these two split-plot factors was the wafer. The photo-resist baking, however, was done with groups or "lots" of wafers as pictured in Figure 4.13. It would have been very time consuming, tedious, and possibly not representative of the true process if one wafer were baked at a time. Thus the experimental unit for baking time was a lot of wafers, and baking time is the whole-plot factor.

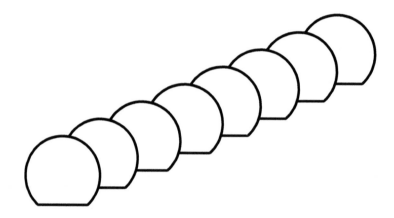

FIGURE 4.13: A "Lot" of Wafers (the Experimental Unit for Baking)

TABLE 4.5
Experimental Factors and Levels for Contact Window Formation
Example

	Levels	
Factor	−	+
X_1 = Development Time (Split-Plot Factor)	30 minutes	60 minutes
X_2 = Exposure Time (Split-Plot factor)	20% under normal	20% over normal
X_3 = Photo-Resist Bake Time (Whole-Plot Factor)	20 minutes	40 minutes

The factors and levels for the experiment are shown in Table 4.5. Each run of the experiment was conducted in the following way:

1. Photo-resist was applied to a lot of wafers.

2. A baking time (20 or 40 min.) was selected at random and all wafers in the lot were baked for the chosen time.

3. Four wafers from the baked lot were selected at random, and each assigned to one of the four development-exposure time conditions (i.e., [30, −20%], [60, −20%], [30,+20%] or [60, +20%]).

4. The four test wafers were exposed and developed according to the individual conditions they were assigned to, and then they were etched. Each wafer represented a run.

5. Five measurements of the line width were made at different locations on the test pattern for each test wafer.

The whole process was repeated with four lots of wafers, two lots on each of two days called reps (short for replicates). Table 4.6 shows the coded and uncoded factor levels. The rep factor (X_4) is treated like a blocking factor, with respect to the whole-plot factor (X_3 = bake time), and the whole-plot standard error of an effect is calculated using the block×whole-plot effect interaction. All the runs in Rep 1 were completed first. Within the rep, a lot was baked for 20 minutes followed by a second lot that was baked 40 minutes. In the second rep, the first lot was baked 40 minutes followed by a second lot that was baked for 20 minutes. The random wafer column, at the right side of Table 4.6, identifies the development and exposure conditions that were applied to the four randomly drawn wafers from each lot. The random run order, shown in the far right column of the figure, resulted from a two-stage randomization: first the order of the baking times within each rep was randomized, and second the order of exposure-development times was randomized within each bake time and rep.

TABLE 4.6
Split-Plot Design with Random Ordering (Contact Window Example)

| | | Coded Factor Levels | | | | Actual Factor Levels | | | | |
Run (Wafer)	X_1	X_2	X_3	X_4	X_1 Develop Time	X_2 Exposure Time	Random Wafer Number	X_3 Bake Time	X_4 Rep	Random Run Order
1	−	−	−	−	30 min	−20%	2	20 min	1	2
2	+	−	−	−	60 min	−20%	4	20 min	1	4
3	−	+	−	−	30 min	+20%	1	20 min	1	1
4	+	+	−	−	60 min	+20%	3	20 min	1	3
5	−	−	+	−	30 min	−20%	2	40 min	1	6
6	+	−	+	−	60 min	−20%	3	40 min	1	7
7	−	+	+	−	30 min	+20%	4	40 min	1	8
8	+	+	+	−	60 min	+20%	1	40 min	1	5
9	−	−	−	+	30 min	−20%	2	20 min	2	14
10	+	−	−	+	60 min	−20%	3	20 min	2	15
11	−	+	−	+	30 min	+20%	1	20 min	2	13
12	+	+	−	+	60 min	+20%	4	20 min	2	16
13	−	−	+	+	30 min	−20%	4	40 min	2	12
14	+	−	+	+	60 min	−20%	1	40 min	2	9
15	−	+	+	+	30 min	+20%	3	40 min	2	11
16	+	+	+	+	60 min	+20%	2	40 min	2	10

Table 4.7 shows the results of the experiment, the response variable Y=$\log_e(s^2)$, and the calculated effects, in the far right columns.[1]

Since the whole-plot factor was randomized to lots within reps, the whole-plot standard error of an effect, s_W, is calculated from the rep×bake time (i.e., the $X_3 X_4$ interaction). This standard error has one degree of freedom. The standard error for the split-plot factors, s_E, is calculated using the other interactions with the rep factor (namely $X_1 X_4$, $X_2 X_4$, $X_1 X_2 X_4$, $X_1 X_3 X_4$, $X_2 X_3 X_4$, and $X_1 X_2 X_3 X_4$). This standard error has 6 degrees of freedom. The calculation of these standard errors is shown below.

$$s_W = \sqrt{E_{34}^2/1} = 0.725,$$

---

[1] Effects are shown in Standard Order, i.e., $\bar{Y}$, X_1, X_2, $X_1 X_2$, X_3, $X_1 X_3$, $X_2 X_3$, $X_1 X_2 X_3$, X_4, $X_1 X_4$, $X_2 X_4$, $X_1 X_2 X_4$, $X_3 X_4$, $X_1 X_3 X_4$, $X_2 X_3 X_4$, $X_1 X_2 X_3 X_4$.

TABLE 4.7
Experimental Results for Contact Window Example

Run	X_1	X_2	X_3	X_4	Replicate Measures of Line Width					s^2	Y $\log_e(s^2)$	Effects
1	−	−	−	−	2.6320	2.6620	2.6210	2.5605	2.4568	0.00661	−5.018	−4.330
2	+	−	−	−	2.4450	2.6380	2.4680	2.6760	2.5794	0.01040	−4.566	0.639
3	−	+	−	−	2.5760	2.5590	2.5964	2.4475	2.6441	0.00530	−5.240	−0.196
4	+	+	−	−	2.5520	2.5470	2.5114	2.6750	2.6880	0.00655	−5.029	0.219
5	−	−	+	−	2.4850	2.4060	2.4735	2.5576	2.6570	0.00912	−4.697	0.923
6	+	−	+	−	2.7206	2.4750	2.6070	2.6450	2.6994	0.00945	−4.662	0.084
7	−	+	+	−	2.4946	2.5586	2.6330	2.5394	2.6559	0.00448	−5.409	0.180
8	+	+	+	−	2.5290	2.6445	2.4032	2.3810	2.3785	0.01366	−4.293	0.168
9	−	−	−	+	2.4420	2.4630	2.5490	2.6855	2.5353	0.00917	−4.692	1.069
10	+	−	−	+	2.4830	2.6290	2.6040	2.7910	2.4863	0.01599	−4.136	0.185
11	−	+	−	+	2.6600	2.5790	2.6285	2.4915	2.5260	0.00486	−5.326	0.061
12	+	+	−	+	2.6930	2.6517	2.5960	2.5970	2.3935	0.01326	−4.323	0.009
13	−	−	+	+	2.5273	2.4960	2.4880	2.3305	2.0803	0.03475	−3.360	0.725
14	+	−	+	+	2.7836	2.8280	2.3250	2.8140	2.3592	0.06567	−2.723	−0.038
15	−	+	+	+	2.3820	2.5190	2.3250	2.6705	2.2109	0.03168	−3.452	0.095
16	+	+	+	+	2.3000	2.2060	2.5160	2.7600	2.9445	0.09540	−2.350	−0.163

$$s_E = \sqrt{[E_{14}^2 + E_{24}^2 + E_{124}^2 + E_{134}^2 + E_{234}^2 + E_{1234}^2]/6},$$

$$s_E = \sqrt{[0.185^2 + 0.061^2 + 0.009^2 + 0.038^2 + 0.095^2 + 0.163^2]/6} = 0.112.$$

The t-ratios for all the effects are shown below. Notice that the t-ratio for bake time (X_3) is computed using the whole-plot standard error of an effect.

$t_1 = 0.639/0.112 = 5.71$
$t_2 = -0.196/0.112 = -1.75$
$t_3 = 0.923/0.725 = 1.27 \Leftarrow$ This is the t-ratio for the whole-plot effect
$t_{12} = 0.219/0.112 = 1.95$
$t_{13} = 0.084/0.112 = 0.75$
$t_{23} = 0.180/0.112 = 1.60$
$t_{123} = 0.168/0.112 = 1.50$

The significance of the effects is judged by comparing the computed t_E to the critical t with the appropriate degrees of freedom (in this case 1 degree of freedom, $t_{1,.05}=12.706$) for the bake time (X_3) effect and 6 degrees of freedom $(t_{6,.05}=2.447)$ for all the others.

As can be seen, the only effect that was significant was X_1 = Development Time, which had a positive effect. That means that increasing development time increases $Y = \log_e(s^2)$, which means it reduces the uniformity of the contact window size. Therefore the most uniform contact windows are formed with a short (30 minute) development time. Notice that the bake time (X_3) effect would be falsely thought to be significant if the wrong standard error had been used in calculating its t-ratio.

A similar experiment on the contact window formation process was conducted at the Bell Laboratories Integrated Circuits Design Capability Laboratory at Murray Hill (MH ICDCL) in 1980 (Phadke et al. [1983]). In three short months of work, these experiments identified improved process parameter settings that accounted for a fourfold reduction in the variance of contact window size and a threefold reduction in defect density (due to un-opened windows). These improvements along with the improved stability and robustness of the process, at the new process parameter settings, gave process engineers added confidence to eliminate a number of in-process checks resulting in a doubling of the total throughput of wafers in the window photo-lithography process.

4.6.2 Designs for Split-Plot 2^k Experiments

The factor effects and interactions for 2^k factorials run in split-plot designs can always be computed using the worksheet method shown earlier in this chapter. However, the rep effects and rep interaction effects that are used to compute the whole-plot standard error of an effect (s_W) and the standard error for the split-plot factors (s_E) cannot be computed using these methods unless the number of replicates is a power of two (i.e., $2^1 = 2$, $2^2 = 4$, etc.). For the more general split-plot 2^k factorial designs, where the number of reps is arbitrary, the analysis and significance tests must be completed using the Analysis of Variance (ANOVA) that will be described in Chapter 5.

There are still many useful 2^k split-plot designs where the number of reps is equal to two. Table 4.8 shows how the computation tables in Appendix B.2 can be used to create and analyze various 2^k split-plot designs with this property. The subscripts for replicate term as well as the terms that should be used to calculate the whole-plot and split-plot errors are shown in this table (i.e., 12 refers to the X_1X_2 interaction, 234 refers to the $X_2X_3X_4$ interaction, etc.).

TABLE 4.8
Two Rep Split-Plot 2^k Designs (with breakdown of error terms)

No. WP Factors	No. SP Factors	Appdx. Table	WP Factors	SP Factors	Rep Factor	WP Error Ints.	SP Error Interactions
1	1	B.2-3	2	1	3	23	13, 123
1	2	B.2-4	3	1, 2	4	34	14, 24, 124, 134, 234, 1234
2	1	B.2-4	2, 3	1	4	24, 34, 234	14, 124, 134, 1234
1	3	B.2-5	4	1, 2, 3	5	45	15, 25, 35, 125, 135, 235, 1235, 145, 245, 345, 1245, 1345, 2345, 12345
2	2	B.2-5	3, 4	1, 2	5	35 45 345	15, 25, 125, 135, 145, 235, 245, 1235, 1245, 1345, 2345, 12345
3	1	B.2-5	2, 3, 4	1	5	25, 35, 45 235, 245 345, 2345	15, 125, 135, 145 1235, 1245 1345, 12345

4.7 Summary

In this chapter we have discussed three additional choices that must be made in planning 2^k type factorial experiments and five additional methods for analyzing data from 2^k experiments. The three choices needed for planning 2^k experiments are: 1) the choice of the number of replicates, 2) the choice of whether or not to run a blocked experiment, and 3) the choice of whether or not to run a split-plot experiment. The choice of the number of replicates depends upon the smallest effect that we do not want to overlook (δ) and the estimate of the standard deviation of a single observation (σ). The choice of whether or not to run a blocked experiment depends on the existence of known background variables. Finally, the choice of whether or not to run a split-plot design usually depends on whether the runs or experiments can be completely randomized to experimental units, or whether there are difficult to vary factors that prevent complete randomization. Figure 4.14 is a flow diagram that illustrates the choices that must be made.

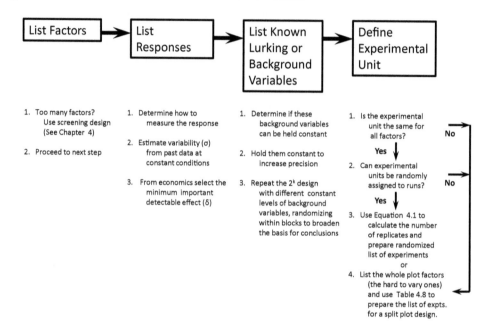

FIGURE 4.14: Flow Diagram of Choices Necessary in the Design of a 2^k Experiment

Additional methods of analysis discussed in this chapter were 1) expressing the results of an experiment in the form of a linear prediction equation that can be used to predict the response at a combination of factor levels not yet tested, 2) including center points in the 2^k design and doing a curvature test to check the adequacy of the linear prediction equation, 3) using contour plots to represent the prediction equation graphically, 4) calculating the standard error of the effect (s_E) from block interaction terms in blocked experiments, and 5) analysis of split-plot experiments that involve calculating the standard error of split and whole-plot factors. In Chapter 8 we will expand on the method of fitting prediction equations to data and present additional methods of checking the adequacy of the models.

4.7.1   Important Equations

$$n_F^* = (8s_P/\delta)^2$$

$$\delta = 8s_P/\sqrt{n_F}$$

$$b_i = [\text{Effect of Factor } X_i]/2$$

$$b_{ij} = [\text{Interaction of Factors } X_i X_j]/2$$

$$s_E = \sqrt{\sum_{i=1}^{m} BI_i^2 / m}$$

4.7.2 Important Terms and Concepts

The following is a list of the important terms and concepts discussed in this chapter.

replicates needed for accuracy
δ, smallest effect we do not want to overlook
α-risk and β-risk
mathematical prediction equation for 2^k factorial
coded factor
uncoded factor
center points
curvature
blocking
known background or nuisance variable
blocking interactions are zero
blocked factorial designs
split-plot factorial designs
whole-plot standard error of an effect, s_W

4.8 Exercises

1. Doty, Gillian, LeBaron, and Sawaya studied the effects of the number of Automatic Guided Vehicles, AGVs (1–3), Speed of AGVs (120/100–180/150 ft./min.), and the pick up and drop off times of AGVs upon the total throughput time in a castings factory. A 2^3 design with replicated center points was used, and the data were generated by the ProModel Simulation program. The data are listed below:

$X_1 =$ Number of AGVs	$X_2 =$ Speed of AGVs	$X_3 =$ Pick/Drop Time	Throughput
$-$	$-$	$-$	180 184 181
$+$	$-$	$-$	202 204 207
$-$	$+$	$-$	199 194 195
$+$	$+$	$-$	206 213 210
$-$	$-$	$+$	190 186 184
$+$	$-$	$+$	208 208 209
$-$	$+$	$+$	196 203 204
$+$	$+$	$+$	215 210 195
0	0	0	194 201 195
			204 197 203

(a) Calculate the mean and variance, s^2, of the replicate responses for each run, and calculate the pooled standard deviation, s_p.

(b) Do a graphical analysis by plotting the mean responses at each run on a cube.

(c) Calculate the main effects, two-way interactions, the three-way interaction, and the curvature effect.

(d) Calculate the standard error of an effect, s_E, and the standard error of the curvature effect, s_C.

(e) Determine which effects and interactions are statistically significant. Is curvature significant?

2. A study was performed to investigate the effects of iron deficiency in diets of rats on their activity. The measure of activity, Y, was the time spent in vertical movement (seconds per 15-minute interval). The response was measured automatically and continuously using a rat cage with infrared beams connected to a computer monitoring system. The response recorded was the average of 40 15-minute readings taken during 10 hours of light or 10 hours of darkness (with 2 hours in between). The factors studied were:

X_1:	Iron in diet	$-$ = Deficient	$+$ = Adequate
X_2:	Age of rat	$-$ = 3 weeks old	$+$ = 6 weeks old
X_3:	Time of day	$-$ = day (light)	$+$ = night (dark)

Four rats were used for each set of conditions, each from a different litter of rats. The data collected are given below.

Iron in Diet X_1	Age, weeks X_2	Day or Night X_3	Response, Y Litter of Rats 1	2	3	4
-1	-1	-1	4.1	9	3.7	5.2
1	-1	-1	2.6	8.4	4.3	7.5
-1	1	-1	4.9	7.2	3.8	3.3
1	1	-1	3.1	8.8	4.8	4.9
-1	-1	1	10.3	14.2	12.7	12.8
1	-1	1	10.9	18.5	12.8	15
-1	1	1	5.6	12.1	4.8	6.3
1	1	1	10.2	15.9	12.8	10.3

(a) Treating the rats as pure replicates, analyze the data to determine which factors and interactions are statistically significant. Summarize your results in equation form.

(b) Treating the rat litters as blocks, reanalyze the data to determine which factors and interactions are statistically significant. Note: Use the interactions of blocking variables with the factors studied to estimate s_E. Summarize your results in equation form.

(c) What was the impact of treating the litters as blocks on your analysis?

Chapter 5

General Factorial Experiments and ANOVA

5.1 Introduction

In Chapters 3 and 4 we presented the simplest of factorial experiments, namely those with only two levels for each of the factors under study. Examples of two-level factorials were shown when the factors were quantitative such as moisture %, or qualitative such as thread type (rolled or tapped). It makes sense to use two levels for quantitative factors if we are interested in linear relationships. However, two levels are only appropriate for qualitative factors when there are just two alternatives. In many practical situations there may be more than two alternatives for qualitative factors, such as component type A, B, C, or D in a prototype, or material types 1, 2, or 3. Also, it may be desirable to study nonlinear relationships for quantitative factors. Chapters 9–11 of this book present experimental designs specifically for studying nonlinear relationships when only quantitative factors are present. In this chapter we study an extension of the factorial experiments that we have already presented in Chapters 3 and 4.

The factorial experiments we present in this chapter will accommodate more than two levels for each factor. This will allow us to study quadratic relationships for quantitative factors and use more than two alternatives for qualitative factors. Quantitative and qualitative factors can be used simultaneously in these designs. The only problem with factorial experiments that contain more than two levels for each factor is that the total number of experiments is greatly increased. For example a 2^3 factorial experiment with two replicates for each treatment combination requires 16 runs. However, a $3 \times 4 \times 3$ factorial with two replicate per treatment combination requires 72 runs. For this reason, large multilevel factorial designs are usually avoided in early stages of research, but they can be quite useful when the important factors have already been identified and the goal is to identify optimum treatment combinations among multiple alternative levels of a few factors.

We will introduce an alternate form of the mathematical model for representing the data and making predictions in multilevel factorials, and we will use Analysis of Variance (or ANOVA that was introduced in Section 2.5.3) for determining which factors and interactions are significant. Orthogonal contrasts will be introduced as a tool for studying specific differences among factor levels, and we will show that the calculation worksheet used in Chapter 3 was a special case of orthogonal contrasts. Finally in this chapter we will show how blocking and split-plots can be used in multiple level factorials, and how to analyze the data from these experiments.

5.2   Multiple Level Factorial Designs

Multiple level factorial designs consist of experiments run at all combinations of levels of the factors involved. In Chapter 3, when a three-factor two-level design was presented, the list of experiments consisted of all possible $2\times2\times2 = 8$ treatment combinations. Likewise a factorial design with three factors that have 2, 2, and 3 levels respectively would consist of all $2\times2\times3 = 12$ possible combinations of the three factor's levels. The list of experiments can be written down in standard order by varying the levels of the first factor fastest, the levels of the second factor second fastest, and levels of the third factor slowest; for example, if we designate the levels of the factors by the integer codes 1,2, etc. The twelve runs of a 2x3x2 factorial experiment and the twelve runs of a 3×4 factorial are both shown in Table 5.1. In this table we include a column for the symbolic response.

TABLE 5.1

A $2\times3\times2$ Factorial and a 3×4 Factorial

2×3×2 Factorial					3×4 Factorial			
Run	A	B	C	Y	Run	A	B	Y
1	1	1	1	Y_{111}	1	1	1	Y_{11}
2	2	1	1	Y_{211}	2	2	1	Y_{21}
3	1	2	1	Y_{121}	3	3	1	Y_{31}
4	2	2	1	Y_{221}	4	1	2	Y_{12}
5	1	3	1	Y_{131}	5	2	2	Y_{22}
6	2	3	1	Y_{231}	6	3	2	Y_{32}
7	1	1	2	Y_{112}	7	1	3	Y_{13}
8	2	1	2	Y_{212}	8	2	3	Y_{23}
9	1	2	2	Y_{122}	9	3	3	Y_{33}
10	2	2	2	Y_{222}	10	1	4	Y_{14}
11	1	3	2	Y_{132}	11	2	4	Y_{24}
12	2	3	2	Y_{232}	12	3	4	Y_{11}

These designs can be represented geometrically as shown in Figure 5.1. The effects of changing factor levels can be visualized by placing the average response in each combination of levels of the factors as shown in Figure 5.1.

Main effects and two-factor interactions can be visualized in the same way as they were in the two-level designs. Figure 5.2 is an example of a main effect plot for a four level factor A on the left, and the interaction plot between a four-level factor A and a two-level factor B on the right. In the figure on the left the symbol $\bar{Y}_i$ for $i = 1, 2, 3, 4$ represents the average response in the ith level of factor A. If the main effect were negligible, these averages would fall close to a horizontal line. In this example, the average for level 4 of the factor appears higher than the averages at the other levels indicating non-zero main effects for factor A.

In the figure on the right $\bar{Y}_{ij}$ for $i = 1, 2, 3, 4$ and $j = 1, 2$ represents the average response in the ith level of factor A and the jth level of Factor B. If Factor B had more than two levels, there would be more than two segmented lines on the graph. If there was no interaction between factors A and B the two segmented lines would be parallel within each segment. Since the lines in right half of Figure 5.2 are not parallel (in fact crossing), it indicates that there is an interaction between factor A and Factor B.

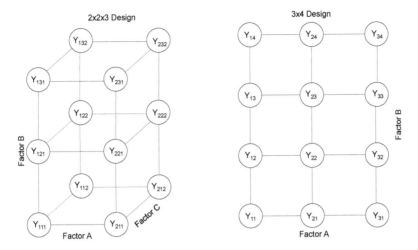

FIGURE 5.1: Geometric Representation of a 2×3×2 Design and a 3×4 Design

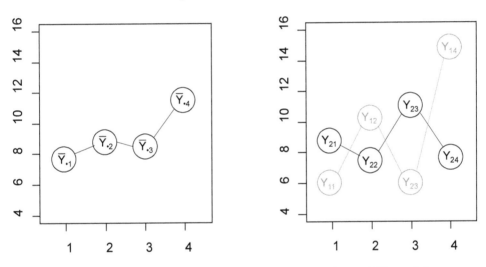

FIGURE 5.2: Main Effect and Interaction Plot for 4×2 Factorial

5.3 Mathematical Model for Multiple Level Factorials

In Chapter 4, a mathematical model for a three-factor factorial was of the form:

$$\hat{Y} = b_0 + b_1 X_1 + b_2 X_2 + b_3 X_3 + b_{12} X_1 X_2 + b_{13} X_1 X_3 + b_{23} X_2 X_3 + b_{123} X_1 X_2 X_3 \quad (5.1)$$

where b_i was half the calculated effect for a factor (or interaction), and X_i was the coded factor level (i.e., -1 or $+1$). This model was useful for reporting experimental results or for calculating predicted values. The effects that were used to calculate the b coefficients in this model were the differences, $\bar{Y}_+ - \bar{Y}_-$, in average response between the high ($+$) and the low ($-$) levels for each factor or interaction column in the worksheet. Using this model we can make predictions for any of the eight combinations of factor levels in the 2^3 experiment by setting (X_1, X_2, X_3) to the appropriate combinations of $+1$ and -1. Thus the particular combination of factor levels is identified in model 5.1 by the combination of $+1$ and -1 used for (X_1, X_2, X_3). However, the mathematical model in Equation 5.1 is inappropriate for multilevel factorials since for some factors there are more than two levels. Therefore, we must find another way of identifying a combination of factor levels, and a new definition for the effects and mathematical model.

The notation we use to identify a particular combination of factor levels in a multilevel factorial is the set of subscripts on the symbolic response. For example, in the simple $2 \times 3 \times 2$ factorial in Table 5.1, we can see that Y_{111} is the response in the first level of factor A, the first level of factor B, and the first level of factor C. Likewise Y_{132} is the response in the first level of factor A, third level of factor B, and second level of factor C. Thus the set of subscripts, ijk, on the symbolic response Y_{ijk} identifies it to be the response from the ith level of factor A, jth level of factor B, and kth level of factor C.

With this notation for a combination of treatment factors in mind, we will define K estimated effects for a factor, F, that has K levels. The first effect, $\hat{F}_1$, is defined to be the difference in the average response in the first level of factor F minus the overall average response. In general, we define the kth effect, $\hat{F}_k$, for $k = 1, \dots, K$ to be the difference of the average response at the kth level of factor F minus the overall average. If F is the third factor in a three-factor design and the first two factors have I and J levels, then we write $\hat{F}_k = \bar{Y}_{..k} - \bar{Y}_{...}$, where the dots replacing subscripts indicate we have averaged the response over all possible values of the subscript in that position, i.e.,

$$\bar{Y}_{..k} = \sum_{i=1}^{I} \sum_{j=1}^{J} Y_{ijk} / IJ.$$

We will illustrate this definition of effects using the following table of data from a simple 2×3 factorial experiment that is presented in a standard order, where the subscripts on the first factor A change fastest, and the subscripts on the second factor B change slowest.

TABLE 5.2
Example Data from 2×3
Factorial

| Factor Levels | | Response |
A	B	Y
1	1	11.5
2	1	14.5
1	2	10
2	2	13
1	3	6
2	3	18.5

Consider the second factor B. The average response at each level of B are:

$$\bar{Y}_{.1} = (11.5 + 14.5)/2 = 13.0,$$

$$\bar{Y}_{.2} = (10 + 13)/2 = 11.5,$$

$$\bar{Y}_{.3} = (6 + 18.5)/2 = 12.25,$$

and the overall average response is:

$$\bar{Y}_{...} = (13 + 11.5 + 12.25)/3 = 12.25.$$

Therefore, the three effects of factor B are:

$$\hat{B}_1 = 13.0 - 12.25 = 0.75,$$

$$\hat{B}_2 = 11.5 - 12.25 = -0.75,$$

$$\hat{B}_3 = 12.25 - 12.25 = 0.00.$$

The two effects for the two level factors A can be calculated in a similar fashion.

The mathematical model for the multilevel factorial (assuming no interaction effects) can be written in terms of the effects as follows:

$$\hat{Y}_{ij} = \hat{\mu} + \hat{A}_i + \hat{B}_j \tag{5.2}$$

where $\hat{Y}_{ij}$ is the predicted value in the ijth combination of levels of factors A, and B, $\hat{\mu}$ is the overall average response, and $\hat{A}_i$ and $\hat{B}_j$ are the effects (or deviations from the overall average) that result from being in the ith level of factor A and the jth level of factor B. This is often called the *additive model* because predictions can be made from this model by adding the appropriate effects for each factor to the overall average. It will yield accurate predictions if there is no interaction between the two factors. The additive model will not produce accurate predictions if there is an interaction between factors. A more flexible model can be obtained by adding interaction effects to the additive model (5.2), just as they were added to model (5.1). For example the model for a two-factor factorial with r replicates per cell and an interaction can be written as:

$$Y_{ijk} = \hat{\mu} + \hat{A}_i + \hat{B}_j + \hat{AB}_{ij} + \epsilon_{ijk}, \tag{5.3}$$

where i is the level of factor A, j is the level of factor B, and k is the replicate number.

The interaction effects $\hat{AB}_{ij}$ are defined as the average responses in the ijth combination of factor levels minus the predicted values from the additive model in the corresponding combination of factor levels. These can be written as:

$$\hat{AB}_{ij} = Y_{ij} - (\hat{\mu} + \hat{A}_i + \hat{B}_j). \tag{5.4}$$

If there are r replicate experiments in each of the IJ combinations of factor levels, then the Y_{ij} in Equation 5.4 is changed to the average of the r replicates, $\bar{Y}_{ij.}$.

The interaction effects can be visualized in Figure 5.3 where the solid dots represent the Y_{ij} (or $\bar{Y}_{ij.}$ if there are replicates), and the open dots represent the predicted values from the additive model, $\hat{Y}_{ij}$. Notice the line segments joining the predicted values from the additive model are parallel which indicates the predictions from this model assume that there is no interaction. The figure shows the interaction effects $\hat{AB}_{ij}$ as the differences in the Y_{ij} (or $\bar{Y}_{ij.}$) and the predicted values $\hat{Y}_{ij}$.

The model for a three-factor factorial with two-factor interactions and the three-factor interaction can be written as:

$$Y_{ijkl} = \hat{\mu} + \hat{A}_i + \hat{B}_j + \hat{C}_k + \hat{AB}_{ij} + \hat{AC}_{ik} + \hat{BC}_{jk} + \hat{ABC}_{ijk} + \epsilon_{ijkl}, \tag{5.5}$$

where the two-factor interaction effects $\hat{AC}_{ik}$, $\hat{BC}_{jk}$ are defined similar to the $\hat{AB}_{ij}$ effects in Equation 5.4, and the three-factor interaction effects are defined by

$$\hat{ABC}_{ijk} = \bar{Y}_{ijk.} - (\hat{\mu} + \hat{A}_i + \hat{B}_j + \hat{C}_k + \hat{AB}_{ij} + \hat{AC}_{ik} + \hat{BC}_{jk}).$$

Models for factorial experiments in four or more factors can be written in a similar way by adding all possible interaction effects. However, if there are more than two levels for multiple factors, the number of experiments required for a multilevel factorial can become very large.

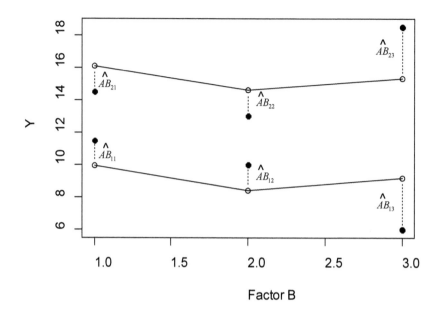

FIGURE 5.3: Interaction Effects for 2×3 Experiment

5.4   Testing the Significance of Main Effects and Interaction Effects

Since there is more than one effect for each factor and interaction in a multilevel factorial, simple t-tests cannot be used to test their statistical significance. Instead, an Analysis of Variance is used, and F-ratios are computed that compare the variability of effects (for each factor or interaction) to the variability of replicate responses in each combination of factor levels. This is similar to the ANOVA that was described in Section 2.5.3. The ANOVA sums of squares for a two-factor factorial with I levels of factor A, J levels of factor B, and r replicate experiments in each combination of factor levels, can be written in terms of the effects that were defined above as follows:

$$ssA = \sum_{i=1}^{I} rJ(\hat{A}_i)^2 \tag{5.6}$$

$$ssB = \sum_{j=1}^{J} rI(\hat{B}_j)^2 \tag{5.7}$$

$$ssAB = \sum_{i=1}^{I} \sum_{j=1}^{J} r(\hat{AB}_{ij})^2 \tag{5.8}$$

$$ssE = \sum_{i=1}^{I} \sum_{j=1}^{J} \sum_{k=1}^{r} (Y_{ijk} - \hat{Y}_{ij.})^2 \tag{5.9}$$

where $\hat{Y}_{ij} = Y_{...} + \hat{A}_i + \hat{B}_j + \hat{AB}_{ij} = \bar{Y}_{ij}$.

The Analysis of Variance table is constructed as shown in Table 5.3.

TABLE 5.3

Analysis of Variance Table for Two-Factor Factorial

Source	df	SS	Mean Square	F-ratio
A	$I-1$	ssA	$msA = ssA/(I-1)$	msA/msE
B	$J-1$	ssB	$msB = ssB/(J-1)$	msB/msE
AB	$(I-1)(J-1)$	$ssAB$	$msAB = ssAB/(I-1)(J-1)$	$msAB/msE$
Error	$(r-1)IJ$	ssE	$msE = ssE/(r-1)IJ$	

If the calculated F-ratio for main effect A, B or the two-way interaction AB exceeds the critical limit (as discussed in Section 2.5.3) it indicates that there are significant differences in the average responses at the different levels of the main effect, or significant differences among the cell means for the interaction. Similar to the ANOVA discussed in Section 2.5.3, the F-statistics are valid if the expected variances of the response within the cells (or combinations of factor levels) are approximately equal and the response data are normally distributed. If these conditions are not satisfied the response data can be transformed (as illustrated in Section 3.9.1) to achieve approximate validity of the F-statistics.

Although it is possible to compute the quantities in Table 5.3 using a spreadsheet program (Excel© Analysis ToolPak add-in has a TWOWAYANOVA function and also the Excel free real statistics add-in [RealStat, 2015]), it is much more convenient to construct ANOVA tables using a statistical program.

5.5 Example of a Multilevel Two-Factor Factorial

In their tutorial on the Analysis of Variance method Stahle and Wold [1989] show the data in Table 5.4 for a two-factor factorial. Levels of the first factor (Reagent) represented four different reagents that could be used in a chemical reaction, and the levels of the second factor (Catalyst) represent three different catalysts that could be used. There are two replicate runs conducted for each combination of reagent type and catalyst, and the response is the yield of the experiment. The format of the table is the normal way that data should be entered into a statistical program for analysis, with separate columns for the levels of the factors and the response variable.

Table 5.5 shows the Analysis of Variance table produced by the `Multi-way ANOVA` menu in the `Rcmdr`[Fox, 2005] GUI interface for the open source statistical program R. An example of the data set creation and menu choices to produce Table 5.5 are shown on the website for the book. Both main effects are highly significant, which means there are significant differences among the marginal means for the reagent and catalyst. Figure 5.4 shows the main effect plots for both the the Catalyst and Reagent main effects. There it can be seen that the Yield is higher on the average using Catalyst 2 and Reagent C. However, since there is a significant interaction, this does not tell the whole story.

Figure 5.5 shows the interaction plot. There are differences in the slopes of the lines connecting the means for each reagent among the three catalysts, and it can be seen from the graph, or the cell means in Table 5.5, that Reagent B in combination with Catalyst 2, or Reagent D with Catalyst 1 or 2, or Reagent C with any one of the three catalysts, should all result in high yields.

TABLE 5.4

Data from 4×3 Factorial
Experiment in Standard Order

Reagent	Catalyst	Yield
A	1	4
A	2	11
A	3	5
A	1	6
A	2	7
A	3	9
B	1	6
B	2	13
B	3	9
B	1	4
B	2	15
B	3	7
C	1	13
C	2	15
C	3	13
C	1	15
C	2	9
C	3	13
D	1	12
D	2	12
D	3	7
D	1	12
D	2	14
D	3	9

The `Rcmdr` can also produce diagnostic plots, similar to those that will be presented in Section 8.6, to determine if the equal variance and normality conditions justifying the F-statistics are satisfied.

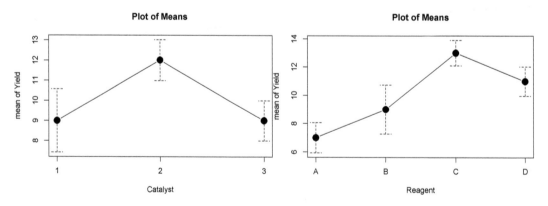

FIGURE 5.4: Main Effect Plots

TABLE 5.5
ANOVA results from a Statistical Program

```
Anova Table (Type II tests)

Response: Yield
                 Sum Sq Df F value   Pr(>F)
Catalyst             48  2     6.0 0.015625 *
Reagent             120  3    10.0 0.001386 **
Catalyst:Reagent     84  6     3.5 0.030802 *
Residuals            48 12
---
Signif. codes:  0 '***' 0.001 '**' 0.01 '*' 0.05 '.' 0.1 ' ' 1

Cell Means
   A  B  C  D
1  5  5 14 12
2  9 14 12 13
3  7  8 13  8

Cell Standard Deviations
        A        B        C        D
1 1.414214 1.414214 1.414214 0.000000
2 2.828427 1.414214 4.242641 1.414214
3 2.828427 1.414214 0.000000 1.414214
```

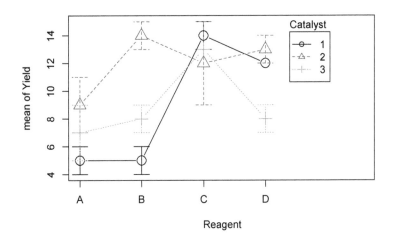

FIGURE 5.5: Catalyst by Reagent Interaction

5.6 Comparison of Means after the ANOVA

Although the F-statistic in the Analysis of Variance tells us there are significant differences among the cell or marginal means for the factor or interaction tested, it does not tell us specifically which of the means, for a factor or interaction in question, are significantly different than each other. In the example presented in the last section, the F-statistic for the interaction shows there are significant differences among the cell means, but it does not say, for example, if there is a significant difference between the mean yield of 14 with Reagent C and Catalyst 1 and the mean yield of 12 with Reagent C and Catalyst 2. This question must be answered by comparing the specific means after the ANOVA.

5.6.1   Least Significant Difference (LSD) Method

One way to determine which means are significantly different is to do a direct comparison with a t-statistic. The t-statistic for comparing two cell means would be:

$$t_{IJ(r-1)} = \frac{\bar{Y}_{ij.} - \bar{Y}_{i'j'.}}{\sqrt{msE\left(\frac{1}{r} + \frac{1}{r}\right)}}, \tag{5.10}$$

where $\bar{Y}_{ij.}$ is the mean in the ij treatment combination, $\bar{Y}_{i'j'.}$ is the mean in any other treatment combination, msE is the mean square error from the ANOVA table, and r is the number of replicates in each treatment combination. For comparing marginal means for a significant main effect replace the cell means $\bar{Y}_{ij.}$ in Equation 5.10 with the marginal means for the factor you want to test, i.e., $\bar{Y}_{i..}$ for the first factor, and change the denominator to $\sqrt{msE\left(\frac{1}{Jr} + \frac{1}{Jr}\right)}$.

In general, whenever two means at different levels of a significant factor or two cell means in different combinations of levels of a significant interaction differ by more than

$$LSD = t_{\nu,1-\alpha}\sqrt{msE\left(\frac{1}{n} + \frac{1}{n}\right)}, \tag{5.11}$$

they can be declared significantly different, where ν is the error or residual degrees of freedom in the ANOVA table, α is the desired significance level, and n is the number of values averaged. This is called the LSD method of comparing means after the ANOVA. As an illustration of this, consider comparing cell means in the chemical reaction experiment presented in the last section where $msE = 4.0$.

$$LSD = t_{12,0.95}\sqrt{msE\left(\frac{1}{2} + \frac{1}{2}\right)} = 2.179\sqrt{4.0(1)} = 4.358$$

Therefore, the difference, in the mean for Reagent C and Catalyst 1 and the mean for Reagent C and Catalyst 2, $(14-12) < 4.358$ and is not significantly different at the $\alpha = 0.05$ level of significance.

The LSD method can be used if one pre-selected comparison of means is to be made after the ANOVA. However, the confidence level is decreased (or the chance of a false positive is increased) when using the LSD method if a number of comparisons are made, or if the means to be compared are chosen after examining their values. This will result in a higher rate of false positives. There are $I \times (I-1)/2$ possible comparisons of means for a factor with I levels, and $I \times J \times (IJ-1)/2$ possible comparisons of means in an interaction with $I \times J$ combinations of levels. If we make one comparison of means at confidence level $1-\alpha$, the probability of a Type I error for the test will be α, but if we make k comparisons at confidence level $1-\alpha$, the probability of a Type I error will be increased to $1-(1-\alpha)^k$. For example, if all six possible comparisons of marginal means are made for a four level factor at the $\alpha = .05$ significance level, the probability of declaring at least one pair significantly different by chance is $1 - .95^6 = 1 - 0.735 = 0.265$, not 0.05!

5.6.2 Tukey's Method of Comparing Means after the ANOVA

In order to control the probability of a Type I error when making many comparisons of means following the ANOVA, the value of the LSD must be increased. Tukey [1949] developed a method that does this by referring to the distribution of the studentized range rather than the t-distribution. Using Tukey's method we declare a difference of two means to be significant if it is greater than

$$HSD = \frac{q(1-\alpha, \nu, k)}{\sqrt{2}} \sqrt{msE \left(\frac{1}{n} + \frac{1}{n}\right)}, \tag{5.12}$$

where HSD stands for Tukey's honestly significant difference, $q_{(1-\alpha, \nu, k)}$ is the $1-\alpha$ percentile of the studentized range statistic with ν degrees of freedom for error, and k means being compared. Finally n is the number of values averaged in any two means being compared. Table A.5 in Appendix A tabulates the 95% confidence values for the studentized range. These are used like the critical values for the t-statistic in Appendix Table A.3.

Figure 5.6 shows a worksheet for comparing all means in the Reagent × Catalyst interaction in the Chemical yield experiment presented in Section 5.5. The means in each of the 12 combinations of Reagent and Catalyst are entered into row 3 columns D–O, and in column C rows 4–15. The absolute value of the differences in means is presented in the body of the worksheet. For example cell F5 contains the formula = ABS(F3-$C5).

	A	B	C	D	E	F	G	H	I	J	K	L	M	N	O
1			Reagent	A	A	A	B	B	B	C	C	C	D	D	D
2			Catalyst	1	2	3	1	2	3	1	2	3	1	2	3
3	Reagent	Catalyst	Mean	5	9	7	5	14	8	14	12	13	12	13	8
4	A	1	5		4	2	0	**9**	3	**9**	7	**8**	7	**8**	3
5	A	2	9			2	4	5	1	5	3	4	3	4	1
6	A	3	7				2	7	1	7	5	6	5	6	1
7	B	1	5					**9**	3	**9**	7	**8**	7	**8**	3
8	B	2	14						6	0	2	1	2	1	6
9	B	3	8							6	4	5	4	5	0
10	C	1	14								2	1	2	1	6
11	C	2	12									1	0	1	4
12	C	3	13										1	0	5
13	D	1	12											1	4
14	D	2	13												5
15	D	3	8												
16															
17			Q(0.05,12,12)= 5.62			HSD=(5.62/1.414)(sqrt(4(1))=7.95									

FIGURE 5.6: Worksheet for Reagent by Catalyst Means using HSD Method

The upper percentile of studentized range statistic can be calculated using the QCRIT function in the real stats add-in for Excel© [RealStat, 2015], but for convenience the 95th percentiles are tabulated in Appendix Table A.5. To compare two cell means, we use $k = 12$ (there are 12 cell means), $df = 12$ (the error degrees of freedom), and find $q_{(0.95,12,12)} = 5.62$ from the appendix table. The value of HSD is calculated in the worksheet as a function of $q_{(0.95,12,12)}$. The means that differ by more than HSD are bolded. By examining the bolded pairs, it can be seen that the mean yield for Reagents A and B in combination with

Catalyst 1 are significantly less than the mean yields with Reagent B and Catalyst 2, or Reagent C with Catalyst 1 or 3, or Reagent D with Catalyst 2. These are the only significant differences in cell means that can be detected with Tukey's method.

The marginal means for Reagent and Catalyst can also be compared in a similar way after the ANOVA. However, for this data these comparisons would be less informative because a significant interaction was found.

5.6.3   Orthogonal Contrasts

Another way to compare cell means or marginal means without increasing the probability of a type I error, or false positive, is to partition the sums of squares for a factor in the ANOVA using orthogonal contrasts. A contrast is a linear combination of marginal means or cell means of the form:

$$C_c = \sum_i c_i \bar{Y}_i$$

where the sum of the contrast coefficients is zero, $\sum_i c_i = 0$. Two contrasts $C_c = \sum_i c_i \bar{Y}_i$ and $C_d = \sum_i d_i \bar{Y}_i$ are orthogonal if the sum of products of their coefficients, $\sum_i c_i \times d_i = 0$, is zero. Any sum of squares (i.e., ssA, or $ssAB$) in the ANOVA table with ν degrees of freedom can be partitioned into ν independent single degree of freedom sums of squares by choosing ν orthogonal contrasts of the marginal means (for a main effect) or cell means (for an interaction). The sums of squares associated with a contrast $C_c = \sum_i c_i \bar{Y}_i$ are

$$SS(C_c) = n \times \frac{(\sum_i c_i \bar{Y}_i)^2}{(\sum_i c_i^2)}$$

where n is the number of responses averaged to get $\bar{Y}_i$.

To illustrate this, contrast coefficients for the Catalyst and Reagent factors in the chemical reaction experiment presented in the last section are shown in Table 5.6.

TABLE 5.6
Contrast Coefficients for Catalyst and Reagent

Catalyst	Mean	c_1	c_2	Reagent	Mean	c_1	c_2	c_3
1	9	-1	-1	A	7	-1	-1	-1
2	12	2	0	B	9	-1	-1	1
3	9	-1	1	C	13	3	0	0
				D	11	-1	2	0

For the two sets of contrast coefficients for Catalyst, it can be seen that the coefficients sum to zero

$$\sum_i c_{1i} = -1 + 2 - 1 = 0$$

$$\sum_i c_{2i} = -1 + 0 + 1 = 0$$

and are orthogonal.

$$\sum_i c_{1i} \times c_{2i} = -1 \times -1 + 2 \times 0 + -1 \times 1 = 0.$$

This can also be verified for the contrasts of Reagent. Examination of the contrast coefficients shows the first contrast for Catalyst will compare twice the mean response for Catalyst 2 to the sum of the mean responses for Catalyst 1 and 3, i.e.,

$$\sum_{i=1}^{3} c_{1i}\bar{Y}_i = -1(9) + 2(12) - 1(9) = 6.$$

The second contrast for Catalyst compares the mean response for Catalyst 1 and Catalyst 3, i.e.,

$$\sum_{i=1}^{3} c_{2i}\bar{Y}_i = -1(9) + 0(12) + 1(9) = 0.$$

The first contrast for Reagent compares the average response for Reagent C to the other reagents, etc. The sums of squares for Catalyst in Table 5.6 can then be partitioned as follows:

$$SS_{Catalyst} = 48 = n \times \frac{(\sum_{i=}^{3} c_{1i}\bar{Y}_i)^2}{(\sum_{i=1}^{3} c_{1i}^2)} + n \times \frac{(\sum_{i=}^{3} c_{2i}\bar{Y}_i)^2}{(\sum_{i=1}^{3} c_{2i}^2)}$$

$$= (8)\frac{6^2}{-1^2 + 2^2 + -1^2} + (8)\frac{0^2}{-1^2 + 0^2 + 1^2}$$

$$= 48 + 0.$$

Thus, it can be seen the sums of squares for Catalyst are due entirely to the difference between the mean response for Catalyst 2 compared to the mean response for the other two catalysts. This is an extreme case. Usually each contrast will contribute something to the total.

The sums of squares for Reagent can be similarly partitioned, as

$$SS_{Reagent} = SS_{c_1} + SS_{c_2} + SS_{c_3} = 72 + 36 + 12,$$

and it can be seen most of the sums of squares come from the first contrast, that compares Reagent C to the others, and second contrast that compares Reagent D to Reagents A and B.

Since the partitions of the sums of squares each have one degree of freedom, the ratio of the sums of squares for each contrast to the error means square in the ANOVA table follows the F-distribution, i.e.,

$$F_{1,12} = \frac{SS_{c_i}}{msE}.$$

The sums of squares and F-ratios can be computed conveniently in a spreadsheet. For example Figure 5.7 shows the computations for the chemical experiment. Columns A and B in the spreadsheet show the indicators for the levels of Reagent and Catalyst. Column C contains the cell means obtained from Table 5.6.

Columns D through H contain the contrast coefficients for reagent (labeled R1 -R3) in the worksheet, and catalyst (labeled C1 - C2). Notice that the contrast coefficients for reagent from Table 5.6 are repeated for each level of the catalyst, and likewise the contrast coefficients for catalyst from Table 5.6 are repeated for each level of the reagent. This is due to the fact that column C in the spreadsheet contains cell means and not the marginal means.

The interaction Catalyst×Reagent has 6 degrees of freedom, and 6 orthogonal contrasts

	A	B	C	D	E	F	G	H	I	J	K	L	M	N
1	Reagent	Catalyst	Mean Yield	R1	R2	R3	C1	C2	R1C1	R1C2	R2C1	R2C2	R3C1	R3C2
2	A	1	5	-1	-1	-1	-1	-1	1	1	1	1	1	1
3	A	2	9	-1	-1	-1	2	0	-2	0	-2	0	-2	0
4	A	3	7	-1	-1	-1	-1	1	1	-1	1	-1	1	-1
5	B	1	5	-1	-1	1	-1	-1	1	1	1	1	-1	-1
6	B	2	14	-1	-1	1	2	0	-2	0	-2	0	2	0
7	B	3	8	-1	-1	1	-1	1	1	-1	1	-1	-1	1
8	C	1	14	3	0	0	-1	-1	-3	-3	0	0	0	0
9	C	2	12	3	0	0	2	0	6	0	0	0	0	0
10	C	3	13	3	0	0	-1	1	-3	3	0	0	0	0
11	D	1	12	-1	2	0	-1	-1	1	1	-2	-2	0	0
12	D	2	13	-1	2	0	2	0	-2	0	4	0	0	0
13	D	3	8	-1	2	0	-1	1	1	-1	-2	2	0	0
14	SUMPRODUCT =			36	18	6	24	0	-36	-4	-9	-13	9	1
15	SUMSQ =			36	18	6	24	8	72	24	36	12	12	4
16			SS	72	36	12	48	0	36	1.3333	4.5	28.167	13.5	0.5
17			F	18	9	3	12	0	9	0.3333	1.125	7.0417	3.375	0.125
18	ANOVA	msE =		4	Critical F(1,12) =	4.747								

FIGURE 5.7: Worksheet for Partitioning Sums of Squares

to partition the interaction sums of squares are shown in the worksheet columns I through N. The coefficients in these contrasts were obtained by calculating the products of the main effect contrasts. In other words, the entry in cell I2 is = D2*G2, and this formula is copied down the column into cells I3 through I13. A similar thing was done to calculate the contrast coefficients R1C2 through R3C2.

The contrasts $\sum_i c_i \bar{Y}_i$ are then calculated by the SUMPRODUCT formula in row 14, similar to the way Effects were calculated for two-level factorials in Section 3.7.2. The cell D14 contains the formula = SUMPRODUCT($C2:$C13,D2:D13) and this formula was copied across the columns from cell E14 to N14. Row 15 in the spreadsheet contains the sums of squares of the contrast coefficients $(\sum_{i=1} c_i^2)$ that is calculated with the SUMSQ formula. Cell D15 contains the formula = SUMSQ(D2:D13), and this formula was copied across the columns to cells E15 through N15. Finally, the sums of squares for each contrast that was given by $(n \times (\sum_{i=} c_i \bar{Y}_i)^2 / (\sum_{i=1} c_i^2))$ are computed in row 16 by dividing the results of 2 times row 14 squared by the results in row 15. In other words, the content of cell D16 is the formula = 2*D14^2/D15 and this formula was copied into cells E16 through N16. In this case, the number of values averaged to get the cell means is $n = 2$. The error mean square $(msE = 4.0)$ was obtained from the ANOVA Table 5.5 by dividing residuals sum of squares by the residuals degrees of freedom, and it was typed into cell C18, and the F-ratios for each contrast are calculated in row 17 by dividing each contrast sum of squares by the error mean square. Each F-ratio has one degree of freedom for the numerator sums of squares and 12 degrees of freedom for the denominator (from Table 5.5).

The critical value for the F distribution with 1 and 12 degrees of freedom is calculated in cell G18 with the formula = FINV(0.05,1,12). Therefore the only contrasts that are significant at the $\alpha = 0.05$ level of significance are R1, R2, C1, R1C1, and R2C2, and these represent independent tests that will not raise the significance level. R1 compares the mean yield for Reagent C to the mean yield for the other reagents, and R2 compares the mean yield for Reagent D to the mean yield for Reagents A and B. R1C1 compares the mean yield for Reagent A, B, and D with Catalyst 2, and Reagent C with Catalyst 1 or 3, to the mean yield for the other seven combinations of reagents and catalysts. Finally, R2C2 compares the difference in average yields for Reagents A and B to the average yield for Reagent D between Catalysts 1 and 3. These significant differences can be visualized in the interaction graph (Figure 5.5).

5.6.4 Other Applications of Orthogonal Contrasts

5.6.4.1 Orthogonal Polynomial Contrasts

Experimenters sometimes use more than two levels for quantitative factors if they are interested in quadratic or other higher order nonlinear effects. This would be the case if the goal were to find the factor value that produced a maximum or minimum response within the range of factor values under study. When the factor levels are equally spaced values of a quantitative factor, orthogonal polynomial contrast coefficients can provide a quick set of comparisons that will allow the experimenter to determine if the response is related to the factor levels linearly, quadratically, or according to a higher level polynomial. Table A.4 in Appendix A shows the coefficients for these contrasts for three-, four-, or five-level factors. More extensive tables can be found in the Biometrika Tables for Statisticians.

As an example of the use of these tables, consider the worksheet shown in Figure 5.8. Column B in the worksheet contains the marginal mean yields for five equally spaced temperature levels in a 3×5 factorial experiment. Columns C–F show the orthogonal polynomial contrasts copied from Appendix Table A.4 for a five-level factor.

	A	B	C	D	E	F
1	Temperature	Mean	linear	quadratic	cubic	quartic
2	1	1.3	-2	2	-1	1
3	2	3.05	-1	-1	2	-4
4	3	3.55	0	-2	0	6
5	4	3.45	1	-1	-2	-4
6	5	2.95	2	2	1	1
7	SUMPRODUCTS		3.7	-5.1	0.85	-0.45
8	SUMSQ		10	14	10	70
9	SS		8.214	11.14714	0.4335	0.01736
10	F		4.8748	6.615515	0.2573	0.0103
11	ANOVA msE =		1.685			
12	Critical F(1,15) = 4.54					

FIGURE 5.8: Worksheet for Partitioning Sums of Squares into Orthogonal Polynomials

The sums of squares for the orthogonal contrasts are computed in row 9 in the same way shown in the last section. The mean square from the ANOVA table is shown in cell C11, and the F-ratios for the contrasts are shown in row 10. From these it can be seen that the linear and quadratic contrasts for the temperature factor are significant.

Since the three-level main effect and the interaction in the ANOVA were not significant, the results can be summarized by Equation 5.13 and Figure 5.9.

$$\text{Mean Yield} = -0.80 + 2.56(\text{Temperature}) - 0.36(\text{Temperature})^2. \qquad (5.13)$$

Use of the orthogonal polynomial contrasts does not produce the coefficients in the quadratic equation, like that shown Equation 5.13, nor does it determine the temperature that is predicted to result in the highest yield. Rather it just provides us with a quick way to determine the appropriate degree (i.e., linear, quadratic, etc.) of the equation that should be fit. To determine the prediction equation, and the temperature predicted to give the highest yield, the least squares method for fitting a quadratic equation should be used. That method will be described in Chapter 8.

5.6.4.2 Analysis of Unreplicated Multilevel Factorials

When there are multiple levels of factors in a factorial design, the total number of combinations of levels can be quite large. One way to reduce the total number of experiments is to eliminate replicate experiments in each combination of treatment levels. However, when

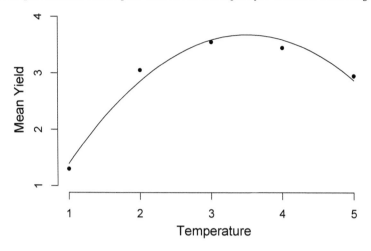

FIGURE 5.9: Quadratic Polynomial Fit to Mean Yield

replicate experiments are eliminated the error mean square in the ANOVA table cannot be computed nor F-statistics to test the significance of main effects and interactions.

In order to determine which effects and interactions are significant graphical techniques can be used like those illustrated in Section 3.10. In that section, a Normal plot and a Pareto Diagram of the effects from an unreplicated 2^4 factorial experiment were made in order to identify which effects were significant. The effects in a 2^k factorial are contrasts of the cell means (or individual responses in an unreplicated design) where the coefficients for the contrasts (-1, and $+1$) are the same for each main effect and interaction calculated.

In a multilevel factorial the contrast coefficients c_i are not the same for each single degree of freedom for main effects or interactions, but if they are standardized by dividing each contrast by the square root of the sum of squares of the contrast coefficients, i.e.,

$$\text{Standardized Contrast} = \frac{\sum_i c_i \bar{Y}_i}{\sqrt{\sum_i c_i^2}},$$

they can be examined graphically to determine which are significant. Consider the following example to illustrate how this can be done.

During the 1960s and 1970s Japanese automobiles were well known for their rust problems. Toyota Corporation's competitive position and future profitability in European markets was dependent on a solution to this problem. The following experiment[1] represents one of a series of experiments that were run by Toyota employees during the course of solving this problem. In this experiment, the response (scab test) was the diameter of a scab corrosion spot after a specified number of cycles of exposure to salt spray. The factors studied in the experiment are shown in Table 5.7. Notice that none of the factors are continuous in nature. The particular choice of levels resulted in a $3^2 \times 2^2$ factorial.

The data are shown in columns A–E of the worksheet shown in Figure 5.10. Here it can be seen that there are no replicates of any of the experimental conditions. To get a preliminary idea of what factors are significant, orthogonal contrasts were constructed for

---

[1]The data presented herein are not the actual experimental data but were fabricated to match the results of an experiment described in *Implementation Manual for Three Day QFD Workshop*, American Supplier Corporation, Dearborn, Michigan.

TABLE 5.7

Rust Test Factors and Levels

Factors	Levels
A- Electro-coating Method	1=Anion, 2=Super Anion, 3=Cation
B- Treatment Method	1=Spray, 2=Dip
C- Number of Coats	1=2 coats, 2= 3 coats
D- Treatment Type	1=A, 2=B, 3=C

each of the factors. The set of contrasts for the first two factors is listed below, and the contrasts for the second two factors are defined similarly.

Electro-coating Contrasts

$$A1 = -2 \times \hat{A}_1 + \hat{A}_2 + \hat{A}_3$$

$$A2 = 0 \times \hat{A}_1 - \hat{A}_2 + \hat{A}_3$$

Treatment Method Contrast

$$B1 = -1 \times \hat{B}_1 + \hat{B}_2$$

The contrast coefficients for the main effects are listed in the worksheet columns F through K. Contrasts for the interaction degrees of freedom (for example A1B1 in column L) were obtained by multiplying the respective contrasts for the main effects together in the same way shown in Figure 5.7. In this example, there were two-way interaction contrasts, three-way interaction contrasts (such as A1B1C1), and four-way interaction contrasts (such as A1B1C1D1), but only the first 7 two-way interaction contrasts are shown in the figure (so that it can be displayed on one page).

The contrasts were calculated in row 38 of the worksheet using the spreadsheet `SUMPRODUCT` formula. The `= SUMSQ` function was used in row 39 to get the sums of squares of the contrast coefficients, and the standardized contrasts in row 40 were calculated by dividing the elements of row 38 by the square root of the elements in row 39.

The normal probability plot of the standardized contrasts is shown in Figure 5.11. Here it can be seen that the first contrast for factors A, B, C, and D, the second contrast for factors A and D, and one contrast from the AB interaction and one from the ABD interaction all appear to be significant. Since none of the contrasts from the four-way interaction, and all the contrasts from interactions involving factor C, do not appear to be significant, these interactions can be eliminated from the model:

$$Y_{ijklm} = \mu + A_i + B_j + C_k + AB_{ij} + AD_{il} + BD_{jl} + ABD_{ijl} + \epsilon_{ijklm} \tag{5.14}$$

leaving degrees of freedom for the error mean square. Although no contrasts from the AD and BD interactions were identified on the normal plot of contrasts, these two interactions were included in the model to preserve effect heredity.

	A	B	C	D	E	F	G	H	I	J	K	L	M	N	O	P	Q	R	S
1	A	B	C	D	ScabD	A1	A2	B1	C1	D1	D2	A1B1	A2B1	A1C1	A2C1	A1D1	A2D1	A1D2	A2D2
2	1	1	1	1	6.90	-2	0	-1	-1	-2	0	2	-0	2	-0	4	-0	-0	0
3	2	1	1	1	5.09	1	1	-1	-1	-2	0	-1	-1	-1	-1	-2	-2	0	0
4	3	1	1	1	5.66	1	-1	-1	-1	-2	0	-1	1	-1	1	-2	2	0	-0
5	1	2	1	1	6.22	-2	0	1	-1	-2	0	-2	0	2	-0	4	-0	-0	0
6	2	2	1	1	5.21	1	1	1	-1	-2	0	1	1	-1	-1	-2	-2	0	0
7	3	2	1	1	3.17	1	-1	1	-1	-2	0	1	-1	-1	1	-2	2	0	-0
8	1	1	2	1	7.66	-2	0	-1	1	-2	0	2	-0	-2	0	4	-0	-0	0
9	2	1	2	1	6.62	1	1	-1	1	-2	0	-1	-1	1	1	-2	-2	0	0
10	3	1	2	1	6.77	1	-1	-1	1	-2	0	-1	1	1	-1	-2	2	0	-0
11	1	2	2	1	6.60	-2	0	1	1	-2	0	-2	0	-2	0	4	-0	-0	0
12	2	2	2	1	5.52	1	1	1	1	-2	0	1	1	1	1	-2	-2	0	0
13	3	2	2	1	4.28	1	-1	1	1	-2	0	1	-1	1	-1	-2	2	0	-0
14	1	1	1	2	7.60	-2	0	-1	-1	1	-1	2	-0	2	-0	-2	0	2	-0
15	2	1	1	2	4.92	1	1	-1	-1	1	-1	-1	-1	-1	-1	1	1	-1	-1
16	3	1	1	2	5.74	1	-1	-1	-1	1	-1	-1	1	-1	1	1	-1	-1	1
17	1	2	1	2	3.97	-2	0	1	-1	1	-1	-2	0	2	-0	-2	0	2	-0
18	2	2	1	2	5.03	1	1	1	-1	1	-1	1	1	-1	-1	1	1	-1	-1
19	3	2	1	2	2.72	1	-1	1	-1	1	-1	1	-1	-1	1	1	-1	-1	1
20	1	1	2	2	7.99	-2	0	-1	1	1	-1	2	-0	-2	0	-2	0	2	-0
21	2	1	2	2	5.25	1	1	-1	1	1	-1	-1	-1	1	1	1	1	-1	-1
22	3	1	2	2	6.83	1	-1	-1	1	1	-1	-1	1	1	-1	1	-1	-1	1
23	1	2	2	2	6.99	-2	0	1	1	1	-1	-2	0	-2	0	-2	0	2	-0
24	2	2	2	2	5.48	1	1	1	1	1	-1	1	1	1	1	1	1	-1	-1
25	3	2	2	2	2.57	1	-1	1	1	1	-1	1	-1	1	-1	1	-1	-1	1
26	1	1	1	3	5.30	-2	0	-1	-1	1	1	2	-0	2	-0	-2	0	-2	0
27	2	1	1	3	4.90	1	1	-1	-1	1	1	-1	-1	-1	-1	1	1	1	1
28	3	1	1	3	3.98	1	-1	-1	-1	1	1	-1	1	-1	1	1	-1	1	-1
29	1	2	1	3	5.92	-2	0	1	-1	1	1	-2	0	2	-0	-2	0	-2	0
30	2	2	1	3	3.76	1	1	1	-1	1	1	1	1	-1	-1	1	1	1	1
31	3	2	1	3	1.56	1	-1	1	-1	1	1	1	-1	-1	1	1	-1	1	-1
32	1	1	2	3	5.62	-2	0	-1	1	1	1	2	-0	-2	0	-2	0	-2	0
33	2	1	2	3	4.85	1	1	-1	1	1	1	-1	-1	1	1	1	1	1	1
34	3	1	2	3	5.72	1	-1	-1	1	1	1	-1	1	1	-1	1	-1	1	-1
35	1	2	2	3	5.19	-2	0	1	1	1	1	-2	0	-2	0	-2	0	-2	0
36	2	2	2	3	3.91	1	1	1	1	1	1	1	1	1	1	1	1	1	1
37	3	2	2	3	2.10	1	-1	1	1	1	1	1	-1	1	-1	1	-1	1	-1
38				Contrasts		-40.3	9.44	-27.2	12.3	-21.5	-12	-8.66	15.58	-0.12	-2.72	-2.96	1.76	1.28	1.24
39				Contrast SS		72	24	36	36	72	24	72	24	72	24	144	48	48	16
40		Standardized Contrast				-4.75	1.93	-4.53	2.05	-2.53	-2.51	-1.021	3.18	-0.01	-0.555	-0.25	0.254	0.185	0.31

FIGURE 5.10: Worksheet for Calculating Standardized Contrasts

A portion of the ANOVA table produced by the Minitab ANOVA-General Linear Model menu, and the model in Equation 5.14, is shown in Table 5.8. The complete output also shows the individual effects for the terms in the model (like those shown in Figure 5.10). The ANOVA table confirms exactly what was seen in the normal plot of contrasts.

All of the main effects appear to be significant as well as the AB interaction (Coating Method × Treatment Method) and the three-factor interaction ABD (Coating Method × Treatment Method × Treatment Type). Significance of the three-factor interaction means that the effect of factor A (Coating Method) will be different depending upon which levels

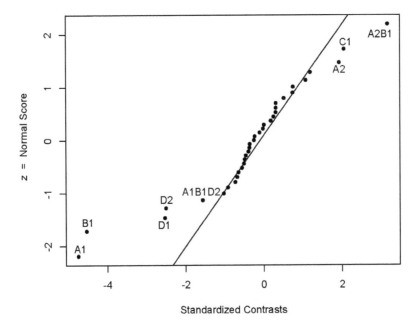

FIGURE 5.11: Normal Plot of Standardized Contrasts

TABLE 5.8
Minitab ANOVA Table for Model 5.14

```
Analysis of Variance

Source   DF   Adj SS   Adj MS   F-Value   P-Value
  A       2   26.2475  13.1237   37.21     0.000
  B       1   20.5511  20.5511   58.27     0.000
  C       1    4.2025   4.2025   11.91     0.003
  D       2   12.7034   6.3517   18.01     0.000
 A*B      2   11.1556   5.5778   15.81     0.000
 A*D      4    0.2556   0.0639    0.18     0.945
 B*D      2    0.7855   0.3928    1.11     0.351
A*B*D     4    3.5536   0.8884    2.52     0.080
Error    17    5.9961   0.3527
Total    35   85.4510
```

of factors B (Treatment Method) and D (Treatment Type) are used. The interaction plot in Figure 5.12 shows the dependency.

The goal was to reduce the diameter of a scab corrosion spot; therefore the treatment method = dip, in combination with the Cation electro-coating method, was the optimal combination.

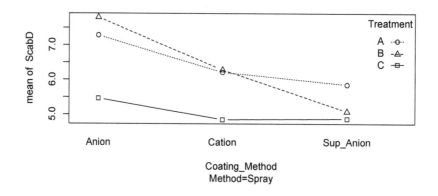

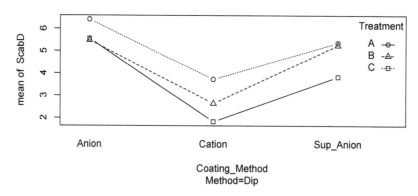

FIGURE 5.12: Interaction Plot A×B×D

5.7 Analysis of Blocked and Split-Plot Factorial Experiments with Multilevel Factors

In Section 4.5 and 4.6 of Chapter 4, the rationale for blocked and split plot factorials was explained, but the method of analysis shown was restricted to the case where all factors have only two-levels ($-$ and $+$) and the number of blocks or reps in the design is restricted to be a power of 2 (i.e., 2, 2^2, etc.). Now that the technique of ANOVA for factorial designs has been introduced, we will illustrate how it can be used to analyze blocked or split plot factorials with an arbitrary number of factor levels for each factor, and an arbitrary number of blocks or reps.

5.7.1 Analysis of a Blocked Factorial by ANOVA

In the example analysis of a blocked factorial in Section 4.5.1, it was shown that all interactions with the block term were pooled to form an error. In the Analysis of Variance context, the pooling can be accomplished by simply leaving those interactions with block terms out of the ANOVA model (like the AC, BC, and $ABCD$ interactions were left out of the model 5.14 in the last section).

For example, consider the data from the four-block pesticide degradation experiment presented in Table 4.4. The data in the columns for Temp, Moisture, Block, and Half-Life in Table 4.4 can be entered into a statistical program (re-coding the − and + levels for Temp and Moist as 1 and 2). Leaving the block interaction terms out, we write the model as:

$$\ln(\text{Half-Life}) = \mu + \text{Temp}_i + \text{Moisture}_j + \text{Temp} \times \text{Moisture}_{ij} + \text{Block}_k + \epsilon_{ijk},$$

and the sums of squares for interactions left out of the model are combined to form the error sums of squares in the ANOVA table. With the model above, a statistical program can be used to produce an ANOVA table like that shown in Table 5.9.

TABLE 5.9
Analysis of Variance for Blocked Factorial in Table 4.4

Source	df	Sum Squares	Mean Squares	F-ratio	P-value
Temp	1	166,601	166,601	7.83	0.021
Moisture	1	55,428	55,428	2.60	0.141
Temp×Moisture	1	28,499	28,499	1.34	0.277
Block	3	164,735	54,911		
Error (Block interactions)	9	191,515	29,279		

5.7.2 Analysis of a Split-Plot Factorial by ANOVA

Data from split plot factorial experiments can be analyzed in a similar fashion. Recall that in Section 4.6.2 of Chapter 4, it was shown that the whole plot error term was formed by pooling all interactions involving whole plot factors and the rep factor, and that the within plot error term was formed by pooling all other interactions involving the rep factor. Again, this pooling can be easily accomplished in the ANOVA context.

Consider the data from the split plot example shown in Section 4.6.1 The whole plot factor C (Bake Time) had two levels, there were two replicate whole plots, and the split plot factors A (Development Time) and B (Exposure Time) each had two levels. An ANOVA table can be constructed using the model:

$$log_e(s^2) = \mu + C_i + Rep_j + C \times Rep_{ij} + A_k + B_l + AC_{ik} + BC_{il} + AB_{kl} + ABC_{ikl} + \epsilon_{ijkl},$$

where the $C \times Rep_{ij}$ interaction is the error term for testing the whole plot factor, and all the other interactions with Rep that have been left out of the model will combine to form the split plot error term ϵ_{ijkl}. The factor levels would be entered into a statistical program in the same format as the right side of Table 4.6, and an ANOVA table similar to Table 5.10 can be produced.

The ANOVA program produces the correct F-statistics for the split plot factors (A = Development Time, B = Exposure Time, and interactions AB, AC, BC, and ABC). These F ratios have 1 degree of freedom for the numerator and 6 degrees of freedom for the denominator. It can be shown that the square root of a random variable that follows and F-distribution with 1 degree of freedom in the numerator follows the t-distribution. Therefore by taking the square root of the F-ratios in Table 5.10 for A, B, and interactions AB, AC, BC, and ABC, the result will be exactly the same t-ratios shown in Section 4.6.1.

The ANOVA program does not produce the correct F-ratio for the whole plot factor C = Bake Time. The denominator of the F-ratio for the whole plot factor should be the mean square for Bake Time × Replicate. This can be easily calculated as:

TABLE 5.10
ANOVA Table for Split-Plot Design from Data in Table 4.6

Source	df	Sum Squares	Mean Square	F-ratio
Bake Time	1	3.406	3.406	67.974
Replicate	1	4.572	4.572	91.213
Bake Time × Replicate	1	2.103	2.103	42.060
Development Time	1	1.633	1.633	32.589
Exposure Time	1	0.154	0.154	3.062
Dev × Exp	1	0.192	0.192	3.836
Dev×Bake	1	0.028	0.028	0.557
Exp × Bake	1	0.013	0.013	2.596
Dev× Exp × Bake	1	0.112	0.112	2.241
Error	6	0.300	0.0501	

$$F_{1,1} = \frac{3.406}{2.103} = 1.619,$$

which again is the square of the t-ratio calculated for the whole plot effect in Section 4.6.1.

The advantage of analyzing the data from a split-plot experiment by ANOVA is that, of course, the number of levels of the factors is not restricted to be 2, and the number of whole plots or replicates does not have to be a power of 2.

5.8   Summary

In this chapter we discussed factorial experiments with more than two levels for the factors. We showed how the Analysis of Variance (first described in Chapter 2) can be extended to test the significance of differences in factor level means and interaction means in multilevel factorial experiments. We also discussed methods for comparing specific means after the Analysis of Variance, so that the significance level will not be increased, and methods to analyze unreplicated multilevel factorials with graphical techniques such as normal plots. Finally, we explained how the Analysis of Variance can be used for blocked and split-plot factorials with more than two levels for the factors.

5.8.1   Important Equations

$$Y_{ijk} = \hat{\mu} + \hat{A}_i + \hat{B}_j + \hat{AB}_{ij} + \epsilon_{ijk}$$

$$Y_{ijkl} = \hat{\mu} + \hat{A}_i + \hat{B}_j + \hat{C}_k + \hat{AB}_{ij} + \hat{AC}_{ik} + \hat{BC}_{jk} + \hat{ABC}_{ijk} + \epsilon_{ijkl}$$

$$LSD = t_{\nu,1-\alpha}\sqrt{msE\left(\frac{1}{n} + \frac{1}{n}\right)}$$

$$HSD = \frac{q(1-\alpha,\nu,k)}{\sqrt{2}}\sqrt{msE\left(\frac{1}{n} + \frac{1}{n}\right)}$$

$$C_c = \sum_i c_i \bar{Y}_i$$

$$SS(C_c) = n \times \frac{(\sum_i c_i \bar{Y}_i)^2}{(\sum_i c_i^2)}$$

$$\text{Standardized Contrast} = \frac{\sum_i c_i \bar{Y}_i}{\sqrt{\sum_i c_i^2}}$$

5.8.2 Important Terms

The following is a list of the important terms and concepts covered in this chapter.

multiple level factorial
mathematical model for multilevel factorial
LSD method
Tukey's HSD method
orthogonal contrasts
standardized contrasts

5.9   Exercises

1. Tannin acyl hydrolase that is commonly referred to as tannase is one of the important hydrolytic microbial enzymes. It has applications in food, animal feed, pharmaceutical, chemical, and leather industries. Aboubakretal et al. [2013] carried out a split-plot design to study the effects of fermentation temperature and fermentation time on tannase production. The response was the extracellular tannase (U/50 mL). The whole-plot factor was the fermentation temperature, and the split-plot factor was the fermentation time. The data for the three reps (shown in three columns) are in the Table below.

   (a) Enter the data into a statistical program and create the ANOVA table.

   (b) Determine the appropriate F-statistics and P-values for the whole-plot factor, and the split-plot factor and whole-plot by split-plot interaction.

   (c) Create appropriate graphs or tables to illustrate significant differences in the cell or marginal means and identify the Temperature and Fermentation Time that will produce the maximum tannase production.

   (d) Use Tukey's HSD method to determine which means in the combinations of Temperature and Fermentation Time are NOT significantly different at the $\alpha = 0.05$ significance level from the combination that produces the maximum.

TABLE 5.11
Data from Split-Plot Experiment

Temp (°C)	Ferm Time (min)	Rep 1	Rep 2	Rep 3
20	24	106.6	106.6	106.6
25	24	126.6	126.6	126.6
30	24	153.9	153.9	153.9
35	24	158.5	158.5	158.5
40	24	115.5	115.5	115.5
20	48	139.3	139.3	139.3
25	48	171.8	171.8	171.8
30	48	213.5	213.5	213.5
35	48	220.2	220.2	220.2
40	48	181.2	181.2	181.2
20	72	176.5	176.5	176.5
25	72	222.8	222.8	222.8
30	72	255.6	255.6	255.6
35	72	271.9	271.9	271.9
40	72	195.6	195.6	195.6
20	96	207.2	207.2	207.2
25	96	258.8	258.8	258.8
30	96	283.3	283.3	283.3
35	96	312.7	312.7	312.7
40	96	256.2	256.2	256.2
20	120	248.0	248.0	248
25	120	265.6	265.6	265.6
30	120	284.6	284.6	284.6
35	120	304.2	304.2	304.2
40	120	288.8	288.8	288.8

Chapter 6

Variance Component Studies

6.1 Introduction

All data collected in engineering, manufacturing, or scientific investigations are subject to variability. Variability in data makes decisions more difficult. However, precise results are rare in process research, and variability is a fact of life. We must learn to make informed decisions in the face of uncertain or variable information. Already in this book we have discussed much about variability, and decision making in the presence of variability. In Chapter 2 we discussed some ways of summarizing data (histograms, sample variances, etc.) to reveal the amount of variability present. We also discussed some of the theory for explaining the cause for variability.

In Chapters 2 and 3 we learned to judge the significance of the effects of experimental factors by comparing their magnitude to experimental noise variability. In this chapter we will explain how variability can be classified or broken down into its component sources. This will help us to focus efforts on the major sources when attempting to identify the causes for variability using techniques like those discussed in Chapter 3.

We will show how to partition estimates of the component sources of variabilty, that normally are not measured directly, from an overall estimate of variance. For example, the total variance in assays of a chemical product may be apportioned into variance due to product variability and laboratory measurement error.

6.2 Additivity of Variances

If a random variable (Y) is actually the sum of more than one independent random variable, such as adding (X) the result of flipping the spinner in Figure 2.12 to (Z) the result obtained from the quincunx in Figure 2.17, then

$$Y = X + Z. \tag{6.1}$$

In that case the variance, σ_Y^2, of the random variable Y can be shown to be the sum of the variances of X and Z.

$$\sigma_Y^2 = \sigma_X^2 + \sigma_Z^2. \tag{6.2}$$

6.3 Simple Experiments for Estimating Two Sources of Variability

In general, the variance of a sum of independent random variables is the sum of their individual variances. In this section we will begin to discuss a method of apportioning variability to component sources. We do this by collecting and analyzing data. We will start with the simplest case of two sources of variability. Suppose we are studying a batch chemical process where two reactants A and B are combined in a pot and stirred until the reaction is complete in forming a product C. Lab assays show there is variability over time in yields of the product C. Suppose we want to determine what part of the variability in measured yields is due to uncontrollable operating condition changes (such as ambient temperature, and purity of reactants that occur between batches), and what part is due to measurement error (such as sampling and analytical lab error). We can think of the measured response (yield) for the jth measurement of the ith operating condition or batch as the sum of two sources of variability. We can write a mathematical model for yield as:

$$Y_{ij} = \mu + B_i + M_{ij} \tag{6.3}$$

where μ represents the overall mean yield, B_i represents the deviation from the mean caused by the ith operating condition or batch, and M_{ij} is the deviation from $\mu + B_i$ caused by the jth measurement.

Model 6.3 in form looks exactly like Model 2.18 in Chapter 2 with the τ_i's replaced by B_i's and the ϵ_{ij}'s replaced by M_{ij}'s , but the interpretation is different. In the experiments we discussed in Chapter 2, deliberate changes were made and the effects, τ_i, represented the difference between the average response in the ith level of Factor T and the overall average of the experiment. In statistical terminology the τ_i's are called fixed effects. In that situation, we were interested in the values of the fixed effects, τ_i, as measures of the effects of the deliberate changes. On the other hand, in Model 6.3, the B_i's represent the differences from the overall average caused by random changes that we have no control over. In statistical terminology the B_i's are called random effects. We are not specifically interested in the values of the B_i's themselves, but rather the variance of the B_i's which is σ_B^2. The random effects M_{ij} in Model 6.3 represent the effects of measurement error such as sampling and laboratory analytical error. We are specifically interested in the variance of the M_{ij}'s which is σ_M^2

Y_{ij} is a sum of a constant μ, and two random variables B_i and M_{ij}. Therefore the variance of Y_{ij} , σ_Y^2 , is equal to the sum of the variances of the two random variables. In other words,

$$\sigma_Y^2 = \sigma_B^2 + \sigma_M^2. \tag{6.4}$$

We call these two variances the variance components. μ contributes nothing to the variance since it is a constant. We can estimate the two variance components separately if a set of data is collected similar to Table 2.7 in Chapter 2. We will have to make n_B of batches (using the same reactor, source of material(s), and operator(s)) and then measure the yield of each batch n_M of times. The experimental design or sampling design is shown schematically in Figure 6.1 and reproduced in the top of Table B.7.1-1. The only decisions to be made are the selection of the number of batches to make, n_B (denoted as n_A in the appendix table), and the number of measurements per batch, n_M. It is recommended that n_B be chosen to be as large as is feasibly possible, normally 25 to 30, in order to get a reliable estimate of σ_B^2. The recommended value for n_M is 2, which is the value shown in Figure 6.1. If more than two measurements were made for each batch, the design would be very "unbalanced" in the sense that there would be much more information available to estimate σ_M^2 than there would be to estimate σ_B^2.

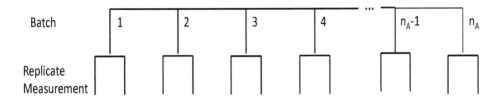

FIGURE 6.1: Simple Experimental Design for Estimating Two Sources of Variability

6.4 Estimation of Variance Components Using ANOVA

One simple way to estimate the two variances from the experimental data is by the Analysis of Variance (ANOVA) method. The calculations needed to complete the ANOVA table are exactly the same as those shown in Table 2.8, the ANOVA for a one-factor experiment. However, additional information is needed to make the ANOVA table useful for estimating the components of variance. This additional information is the expected values of the mean squares. The mean squares, like sample variances, are random variables. Their expected values are linear combinations of the variance components. For the situation with two sources of variability, the complete ANOVA table with expected mean squares is shown symbolically in Table 6.1. Derivation of the expected mean square values is beyond the scope of this book, but can be found in Graybill [1976]. To estimate the variance components, set up two linear equations by equating the mean squares (which are numerical constants after the ANOVA calculations are complete) to their expected values, and solving simultaneously.

TABLE 6.1
Analysis of Variance Table

Source	df	Sum of Squares	Mean Squares	F-ratio	Expected Mean Squares
Treatment	$t-1$	ssT	$msT = ssT/(t-1)$	$F = msT/msE$	$\sigma_M^2 + 2\sigma_B^2$
Error	$n-t$	ssE	$msE = ssE/(n-t)$		σ_M^2
Total	$n-1$	$ssTotal$	$msTotal$		

We will illustrate estimating variance components for the chemical batch yield example using the data in Table 6.2. In this table $n_B = 10$ for illustrative purposes, although in practice a larger number of batches (>25) should be sampled to get a reasonable estimate of σ_B^2. The number of measurements of each batch is $n_M = 2$ (the recommended value).

TABLE 6.2

Yields of Batch Chemical Process

i	j	y_{ij}	$\bar{y}_{i\cdot}$	$\sum_{j=1}^{2}(y_{ij}-\bar{y}_{i\cdot})^2=(y_{i1}-y_{i2})^2/2$
1	1	95.24		
1	2	95.20	95.220	0.0008
2	1	95.00		
2	2	95.03	95.066	0.00045
3	1	95.05		
3	2	95.05	95.050	0.00
4	1	95.13		
4	2	95.14	95.135	0.00005
5	1	94.70		
5	2	94.68	94.690	0.0002
6	1	94.97		
6	2	94.97	94.970	0.000
7	1	95.14		
7	2	95.13	95.135	0.00005
8	1	94.66		
8	2	94.68	94.670	0.0002
9	1	95.16		
9	2	95.17	95.165	0,00005
10	1	94.95		
10	2	94.95	94.950	0.000
Totals				$ssE = 0.0018$

In the table sample data are given, along with the batch averages, and within batch sum of squares. These summary statistics can be easily computed with a spreadsheet program and used to calculate the sums of squares in the ANOVA. Alternatively the ANOVA table can be created with the spreadsheet functions described in Section 2.5.3. The ANOVA table calculated with the data in Table 6.2 is shown in Figure 6.2.

	A	B	C	D	E
1	ANOVA				
2	*Source of Variation*	*SS*	*df*	*MS*	*F*
3	Between Batches	0.6464	9	0.071822	399.0123457
4	Within Batches	0.0018	10	0.00018	
5					
6	Total	0.6482	19		

FIGURE 6.2: Analysis of Variance for Data in Table 6.2

To estimate the variance components set the mean squares in the ANOVA table equal to their expected values in Table 6.1 and then solve the two simultaneous linear equations:

$$0.00180 = \sigma_M^2,$$
$$0.071822 = \sigma_M^2 + 2\sigma_B^2.$$

This is an easy set of equations to solve since the first equation is already solved for σ_M^2. If we substitute the solution to the first equation into the second we get:

$$0.071822 = 0.000180 + 2\sigma_B^2.$$

The solution to both equations is then:

$$\hat{\sigma}_M^2 = 0.000180$$

$$\hat{\sigma}_B^2 = 0.035821$$

where the $\hat{}$'s over the symbols indicate they are estimated values. The estimated variance of a single measurement on one batch of material is:

$$\hat{\sigma}^2 = \hat{\sigma}_B^2 + \hat{\sigma}_M^2 = 0.035821 + 0.00018 = 0.036001$$

and the estimated proportion of variation that is due to changes in operating conditions is:

$$\frac{\hat{\sigma}_B^2}{\hat{\sigma}^2} = \frac{0.035821}{0.036001} = 0.995.$$

The measurement error due to sampling and laboratory error is a very small proportion (i.e., $1.0 - 0.995 = 0.005$) of the total variability and, in this case, designed experiments could be used effectively to determine effects of various process operating factors upon the changes in yield.

One problem with the ANOVA method of estimating the variance components is that occasionally the mean square for measurement error is larger than the mean square for batch in the ANOVA table. This will result in a negative estimate for σ_B^2 when the two linear equations involving the expected mean squares are solved for the variance components. If this occurs, the estimate, σ_B^2, is set equal to zero, since it is impossible, by definition, to have a negative variance.

6.5 Graphical Representation of Data from Simple Experiments for Estimating Two Sources of Variability

One odd or unrepresentative value can have a major influence on variance estimates. When the variability in measurement error calculated from a simple experiment for estimating two sources of variability is large ($> 25\%$ of the total) it is useful to know whether this measurement error is approximately equal in all batches or parts measured, or whether a majority of the measurement error occurred on one batch or part. If the majority of measurement error occurred on one batch or part, it may be possible to repeat the lab analysis to determine if something out of the ordinary has occurred. Graphical displays [Snee, 1983] allow us to quickly view data from a simple experiment to determine if there are any strange values or outliers. A histogram of all the data, possibly overlaid with the batch averages, will allow us to detect clusters of atypical results and peculiar distribution shapes. For example, Figure 6.3 shows a histogram of the data in Table 6.2. From this graph we can see that the distribution shape is not normal, and that the majority of the variability among batches is due to the two batches with low yields around 94.7.

Dot frequency diagrams in which all data points are shown in a single plot with notations indicating the sources of variation are useful for visualizing all the sources of variation in one diagram. For example, the data of Table 6.2 are graphed in this manner in Figure 6.4. In this figure, boxes are drawn around replicate measurements from the same batch. The vertical

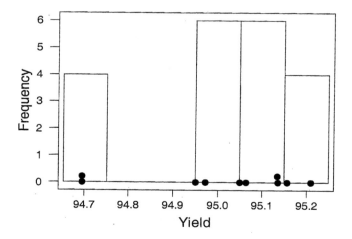

FIGURE 6.3: Histogram of Yield Data with Dot Diagram of Batch Averages

length of the box indicates the measurement variability, and the variability of placement of the boxes around the horizontal line (representing the grand average) shows the variability from batch to batch. The near equality of the box lengths shows the variability within batch is fairly consistent, while the batch-to-batch variability is dominated by batches 5 and 8.

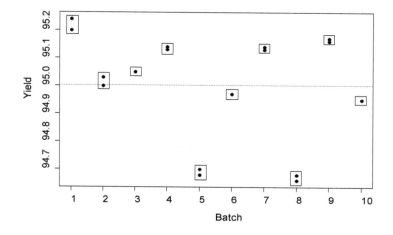

FIGURE 6.4: Dot Frequency Diagram of Yield Data from Table 6.2

To visualize the effect of one faulty measurement on variance component estimates, consider changing the value of the second measurement on the second batch in Table 6.2, from 95.03 to 93.10. The estimate of measurement error would change from $\hat{\sigma}_M^2 = 0.00018$ (calculated using the ANOVA method) to $\hat{\sigma}_M^2 = 0.18264$; three orders of magnitude! However, Figure 6.5 would easily draw attention to the fact that the majority of the measurement error is occurring within the second batch, and warrant investigation of the possibility of a mistake. In general, graphical analysis should be done in conjunction with any variance component estimation in order to prevent measurement errors or mistakes from influencing estimates.

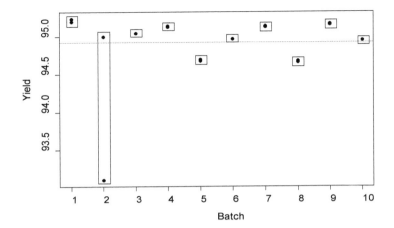

FIGURE 6.5: Dot Frequency Diagram of Yield Data from Table 6.2 with One Faulty Measurement

---

6.6 Components of Variance—Multiple Sources

6.6.1 Introduction

In most industrial data used in research, there are multiple sources of variability that should be understood. The simple experiments described in the last section are only adequate for estimating one source of error variability by separating it from process or piece-to-piece variation. In this section, we will describe nested sampling experiments that are often used in the chemical and process industries for estimating multiple sources of variability. In Section 6.7 we will describe staggered nested designs which allow estimation of the same quantities with less experimentation.

6.6.2 Nested Designs for Estimating Multiple Components of Variance

Tables B.7.1-2 through B.7.1-3 schematically represent nested experimental designs that are used for estimating three, four, or five different sources of variability. At the bottom of the tables are the formulas for the ANOVA calculations used to estimate the variance components.

To illustrate their use, let's consider an example experiment and analysis for three sources of variability. In the semiconductor industry, several integrated circuits, called dies, are manufactured on wafers of silicon. Routine tests are conducted to measure various electrical properties of the individual circuits. For illustrative purposes the data in Table 6.3 represent a typical set of test values collected from the production line during one week. In practice, 25 or 30 days of data would be preferred. Two wafers were chosen randomly from each day's production. Two circuits (or dies) were measured on each wafer. The data in Table 6.3 follow the pattern shown schematically in Table B.7.1-2. Mathematically we can represent the data by the equation:

$$Y_{ijk} = \mu + D_i + W_{ij} + C_{ijk} \tag{6.5}$$

where Y_{ijk} is the measurement on the kth circuit or die on the jth wafer on the ith day. The symbol, μ, represents the overall average measurement, D_i represents the difference from the overall average and the average for the ith day, W_{ij} represents the difference from the ith day average and the jth wafer average, and C_{ijk} represents the difference of the kth circuit or die from the average of the jth wafer in the ith day.

This is called the *nested model* for three sources since the two wafers on each day are unique to that day and the two dies on each wafer are unique to that wafer. The purpose is to estimate the variance of the terms D_i, W_{ij}, and the C_{ijk} in the previous equation, namely σ_D^2, σ_W^2, and σ_C^2 in order to characterize the process variability.

The summary statistics in the right five columns of Table 6.3 can be easily computed with a spreadsheet, and are used for calculating the sums of squares for the ANOVA table.

TABLE 6.3

Experimental Data and Summary Statistics for Semiconductor Measurements

Day i	Wafer j	Die k	y_{ijk}	y_{ijk}^2	$T_{ij\cdot}$	$T_{ij\cdot}^2$	$T_{i\cdot\cdot}$	$T_{i\cdot\cdot}^2$
1	1	1	5.1	26.01				
1	1	2	5.4	29.16	10.5	110.25		
1	2	1	5.5	30.25				
1	2	2	5.9	34.81	11.4	129.96	21.9	479.61
2	1	1	5.1	26.01				
2	1	2	5.7	32.49	10.8	116.64		
2	2	1	5.5	30.25				
2	2	2	5.9	34.81	11.4	129.96	22.2	492.84
3	1	1	4.9	24.01				
3	1	2	5.5	30.25	10.4	108.16		
3	2	1	5.4	29.16				
3	2	2	5.7	32.49	11.1	123.21	21.5	462.25
4	1	1	6.0	36.0				
4	1	2	5.6	31.36	11.6	134.56		
4	2	1	5.8	33.64				
4	2	2	6.3	39.69	12.1	146.41	23.7	561.69
		Totals	89.3	500.39		999.15		1996.39

The formulas for computing the ANOVA sums of squares, given at the bottom of Table B.7.1-2, are illustrated below. The formula for ssA in Table B.7.1-2 is $ssDay$, ssB is $ssWafer$, and ssC is $ssDie$.

$$ssDay = \left(\sum_{i=1}^{4} \frac{T_{i\cdot\cdot}^2}{4} \right) - \frac{T_{\cdots}^2}{4n_A}$$

$$= \frac{479.61}{4} + \frac{492.84}{4} + \frac{462.25}{4} + \frac{561.69}{4} - \frac{(89.3)^2}{4 \cdot 4}$$

$$= \frac{1996.39}{4} - \frac{89.3^2}{16}$$

$$= 499.0975 - 498.405625$$

$$= 0.6919$$

where $T_{\cdots} = 89.3$ is the grand total of all the observed values.

$$ssWafer = \sum_{i=1}^{4} \left(\sum_{j=1}^{2} \frac{T_{ij\cdot}^2}{2} - \frac{T_{i\cdot\cdot}^2}{4} \right)$$

$$= \frac{999.15}{2} - \frac{1996.39}{4}$$

$$= 499.575 - 498.0975$$

$$= 0.4775.$$

$$ssDie = \sum_{i=1}^{4} \sum_{j=1}^{2} \sum_{k=1}^{2} \left(y_{ijk}^2 - \frac{T_{ij\cdot}^2}{2} \right)$$

$$= 500.39 - \frac{999.15}{2}$$

$$= 499.575 - 499.575$$

$$= 0.815.$$

These sums of squares are used to create the ANOVA Table 6.4, where the expected values of the mean squares come from Table B.7.1-2. To estimate the variance components,

TABLE 6.4

Analysis of Variance Table

Source	df	Sum of Squares	Mean Squares	Expected Mean Squares
Days	3	0.6919	0.2306	$\sigma_{Die}^2 + 2\sigma_{Wafer}^2 + 4\sigma_{Day}^2$
Wafer	4	0.4775	0.1194	$\sigma_{Die}^2 + 2\sigma_{Wafer}^2$
Die	8	0.8150	0.1019	σ_{Die}^2

we solve the three equations obtained by equating the calculated mean squares in Table 6.4 to their expected values given in the table.

$$0.2306 = \sigma_{Die}^2 + 2\sigma_{Wafer}^2 + 4\sigma_{Day}^2$$

$$0.1194 = \sigma_{Die}^2 + 2\sigma_{Wafer}^2$$

$$0.1019 = \sigma_{Die}^2.$$

The solution is:

$$\hat{\sigma}_{day}^2 = 0.0278$$

$$\hat{\sigma}_{Wafer}^2 = 0.0088$$

$$\hat{\sigma}_{Die}^2 = 0.1019$$

where the hats over the letters indicate these are estimates calculated from the data. The total variability of measurements is estimated to be:

$$\hat{\sigma}^2 = \hat{\sigma}_{Day}^2 + \hat{\sigma}_{Wafer}^2 + \hat{\sigma}_{Die}^2 = 0.1384.$$

Now it can be seen that most of the variability is due to variation from die to die within a single wafer, i.e.,

$$100\% \times \left(\frac{0.1019}{0.1384} \right) = 73.58\%.$$

The day-to-day and wafer-to-wafer variability is a minor part. This information is very useful. If experimentation were to be conducted to discover the cause of the variability in electrical charactersistics of dies (or individual circuits), the factors should be chosen from among independent variables that could be varied within a single wafer and may cause changes from die to die within a wafer, rather than among independent variables that can be changed from wafer to wafer or day to day.

Again simple graphs like those shown in Section 4.4.1 can help to make sure no odd measurements are influencing the estimates. Figure 6.6 shows a dot frequency diagram of the data in Table 6.3.

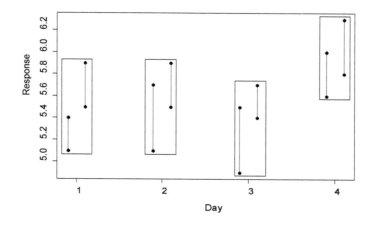

FIGURE 6.6: Dot Frequency Diagram of Yield Data from Table 6.3

Here the boxes contain the four measurements for each day and vertical lines connect the two measurements from the two dies on each wafer. In the figure it can be seen that the variability or difference between the two circuits is fairly uniform from wafer to wafer. There are no problem measurements like those shown in Figure 6.5. The variability from wafer to wafer within a day also appears to be fairly consistent. The only odd characteristic that can be seen is the fact that the measurements were always higher in the second wafer measured each day than they were in the first. Normally in an actual experiment like this, we should have 25 or more days of data in order to have a good estimate of σ^2_{Day}. If we were to see a pattern of the second wafer always being higher with 25 or more days of data, there would be cause for concern. However, with only four days of data we wouldn't be too concerned. After all it is possible to toss four heads in a row while coin tossing.

Next let's consider an example for four sources of variability. The data in Table 6.5 are from Bennett [1954] who describes a sampling study to separate the process variability from three sources of measurement error in lots of metallic oxide.

In this example two samples were taken from each of 13 different lots of material. Two chemists (not necessarily the same for each lot) each made duplicate analysis from each sample. The figures reported in the table are the metal content minus 80. Mathematically we can represent the measured metal content as a sum of five different quantities, i.e.,

TABLE 6.5

Metallic Oxide Data

Lot	Sample	Chemist 1		Chemist 2	
1	1	3.4	3.4	3.6	3.5
1	2	3.7	3.5	3.1	3.4
2	1	4.2	4.1	4.3	4.2
2	2	4.2	4.2	4.3	4.2
3	1	3.5	3.5	4.2	4.5
3	2	3.4	3.7	3.9	4.0
4	1	3.4	3.3	3.5	3.1
4	2	4.2	4.2	3.3	3.1
5	1	3.2	2.8	3.1	2.7
5	2	3.0	3.0	3.2	2.7
6	1	0.2	0.7	0.8	0.7
6	2	0.3	0.4	0.2	-0.1
7	1	0.9	0.6	0.3	0.6
7	2	1.0	1.1	0.7	1.0
8	1	3.3	3.5	3.5	3.4
8	2	3.9	3.7	3.7	3.7
9	1	2.9	2.6	2.8	2.9
9	2	3.1	3.1	2.9	2.7
10	1	3.8	3.8	3.9	3.9
10	2	3.4	3.6	4.0	3.8
11	1	3.8	3.4	3.6	3.8
11	2	3.8	3.6	3.9	4.0
12	1	3.2	2.5	3.0	3.5
12	2	4.3	3.5	3.8	3.8
13	1	3.4	3.4	3.3	3.3
13	2	3.5	3.5	3.2	3.3

$$Y_{ijkl} = \mu + L_i + S_{ij} + C_{ijk} + A_{ijkl}$$

where:

μ represents the grand average,

L_i represents the deviation of the average of Lot i from the grand average,

S_{ij} represents the deviation of the average for the kth sample in Lot i from the average in Lot i,

C_{ijk} represents the deviation of the average for the kth chemist for the jth sample in Lot i from the average in the jth sample in Lot i,

A_{ijkl} represents the deviation of the lth analysis by the kth chemist of the jth sample from Lot i from the average for the kth chemist on the jth sample in Lot i.

This is called a nested model for four sources. The variance of Y_{ijkl}'s, σ^2, can be written as the sum of four different variances, i.e.,

$$\sigma^2 = \sigma_L^2 + \sigma_S^2 + \sigma_C^2 + \sigma_A^2.$$

These variance components can be estimated by the ANOVA method. We can complete the ANOVA calculations by hand using the formulas at the bottom of Table B.7.1-3 as in the last example. There is a spreadsheet illustrating these calculations on the website for the book. However, procedures for making the ANOVA calculations automatically with modern computer software such as MINITAB or the open source program R have made this task much easier. Table 6.6 shows the ANOVA for this data.

TABLE 6.6
Analysis of Variance Table for Metallic Oxide Data

Source	df	Sum of Squares	Mean Squares	Expected Mean Squares
Lot	12	128.8235	10.7353	$\sigma_A^2 + 2\sigma_C^2 + 4\sigma_S^2 + 8\sigma_L^2$
Sample	13	2.5450	0.1958	$\sigma_A^2 + 2\sigma_C^2 + 4\sigma_S^2$
Chemist	26	2.9550	0.1137	$\sigma_A^2 + 2\sigma_C^2$
Analysis	52	1.8700	0.0360	σ_A^2

The estimates of the variance components are:

$$\hat{\sigma}_A^2 = 0.03596$$
$$\hat{\sigma}_C^2 = 0.03885$$
$$\hat{\sigma}_S^2 = 0.02053$$
$$\hat{\sigma}_L^2 = 1.31744.$$

In this example the measurement error is the sum of three components, $\sigma_A^2 + \sigma_C^2 + \sigma_S^2$, and appears to be a small proportion of the total variability, i.e.,

$$\frac{\sigma_A^2 + \sigma_C^2 + \sigma_S^2}{\sigma_A^2 + \sigma_C^2 + \sigma_S^2 + \sigma_L^2} = \frac{0.03596 + 0.03885 + 0.02053}{0.03596 + 0.03885 + 0.02053 + 1.31744} = 0.0675.$$

However, a simple histogram of the data in Figure 6.7 shows that the majority of the lot-to-lot variability can be explained by the fact that lots 6 and 7 have much lower metal content than the other 11 lots.

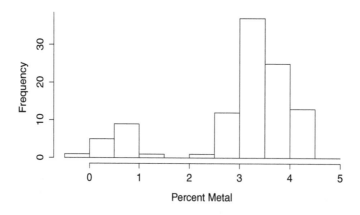

FIGURE 6.7: Histogram of Data from Table 6.5

If an assignable cause can be found for the low values found in these two lots, they should not be included in a study to identify and quantify the relative importance of measurement error. If lots 6 and 7 were removed from the data and the ANOVA calculations redone, the lot-to-lot variance, $\hat{\sigma}_L^2$ would be reduced substantially. This would cause the estimated proportion of the variance due to measurement to increase to over 50% of the total. This is much different than the results from all the data. If a larger study was conducted with 25 or more lots, the conclusions would be less dependent on the results of one or two lots. As in the last section, we always recommend the use of simple graphical displays of the data in conjunction with the variance component estimation.

6.7 Staggered Nested Designs

It can be seen in the ANOVA Table 6.6 from the last examples that the degrees of freedom for the mean squares are unequal or unbalanced when we have more than two components of variance. In the last example, with four components, fully half of the information we collected went into estimating the variance due to repeat analysis by the same chemist (the least important component). This seems like a waste, and it is! If we would have collected 25–30 lots, as recommended, we would have needed to make as many as 30×8=240 lab analyses. In cases like this, there are designs that we can employ to reduce the number of lab analysis by 50% or more and make things more balanced. These designs are called *staggered nested designs* [Smith and Beverly, 1981].

6.7.1 Design and Analysis with Staggered Nested Designs

As with the nested designs for estimating multiple sources discussed in Section 6.6.2, there are tables in Appendix B.7 that give the appropriate staggered nested designs for three, four, and five sources of variation (Tables B.7.2-1, B.7.2-2, and B.7.2-3, respectively). As with the other tables, the formulas for the ANOVA table are included below each design.

Let us revisit the first example discussed in Section 6.6.2 to illustrate the use of the staggered nested designs. The appropriate design corresponding to three sources of variation is given in Table B.7.2-1. It is the same as the nested design already used, except that only one circuit is measured for one of the wafers from each day, while two circuits are measured for the other wafer, as in the full nested design. The reduced set of data are given in Table 6.7. This table looks much like Table 6.3 (for the full nested design) with some of the rows removed. The missing rows correspond to the circuits that would not be measured in the staggered design.

Table 6.7 also has several additional columns of summary statistics that were used to compute the sums of squares in the ANOVA Table 6.8. These columns consist of totals over one or more subscripts, the number of analyses totaled (n's), and the total squared divided by the number of analyses in the total. It should be noted that these are the same summary statistics used for the full nested design, except that the number of analyses in any column was constant for the full nested design. Therefore the step of dividing by the number of analyses in the total was done later. The number of analyses totaled was not even listed in Table 6.3 but was included in the calculations for the ANOVA table given in the appendix. For the staggered nested design, the number of analyses in the total is not constant (except

TABLE 6.7

Semiconductor Measurements—Staggered Nested Design

Day i	Wafer j	Die k	y_{ijk}	y_{ijk}^2	$T_{ij\cdot}$	n_{ij}	$T_{ij\cdot}^2/n_{ij}$	$T_{i\cdot\cdot}$	n_i	$T_{i\cdot\cdot}^2/n_i$
1	1	1	5.1	26.01	5.1	1	26.01			
1	2	1	5.5	30.25						
1	2	2	5.9	34.81	11.4	2	64.98	16.5	3	90.75
2	1	1	5.1	26.01	5.1	1	26.01			
2	2	1	5.5	30.25						
2	2	2	5.9	34.81	11.4	2	64.98	16.5	3	90.75
3	1	1	4.9	24.01	4.9	1	24.01			
3	2	1	5.4	29.16						
3	2	2	5.7	32.49	11.1	2	61.605	16.0	3	85.833
4	1	1	6.0	36.0	6.0	1	36.0			
4	2	1	5.8	33.64						
4	2	2	6.3	39.69	12.1	2	73.205	18.1	3	109.203
		Totals	67.1	377.13	67.1		376.8	67.1		376.036

for $T_{i\cdot\cdot}$ where n_i is always 3), and so n is listed and the step of dividing by n is done after squaring the totals. The totals and counts can be easily computed with a spreadsheet.

The calculations needed to compute the sums of squares and mean squares in the ANOVA shown in Table 6.8 were carried out using the formulas in Appendix B, Table B.7.2-1. For example, the sum of squares for days was calculated as:

$$ssA = 376.036 - (67.1)^2/(3 \times 4) = .8355;$$

the sum of squares for wafers was calculated as:

$$376.8 - 376.036 = .7637;$$

and the sum of squares for circuit was calculated as:

$$377.13 - 376.8 = .3300.$$

From the mean squares, the components of variance were estimated as shown on the next page. It can be seen by comparing Table 6.8 to Table 6.4 that the variance estimate for wafers is a bit different due to the small number of days in this example. For larger studies the variance component estimates obtained from a staggered nested design are normally quite comparable to those that would be obtained from a full nested design with a considerable savings in sampling and measurements. For the three-component study the reduction in number of observations was only 1/4 (12 in Table 6.7, compared to 16 in Table 6.3); however, the savings are much greater for larger studies (i.e., 1/2 for four components and 9/16 for five components).

$$\sigma_C^2 = 0.08250 \ (44.85\%)$$
$$\sigma_W^2 = [0.019092 - 0.0825]/(4/3) = 0.08125 \ (44.17\%)$$
$$\sigma_D^2 = [0.27851 - 0.0825 - (5/3)0.08125]/3 = 0.02199 \ (10.98\%)$$
$$\sigma^2 = \sigma_C^2 + \sigma_W^2 + \sigma_D^2 = 0.18390 \ (100\%).$$

TABLE 6.8

Analysis of Variance Table for Staggered Nested Design

Source	df	Sum of Squares	Mean Squares	Expected Mean Squares
Day	3	0.8355	0.27851	$\sigma_C^2 + (5/3)\sigma_W^2 + 3\sigma_D^2$
Wafer	4	0.7637	0.19092	$\sigma_C^2 + (4/3)\sigma_W^2$
Circuit (Die)	4	0.3300	0.0825	σ_C^2

6.7.2 Staggered Nested Design Example with Four Sources

A dye manufacturer is interested in measuring the strength of a batch of dye so that a consistent strength product can be formulated. The measurement process involves taking a grab sample from a batch of dye, making up a solution with the dye, dyeing a piece of white cloth, and finally measuring the strength of the color of the cloth with a spectrophotometer. Each step in this procedure introduces variability and incurs a cost. In order to get the best assessment of dye strength for a given expenditure we must know how much variability is introduced at each step.

Table 6.9 shows the data from a staggered nested design that can be used to collect data to estimate the sources of variability. In this example, 10 grab samples were taken (but, again, 25 or more batches should be used in a practical application). The design is patterned after Table B.7.2-2. In this table, as with the previous examples, the data were coded (by subtracting 100) to minimize the number of digits needed to be written down. There is great efficiency in this design since only four measurements are made per batch, rather than eight that would be required by a full nested design.

TABLE 6.9
Dye Strengths—Staggered Nested Design

G	S	C	M	y_{ijkl}	y^2_{ijkl}	$T_{ijk\cdot}$	n_{ijk}	$\dfrac{T^2_{ijk\cdot}}{n_{ijk}}$	$T_{ij\cdot\cdot}$	n_{ij}	$\dfrac{T^2_{ij\cdot\cdot}}{n_{ij}}$	$T_{i\cdots}$	n_i	$\dfrac{T^2_{i\cdots}}{n_i}$
1	1	1	1	1.3	1.69	1.3	1	1.69	1.3	1	1.69			
1	2	1	1	3.7	13.69	3.7	1	13.69						
1	2	2	1	4.0	16.00									
1	2	2	2	4.6	21.16	8.6	2	36.98	12.3	3	50.43	13.6	4	46.24
2	1	1	1	7.8	60.84	7.8	1	60.84	7.8	1	60.84			
2	2	1	1	5.4	29.16	5.4	1	29.16						
2	2	2	1	7.0	49.00									
2	2	2	2	6.2	38.44	13.2	2	87.12	18.6	3	115.32	26.4	4	174.24
3	1	1	1	6.1	37.21	6.1	1	37.21	6.1	1	37.21			
3	2	1	1	6.1	37.21	6.1	1	37.21						
3	2	2	1	3.7	13.69									
3	2	2	2	3.7	13.69	7.4	2	27.38	13.5	3	60.75	19.6	4	96.04
4	1	1	1	5.8	33.64	5.8	1	33.64	5.8	1	33.64			
4	2	1	1	2.4	5.76	2.4	1	5.76						
4	2	2	1	1.2	1.44									
4	2	2	2	0.6	0.36	1.8	2	1.62	4.2	3	5.88	10.0	4	25.00
5	1	1	1	8.3	68.89	8.3	1	68.89	8.3	1	68.89			
5	2	1	1	7.1	50.41	7.1	1	50.41						
5	2	2	1	10.5	110.25									
5	2	2	2	9.7	94.09	20.2	2	204.02	27.3	3	248.43	35.6	4	316.94
6	1	1	1	1.2	1.44	1.2	1	1.44	1.2	1	1.44			
6	2	1	1	1.4	1.96	1.4	1	1.96						
6	2	2	1	3.0	9.00									
6	2	2	2	2.8	7.84	5.8	2	16.82	7.2	3	17.28	8.4	4	17.64
7	1	1	1	3.3	10.89	3.3	1	10.89	3.3	1	10.89			
7	2	1	1	6.3	39.69	6.3	1	39.69						
7	2	2	1	6.0	36.00									
7	2	2	2	5.4	29.16	11.4	2	64.98	17.7	3	104.43	21.0	4	110.25
8	1	1	1	7.0	49.00	7.0	1	49.00	7.0	1	49.00			
8	2	1	1	6.4	40.96	6.4	1	40.96						
8	2	2	1	7.5	56.25									
8	2	2	2	6.5	42.25	14.0	2	98.00	20.4	3	138.72	27.4	4	187.69
9	1	1	1	12.7	161.29	12.7	1	161.29	12.7	1	161.29			
9	2	1	1	9.9	98.01	9.9	1	98.01						
9	2	2	1	12.7	161.29									
9	2	2	2	11.9	141.61	24.6	2	302.58	34.5	3	396.75	47.2	4	556.96
10	1	1	1	9.4	88.36	9.4	1	88.36	9.4	1	88.36			
10	2	1	1	8.4	70.56	8.4	1	70.56						
10	2	2	1	10.2	104.04									
10	2	2	2	9.6	92.16	19.8	2	196.02	28.2	3	265.08	37.6	4	353.44
	Totals			246.8	1938.38			1936.18			1916.32			1884.34

The calculations needed to compute the sums of squares and mean squares in the ANOVA shown in Table 6.10 were carried out using the formulas in Appendix B, Table B.7.2-2. For example, the sum of squares for grab samples (ssG) was calculated as:

$$ssG = 1884.34 - (246.8^2)/40 = 361.584;$$

the sums of squares for solutions (ssS) was calculated as:

$$ssS = 1916.32 - 1884.34 = 31.98;$$

the sums of squares for cloth (ssC) was calculated as:

$$ssC = 1936.18 - 1916.32 = 19.86;$$

and finally the sums of squares for measurement (ssM) was calculated as:

$$ssM = 1938.38 - 1936.18 = 2.2.$$

TABLE 6.10
Analysis of Variance Table for Staggered Nested Design

Source	df	Sum of Squares	Mean Squares	Expected Mean Squares
Grab Sample (G)	9	361.584	40.176	$\sigma_M^2 + (3/2)\sigma_C^2 + (5/2)\sigma_S^2 + 4\sigma_G^2$
Solution (S)	10	31.98	3.198	$\sigma_M^2 + (7/6)\sigma_C^2 + (3/2)\sigma_S^2$
Cloth (C)	10	19.86	1.986	$\sigma_M^2 + (4/3)\sigma_C^2$
Measurement (M)	10	2.2	0.22	σ_M^2

The variance components were computed by setting the mean squares equal to their expected values solving the resulting equations as shown below.

$$\hat{\sigma}_M^2 = 0.220 \ (1.9\%)$$

$$\hat{\sigma}_C^2 = [1.986 - 0.220]/(4/3) = 1.324 \ (11.6\%)$$

$$\hat{\sigma}_S^2 = [3.19 - 0.22 - (7/6)(1.324)]/4 = 0.995 \ (8.4\%)$$

$$\hat{\sigma}_G^2 = [40.176 - 0.22 - (3/2)(1.342) - (5/2)(0.955)] = 8.895 \ (78.1\%).$$

From these calculations, it can be seen that most of the variance comes from the samples. If n_G grab samples are taken from a batch, n_S solutions are made from each sample, n_C cloths are dyed from each solution, and n_M measurements are made on each cloth, then according to Equation 2.10 and a generalization of Equation 4.4, the grand average measurement will have standard deviation:

$$\sigma_{AM} = \sqrt{\frac{\sigma_G^2}{n_G} + \frac{\sigma_S^2}{n_G \times n_S} + \frac{\sigma_C^2}{n_G \times n_S \times n_C} + \frac{\sigma_M^2}{n_G \times n_S \times n_C \times n_M}}.$$

Since the cost of taking a grab sample or making a solution from that sample is greater than the cost of dyeing a piece of cloth or measuring the strength of the color of the cloth with a spectrophotometer, the cost and resulting estimated standard deviation of the average strength of color measurement can be calculated for various combinations of numbers of samples, number of solutions from each sample, number of cloths dyed with each solution, and number measurements of each cloth as illustrated in Table 6.11.

In this table the cost of each plan is a sum of the products of the number of samples (etc.) for each plan and the cost of each step in the plan. By having the estimated variance components and calculating the cost and standard deviation of various plans the manufacturer can determine how to best assess the dye strength for a given expenditure. In this table it can be seen that the standard deviation of the average measurement is $\sigma_{AM} = 1.95$

TABLE 6.11

Cost and Standard Deviation of Various Inspection Alternatives

Step in Inspection Alternative	Estimated Variance Component	Cost	Number of Samples			
			Plan 1	Plan 2	Plan 3	Plan 4
Grab Sample (G)	$\hat{\sigma}_G^2 = 8.895$	$5	1	2	3	3
Solution (S)	$\hat{\sigma}_S^2 = 0.955$	$4	1	2	1	1
Cloth (C)	$\hat{\sigma}_C^2 = 1.324$	$1	1	2	1	3
Measurement (M)	$\hat{\sigma}_M^2 = 0.220$	$2	1	2	1	1
		Estimated Standard Dev. $(\hat{\sigma}_{AM})$	3.38	2.21	1.95	1.86
		Total Cost	$12	$24	$22	$24

when three grab samples are taken, one solution is made from each sample, one cloth is dyed with the solution, and one measurement made on the result (three measurements). This is less than the standard deviation of the average measurement when two grab samples are taken, two solutions are made from each sample, two cloths are dyed with each solution, and two measurements are made for each cloth (16 measurements with $\sigma_{AM} = 2.21$). In addition the cost is less ($22 compared to $24) even though the cost of taking a sample is the most expensive step.

6.8 Sequential Experimentation Starting With Components of Variation—Case Study

In most research studies more than one single set of planned experiments (and the associated analysis and interpretation of data) is required in order to acquire sufficient knowledge about a phenomenon under study. In Section 1.6, an example was presented where a screening design was followed by a full factorial and finally a response surface design in order to find operating conditions to maximize the yield of a chemical process.

In many situations there is more than one process step and a large number of potential factors that could be included in experiments. In these situations, it may be best to start a sequence of experiments with a variance component design as illustrated in Figure 1.4. By focusing on steps where the majority of variability exists, the number of potential factors to be included in factorial and screening designs can reduced.

6.8.1 Introduction

In this section we will discuss an example based on research in the BYU Department of Chemistry and Biochemistry. In one lab there is interest in catalyst support materials. The rate of catalyzed reactions can be increased by increasing the number of catalytic sites. To do this catalysts are dispersed on a porous support material. Alumina (Al_2O_3) is one such porous material. It has high thermal stability, high surface area, and a mesoporus nature. This support material is used in automotive catalytic converters (which convert toxic chemicals such as carbon monoxide and unburned hydrocarbons through catalyzed reactions to CO_2 and H_2O), in Fisher–Tropsch synthesis where liquid fuels are produced from natural gas, and various other catalyzed reactions. Alumina is produced in the lab by mixing aluminum isopropoxide, tetraethylorthosilate, and water in a pot. Next the mixture is dried and calcined. The goal is to be able to produce alumina with a high surface area and to control mesopore diameters to match specific applications (different catalytic reactions using different catalysts require different size mesopores).

6.8.2 Staggered Nested Experiment

There were many factors such as the ratios of components, mixing conditions, drying conditions, and calcining conditions that were hypothesized to affect both the surface area and mesopore diameter of alumina. The most obvious factor was the calcination temperature. In order to reduce the list of factors to consider in experiments, a variance component design was first undertaken to determine which process step (mixing or drying-calcining) contributed to the most variability to the final surface area and mesopore diameter. Calcination termperature was held constant in these experiments. A staggered nested design similar to the one shown in Table 6.7 was utilized. Fourteen batches of alumina were produced in the lab. Two samples were taken from each batch, and each sample was dried. The second dried sample from each batch was split again, and the two portions along with the first sample were calcined in different but identical ovens. The data from this staggered nested design are shown in Table 6.13.

The elements in the ANOVA table for the response log(Surface Area) were constructed using the calculation formulas illustrated in Section 6.7.1 and are shown in Table 6.12.

TABLE 6.12
Analysis of Variance Table for Staggered Nested Design

Source	df	Sum of Squares	Mean Squares	Expected Mean Squares
Batch	13	0.269022	0.020694	$\sigma_R^2 + (5/3)\sigma_S^2 + 3\sigma_B^2$
Sample	14	0.048390	0.003456	$\sigma_R^2 + (4/3)\sigma_S^2$
Replicate Oven	14	0.010964	0.000783	σ_R^2

From the mean squares in the table and the expected mean squares, the variance components were estimated to be:

$$\hat{\sigma}_B^2 = 0.0055231 (66.45\%)$$
$$\hat{\sigma}_S^2 = 0.0020049 (24.12\%)$$
$$\hat{\sigma}_O^2 = 0.0007832 (9.42\%).$$

Here it can be seen that the majority of variability in log(Surface Area) occurs between different batches. A smaller proportion of the variability occurs from sample to sample within a batch, and very little of the variability is due to calcining. A similar result was obtained after analysis of the data for pore diameter, and it is left as an exercise.

TABLE 6.13

Semiconductor Measurements—Staggered Nested Design

Batch	Sample	Oven	log(Surf. Area)	Pore Diam.
1	1	1	5.23682	33.4752
1	2	1	5.14813	33.2412
1	2	2	5.10208	32.5057
2	1	1	5.26304	31.5017
2	2	1	5.21170	35.2299
2	2	2	5.24759	36.2005
3	1	1	5.24009	31.4073
3	2	1	5.20672	35.5778
3	2	2	5.21101	34.0814
4	1	1	5.21009	32.2589
4	2	1	5.14559	34.2800
4	2	2	5.14680	34.5266
5	1	1	5.24152	36.6682
5	2	1	5.14286	39.7545
5	2	2	5.13941	38.2177
6	1	1	5.19048	32.1564
6	2	1	5.16443	32.4399
6	2	2	5.18900	31.3504
7	1	1	5.20697	32.0982
7	2	1	5.10396	31.9144
7	2	2	5.09383	31.7503
8	1	1	5.13185	33.5462
8	2	1	5.11872	32.8021
8	2	2	5.21635	32.0952
9	1	1	5.22351	35.4826
9	2	1	5.15600	36.6926
9	2	2	5.10482	37.5519
10	1	1	5.12308	37.5313
10	2	1	5.05064	35.3344
10	2	2	5.04090	36.7813
11	1	1	5.22686	27.6063
11	2	1	5.13627	34.6231
11	2	2	5.14593	34.9339
12	1	1	5.28308	27.5007
12	2	1	5.19549	30.6562
12	2	2	5.19170	30.3087
13	1	1	5.37740	30.5220
13	2	1	5.36597	29.3253
13	2	2	5.33676	30.1663
14	1	1	5.38495	32.4439
14	2	1	5.41341	32.2991
14	2	2	5.34578	31.7875

6.8.3 Follow-up Split-Plot Experiment

Based on the results of the variance component study, follow-up experiments were planned. In these experiments factors that could be changed in the batch mixing process, and in the drying and calcining of batches, were chosen to be studied. The following example is a simplification of what was actually done but illustrates the experimental procedure and results.

Two factors in the batch mixing step and one factor in the batch drying step are shown along with their levels in Table 6.14.

TABLE 6.14

Factors for Follow-up Experiments

Process Step	Factor	(-) Level	(+) Level
Mixing Step	X_2=Excess Alcohol added	No	Yes
Mixing Step	X_3=Mixing Time	15 minutes	30 minutes
Drying Step	X_1=Drying Temperature	Room Temp.	100 Deg. C

Since two of the factors were in the mixing step, and one was in the drying step, a split-plot experiment was carried out. Four batches were mixed in a random order corresponding to all combinations of factors X_2 = Excess Alcohol and X_3 = Mixing Time. Each batch was split and one half was randomly dried for 15 minutes and the other half was dried for 30 minutes. The whole process was repeated in a second replicate. This was done according to the plan in Table 4.8, using two whole plot factors and one split-plot factor. The results are shown in Table 6.15.

TABLE 6.15

Factors for Follow-up Experiments

Batch	Run Order	Dry Temp	Excess Alc	Mix Time	Replicate	log(SurfArea)	Pore Diameter
1	1	−	−	−	−	5.8176	48.3812
1	2	+	−	−	−	5.4868	48.5106
2	4	−	+	−	−	5.8273	12.7832
2	3	+	+	−	−	5.7137	15.8558
3	6	−	−	+	−	5.8854	48.4931
3	5	+	−	+	−	5.5524	47.4195
4	8	−	+	+	−	5.8642	55.3587
4	7	+	+	+	−	5.7514	55.0341
5	10	−	−	−	+	5.8107	37.7041
5	9	+	−	−	+	5.4883	36.406
6	12	−	+	−	+	5.8375	18.8043
6	11	+	+	−	+	5.7701	13.5869
7	13	−	−	+	+	5.7934	44.7954
7	14	+	−	+	+	5.4730	40.6455
8	16	−	+	+	+	5.8107	52.9249
8	15	+	+	+	+	5.7469	52.0090

To analyze the data, all effects and interactions effects were calculated in a spreadsheet (similar to the example in Figure 3.16). Figure 6.8 shows the results for the log(Surface Area) response. The analysis of the Pore Diameter Response is left as an exercise. The whole-plot standard error was computed as the square root of the sum of squared effects (divided by the number of effects summed and squared) as indicated in Table 4.8 for the design with two whole plot factors and one split-plot factor, i.e.,

$$s_W = \sqrt{\frac{E_{24}^2 + E_{34}^2 + E_{234}^2}{3}} = \sqrt{\frac{.023177^2 + (-.03632)^2 + .0005179^2}{3}} = 0.025055.$$

Thus, the formula in cell B24 in the spreadsheet shown in Figure 6.8 is:

```
= SQRT((K20^2+L20^2+P20^2)/3).
```

The split-plot standard error was calculated using the error effects indicated in Table 4.8 as:

$$s_E = \sqrt{\frac{E_{14}^2 + E_{124}^2 + E_{134}^2 + E_{1234}^2}{4}} = \sqrt{\frac{.014507^2 + .009253^2 + .000885^2 + (-.00019)^2}{4}}$$

$$= 0.017231.$$

The t-statistics were calculated by dividing the whole-plot effects by s_W and the split plot effect and interactions by s_E. The whole-plot error had 3 degrees of freedom and the split-plot error had 4 degrees of freedom. The P-values were calculated in the spreadsheet as described in Section 2.5.2. The factors $X_1 =$ Drying Temperature, $X_2 =$ Excess Alcohol, and their interaction were significant at the $\alpha = 0.05$ level of significance.

	Run	Wp	X1 DryM	X2 ExAlc	X3 MxTime	X4 Rep	X1X2	X1X3	X1X4	X2X3	X2X4	X3X4	X1X2X3	X1X2X4	X1X3X4	X2X3X4	X1X2X3X4	logSA
	1	1	-1	-1	-1	-1	1	1	1	1	1	1	-1	-1	-1	-1	1	5.81758
	2	1	1	-1	-1	-1	-1	-1	-1	1	1	1	1	1	1	-1	-1	5.486837
	3	2	-1	1	-1	-1	-1	1	1	-1	-1	1	1	1	-1	1	-1	5.827268
	4	2	1	1	-1	-1	1	-1	-1	-1	-1	1	-1	-1	1	1	1	5.713744
	5	3	-1	-1	1	-1	1	-1	1	-1	1	-1	1	-1	1	1	-1	5.885424
	6	3	1	-1	1	-1	-1	1	-1	-1	1	-1	-1	1	-1	1	1	5.552354
	7	4	-1	1	1	-1	-1	-1	1	1	-1	-1	-1	1	1	-1	1	5.864162
	8	4	1	1	1	-1	1	1	-1	1	-1	-1	1	-1	-1	-1	-1	5.751417
	9	5	-1	-1	-1	1	1	1	-1	1	-1	-1	-1	1	1	1	-1	5.810726
	10	5	1	-1	-1	1	-1	-1	1	1	-1	-1	1	-1	-1	1	1	5.488338
	11	6	-1	1	-1	1	-1	1	-1	-1	1	-1	1	-1	1	-1	1	5.837521
	12	6	1	1	-1	1	1	-1	1	-1	1	-1	-1	1	-1	-1	-1	5.770131
	13	7	-1	-1	1	1	1	-1	-1	-1	-1	1	1	1	-1	-1	1	5.793419
	14	7	1	-1	1	1	-1	1	1	-1	-1	1	-1	-1	1	-1	-1	5.473008
	15	8	-1	1	1	1	-1	-1	-1	1	1	1	-1	-1	-1	1	-1	5.810744
	16	8	1	1	1	1	1	1	1	1	1	1	1	1	1	1	1	5.746908
Sum			-1.66411	1.014207	0.12529	-0.16799	0.949117	0.003982	0.116057	-0.07616	0.185418	-0.29057	0.004682	0.074028	0.007077	0.041431	-0.00153	
Effect			-0.20801	0.126776	0.015661	-0.021	0.11864	0.000498	0.014507	-0.00952	0.023177	-0.03632	0.000585	0.009253	0.000885	0.005179	-0.00019	
t-statistic			-12.0721	5.059985	0.625087		6.885285	0.028886		-0.37996			0.033963					
df			4	3	3		4	4		3			4					
P-value			0.00027	0.014897	0.576232		0.002332	0.978339		0.72928			0.974534					

S$_W$ = 0.025055

S$_E$ = 0.017231

FIGURE 6.8: Split-Plot Analysis of Log(Surface Area)

The analysis of the Pore Diameter response showed that the significant factors were X_2=Excess Alcohol and X_3=Mixing Time, and their interaction. The following equations summarize the results.

$$\log(\text{Surface Area}) = 5.727 - 0.104X_1 + 0.063X_2 + 0.059X_1X_2$$

$$\text{Pore Diameter} = 39.29 - 4.75X_2 + 10.29\left(\frac{\text{Mixing Time} - 22.5}{7.5}\right)$$
$$+ 9.00X_2\left(\frac{\text{Mixing Time} - 22.5}{7.5}\right)$$

where $X_1 = -1$ implies Drying Temperature is room temperature as defined in Table 6.14, etc.

6.8.4   Conclusions

Since there were significant interactions found in the analysis of both responses, interaction graphs can help us interpret the results. Figure 6.9 shows the interaction plot of Excess Alcohol and Drying Temperature that had a significant effect on log(Surface Area). In this figure it can be seen that adding excess alcohol to the mixture causes log(Surface Area) to increase when the batch is dried at 100 °C. However, when the batch is dried at room temperature, adding excess alcohol has little effect and log(Surface Area) is uniformly greater than it is when the batch is dried at 100 °C.

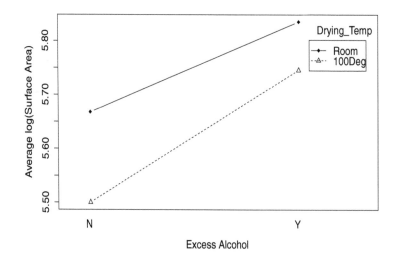

FIGURE 6.9: Interaction Between Excess Alcohol and Drying Method

Figure 6.10 shows the interaction of Excess Alcohol and Mixing Time that had a significant effect on Pore Diameter. This figure shows that increasing the mixing time causes the pore diameter to increase when there is excess alcohol in the mixture, but when there is no excess alcohol in the mixture, increasing the mixing time has a very small effect.

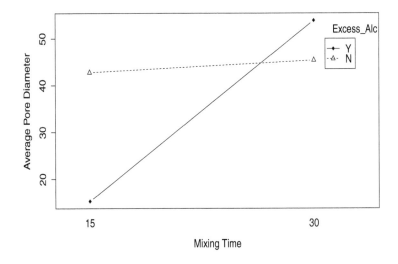

FIGURE 6.10: Interaction Between Excess Alcohol and Mixing Time

The original purpose of experimenting was to discover conditions that would allow making alumina with a high surface area, and the ability to control the pore diameter for various catalytic applications. Since Drying Temperature affected log(Surface Area) but not pore diameter, it should be set to temperature in order to produce alumina with a uniformly high surface area. Next excess alcohol should be added, and the mixing time can be adjusted to produce alumina with the pore diameter needed for specific applications.

This series of experiments started with a variance component study that showed which process steps contributed the most variability, and led to a choice of factors that could be manipulated in those steps. Since factors from two process steps were studied, a split plot experiment was utilized to determine the significant effects and models that allowed discovery of process conditions to satisfy the original objectives.

6.9   Summary

In this chapter we discussed methods for estimating sources of variability. When two random variables are added together the variance of the sum turns out to be the sum of the variances of each of the random variables. Using this fact we described a sampling experiment that can be used to apportion the variance in random data to two sources or variance components. The Analysis of Variance and the expected mean squares from the Analysis of Variance were used to calculate estimates of the variance components. We described some simple graphical methods that can help in understanding why the estimated variance gets apportioned as calculated. Next, we expanded the designs to allow estimation of several sources of variation. Nested designs and time saving staggered nested designs are useful for this purpose. Finally, we showed an example that illustrated how estimating the sources of variation helped to determine the process step where estimating factor effects would be most beneficial in follow-up experiments.

6.9.1   Important Equations

$Y = X + Z$

$\sigma_Y^2 = \sigma_X^2 + \sigma_Z^2$

$Y_{ij} = \mu + B_i + M_{ij}$

$\sigma_Y^2 = \sigma_B^2 + \sigma_M^2$

$\dfrac{\hat{\sigma}_B^2}{\hat{\sigma}^2}$

$Y_{ijk} = \mu + D_i + W_{ij} + C_{ijk}$

$\hat{\sigma}^2 = \hat{\sigma}_D^2 + \hat{\sigma}_W^2 + \hat{\sigma}_C^2$

6.9.2   Important Terms

The following is a list of the important terms and concepts covered in this chapter.

sampling experiment
variance components
nested design
expected mean squares
staggered nested design

6.10   Exercises

1. Finish part dimensions are measured on a micrometer, and the variability in measurements can be assumed to be a sum of the part-to-part variance, the variance due to gauge, and finally the variance due to repeat measurements. Symbolically this is expressed as:

$$\sigma_T^2 = \sigma_P^2 + \sigma_G^2 + \sigma_R^2. \tag{6.6}$$

If the three variance components are, respectively, 2.73, 0.034, and 0.0132:

   (a) Calculate the total variance in the measured dimensions.

   (b) Calculate the percentage or proportion of the total variation in measures that is actually due to part-to-part differences in dimensions.

2. Wernimont [1947] studied the variability in the measured melting point of hydroquinone. There were four different thermometers used in the study, and two repeat experiments were performed with each thermometer. The data are shown below.

Thermometer	Measurement	Melting Point
A	1	174.0
A	2	173.0
B	1	173.0
B	2	172.0
C	1	171.5
C	2	171.0
D	1	173.5
D	2	171.0

(a) Complete the ANOVA table, and compute the variance components for thermometer, and repeat measurement.

(b) Make a dot frequency diagram of the data and spot any outliers.

3. Duncan [1967] presented data from a rubber tread experiment. Two mixes were made from each of six different tread formulas. Two samples were taken from the first slab cured from the first mix and duplicate tests run. Only one sample was taken the second mix and one test run. The data are shown below.

Tread Formula	Mix	Sample	Test Result
1	1	1	98
1	1	2	52
1	2	1	86
2	1	1	75
2	1	2	96
2	2	1	64
3	1	1	34
3	1	2	7
3	2	1	34
4	1	1	32
4	1	2	19
4	2	1	7
5	1	1	138
5	1	2	113
5	2	1	53
6	1	1	102
6	1	2	74
6	2	1	204

(a) Write the model for this data. Define each term in your model.

(b) With the help of Appendix Table B.7.2-1, compute the ANOVA and estimate the three components of variance.

(c) Make a dot frequency diagram of the data, and note any unusual data points and describe how they might affect the estimates.

Chapter 7

Screening Designs

7.1 Introduction

Chapters 3 and 4 discussed two-level factorial designs. These designs are extremely efficient and not only allow one to determine the effects of several factors, but also to determine if interactions exist between those factors. However if the number of factors, k, being examined is large, the number of experimental runs, $n_F = 2^k$, may be excessive, as shown in Table 7.1.

TABLE 7.1

Number of Experiments Needed for a Full 2^k Factorial Design

Number of Factors, k	Number of Experiments, $n_F = 2^k$
2	4
3	8
4	16
5	32
6	64
7	128
8	256
9	512
10	1,024
11	2,048
12	4,096
13	8,192
14	16,384
15	32,768

The preliminary stage of experimentation, where the objective of the experiments is to determine which factors are important from a list of candidates, is called the screening stage. Often cause-and-effect diagrams, described in the next section, are used in brainstorming sessions where candidate factors are chosen; the typical result is a long list of potentially important factors. Therefore, at this stage it is normal to experiment with a large number of factors. The strategic position of this type of experiment is illustrated in Figure 7.1. But even though screening experiments will usually involve a large number of factors, in all likelihood only a few of them will actually have large effects. Further, it is extremely unlikely that the higher order interactions (that can be estimated from 2^k designs) will have large effects. Therefore, it would be a waste of resources to make as many experimental runs as required by a 2^k design at the screening stage. In order to reduce the amount of experimentation in this situation, special experimental designs constructed specifically for screening should be used.

| | | **Present** ⇓ | | **Goal** ⇓ | | |

		Present		**Goal**		

	0%			**Knowledge**		100%
Objective:	Preliminary Exploration	Screening Factors	Effect Estimation		Optimization	Mechanistic Modeling
No. of Factors		5 - 20	3 - 6		2 - 4	1 - 5
Model:	Variance Components	Linear	Linear + Interactions		Linear + Interactions + Quadratics	Mechanistic Model
Purpose:	Identify Sources of Variability	Identify Important Factors	Estimate Factor Effects + Interactions		Fit Empirical Model Interpolate	Estimate Parameters of Theory Extrapolate
Designs:	Nested Designs	Fractional Factorials	Two-level Factorials		RSM	Special Purpose

FIGURE 7.1: Strategic Position of Screening Designs in the Course of Experimentation

Typically when a large number of factors are under study, even efficient researchers with some familiarity with the use of statistical experimental strategies may abandon their thoughts of factorial designs and revert to a seat-of-the-pants approach or one-at-a-time experimentation. The one-at-a-time approach was discussed in some detail in Chapter 1. Remember that the strategy involves performing one control experiment followed by an additional experiment for each factor, varying its setting to a new value while holding all other factors constant at their control value for that experiment. As mentioned in Chapter 1, this is a horribly poor strategy for a number of reasons. In our screening situation, the main shortcoming is that the effect of each factor is determined by the difference in response between only two runs. The result is that there is no averaging of data to reduce the noise, and so the experimental variability may overshadow the factor effects.

Another strategy often used when a large number of factors are under study is to pick a subset of the factors and do a complete factorial design. The choosing of a subset of the factors is done by intuition, hunch, opinion, experience, etc., and it may seem like there is some scientific basis for the process. But in reality, this is even a worse strategy than one-at-a-time. The whole purpose of working with the large number of factors is to determine which subset of the factors is most important. If you already knew which factors were important, there would be no need for the experiments at all. By guessing which factors are of practical significance, it is very likely that one or more of the most important factors will be left out of the design. Then the benefits of knowing the effects of those factors would be lost to you. The penalty for such an oversight is amplified in a competitive world, because the competition may find them. The best strategy is to choose a subset of the experiments required for the full 2^k design. This strategy will be the main focus of this chapter.

7.2 Cause-and-Effect Diagrams

The chances of hypothesizing, or proposing, an adequate solution to a problem are higher if brainstorming sessions are used where the opinions of many individuals who are knowledgeable can be obtained. A good tool for summarizing brainstorming sessions on the cause of a problem, or proposed solutions to a problem, is the cause-and-effect diagram.

The cause-and-effect diagram was originated by Kaoru Ishikawa, the famous Japanese quality control guru, while he was explaining to some engineers the myriad of factors that can affect the variability in a quality characteristic. Cause-and-effect diagrams are best constructed in team meetings, because the thoughts of one individual may stimulate ideas in others. The diagram tends to keep the discussion focused on enumerating causes of a problem, and the process of constructing the diagram usually proves to be very educational for everyone involved.

Cause-and-effect diagrams are constructed by writing the problem or concern in a box at the right. Next, a main arrow is drawn from the left pointing at the box, as shown in Figure 7.2(a). Then, stems representing the main classifications of causes are attached to the main arrow. Ishikawa has suggested organizing causes into four main classifications: Materials, Methods, Machines, and Man. Of course these classifications are not sacred and can be modified to be more appropriate for a particular problem when desired. However, the number of main branches should always be near four, never less than three or more than six, because that would represent poor organization. Adding Ishikawa's four classifications to the diagram results in Figure 7.2(b). Finally, specific things thought to be causes of the problem can be added as twigs to the main stems. For example, in a circuit board assembly plant, a cause-and-effect diagram addressing the problem of faulty circuit boards might appear as Figure 7.3.

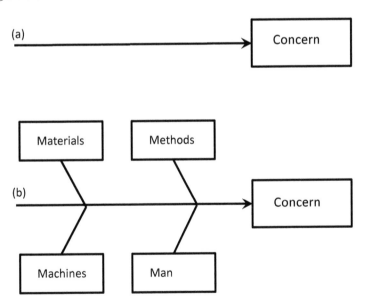

FIGURE 7.2: Cause-and-Effect Diagram Construction

Another way of organizing a cause-and-effect diagram is to use the pattern of a process flow diagram. In this version of the cause-and-effect diagram, the main arrow coming from

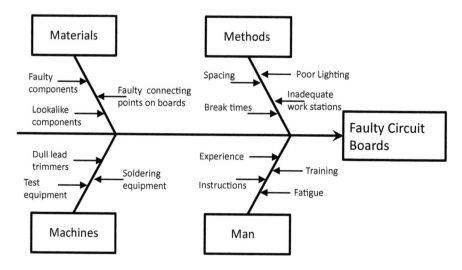

FIGURE 7.3: Cause-and-Effect Diagram for Faulty Circuit Boards

the left shows the major process steps and decision points. The specific causes of the problem are then attached at the process step where they would occur. An example of a cause-and-effect diagram of this type for the hand-assembly portion of a circuit board assembly might appear like Figure 7.4.

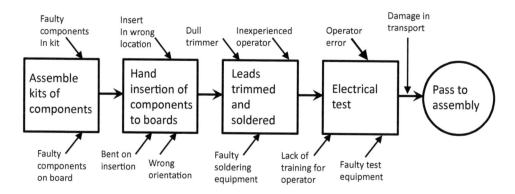

FIGURE 7.4: Process Flow Cause-and-Effect Diagram for Faulty Circuit Boards

Specific causes should be added to the diagram through group brainstorming. One convenient way this can be done is to record an idea from each person in the group until all ideas are exhausted. Using this technique, each person takes a turn and must either add one specific cause to the chart or pass.

Another way to stimulate ideas for specific causes, discussed by Ishikawa, is to ask "Why?" five times. In other words, one could question the brainstorming group with: Why are circuit boards faulty? If one of the group's answers was because some components placed on the board do not function, the next question might be: Why don't some components function? If one of the answers to this question was because some components are not placed properly on the board, a logical next question might be: Why are some components not placed on the board properly? By asking "Why?" five times or until all ideas are exhausted, many detailed causes can be discovered and added to the chart.

Cause-and-effect diagrams for the most part represent opinions or hypotheses. However, if certain causes are known and quantifiable, that information could be added to the chart with footnotes or other symbols. In this way, the cause-and-effect diagram can show the level of process knowledge.

After completion of the cause-and-effect diagram, solutions may be hypothesized for correcting the problem. As these hypothesized solutions are tested through experiments, the resulting information can be used to update the diagram. Causes found not to be important can be crossed out. Others that are demonstrated to be important can be highlighted or footnoted. Demonstrated solutions can be added to the appropriate causes of problems with footnotes, etc. In this way, the cause-and-effect diagram not only serves to collect and organize opinions about problems, but it can be used to guide and record the progress of research efforts aimed at solving the problems.

7.3   Fractionating Factorial Designs

After brainstorming, we typically have a long "laundry list" of possible factors that could have important effects on our system. We now need to sift through (or screen) the factors to separate the ones that have a large impact from those that have a small (or no) impact. The best strategy for screening experiments is to choose a subset or fraction of the experimental runs required for a full 2^k design in all the factors under study. We will discuss two categories of screening designs in this chapter, both obtained by taking a subset of the full 2^k factorial design: (a) fractional factorial designs, and (b) Plackett-Burman designs. The fractional factorial designs are obtained by taking a regular fraction of a 2^k design like $1/2, 1/4, 1/8$ (i.e., where the denominator is a power of 2). For example, to study the effect of six factors, a $1/2$ fractional factorial would consist of $1/2 \times 2^6 = 32$ runs. A $1/4$ fractional factorial would consist of $1/4 \times 2^6 = 16$ runs. The Plackett-Burman designs (named after the two gentlemen who developed them) are irregular fractions of 2^k designs which are constructed in increments of 4 runs. So, for example, you could study six factors using a Plackett-Burman design with 12 runs.

Choosing the runs to be eliminated from a full factorial, in order to obtain a fractional screening design, cannot be done in a haphazard fashion. For example, if the indicated runs in the 2^4 design shown in Table 7.2 were arbitrarily chosen to be eliminated, the resulting design shown in Table 7.3 has Factors 2 and 3 completely confounded. In other words, whenever Factor 2 is at the low level so is Factor 3, and whenever Factor 2 is at the high level so is Factor 3. Therefore, the effect of Factor 2 and the effect of Factor 3 would both be calculated as:

$\text{Effect}_2 = (Y_3+Y_4+Y_7+Y_8)/4 \text{ - } (Y_1+Y_2+Y_5+Y_6)/4$

$\text{Effect}_3 = (Y_3+Y_4+Y_7+Y_8)/4 \text{ - } (Y_1+Y_2+Y_5+Y_6)/4.$

In this case it would be impossible to say whether the difference in these two averages was caused by the change in Factor 2 or by the change in Factor 3. In fact, what we are actually estimating by the difference in the averages above is the sum of the X_2 effect and the X_3 effect. Since we have no way of sorting out how much X_2 contributed to the difference in averages versus how much X_3 contributed, we say that Factors 2 and 3 are confounded.

An alternative, and strategic method for eliminating the runs from a 2^4, is shown in

TABLE 7.2

Full 2^4 Factorial Design with Arbitrarily
Eliminated Runs

Run	X_1	X_2	X_3	X_4	Delete?
1	−	−	−	−	
2	+	−	−	−	
3	−	+	−	−	Delete
4	+	+	−	−	Delete
5	−	−	+	−	Delete
6	+	−	+	−	Delete
7	−	+	+	−	
8	+	+	+	−	
9	−	−	−	+	
10	+	−	−	+	
11	−	+	−	+	Delete
12	+	+	−	+	Delete
13	−	−	+	+	Delete
14	+	−	+	+	Delete
15	−	+	+	+	
16	+	+	+	+	

TABLE 7.3

Runs Remaining from Full 2^4 Factorial
Design after 8 Runs Arbitrarily Eliminated

Run	X_1	X_2	X_3	X_4	Response
1	−	−	−	−	Y_1
2	+	−	−	−	Y_2
3	−	+	+	−	Y_3
4	+	+	+	−	Y_4
5	−	−	−	+	Y_5
6	+	−	−	+	Y_6
7	−	+	+	+	Y_7
8	+	+	+	+	Y_8

Table 7.4. By eliminating the indicated runs in this table, the remaining runs shown in Table 7.5 form a design with no main effects confounded. In fact, we can observe that this design is perfectly orthogonal, by the fact that the factor level combinations for every pair of factors forms a 2^2 full factorial with two replicates at each run. See Table 7.6 for two examples.

7.4   Fractional Factorial Designs

A special class of designs called fractional factorials provide us with a method for strategically picking a subset of runs from a full factorial without confounding main effects. To use this method, we actually start with a full factorial design with the correct number of runs and then add more factors, rather than starting with a full factorial with the correct number

TABLE 7.4

Full 2^4 Factorial Design with
Strategically Eliminated Runs

Run	X_1	X_2	X_3	X_4	Delete?
1	−	−	−	−	
2	+	−	−	−	Delete
3	−	+	−	−	Delete
4	+	+	−	−	
5	−	−	+	−	Delete
6	+	−	+	−	
7	−	+	+	−	
8	+	+	+	−	Delete
9	−	−	−	+	Delete
10	+	−	−	+	
11	−	+	−	+	
12	+	+	−	+	Delete
13	−	−	+	+	
14	+	−	+	+	Delete
15	−	+	+	+	Delete
16	+	+	+	+	

TABLE 7.5

Runs Remaining from Full 2^4 Factorial
Design after 8 Runs Strategically
Eliminated

Run	X_1	X_2	X_3	X_4	Response
1	−	−	−	−	Y_1
4	+	+	−	−	Y_4
6	+	−	+	−	Y_6
7	−	+	+	−	Y_7
10	+	−	−	+	Y_{10}
11	−	+	−	+	Y_{11}
13	−	−	+	+	Y_{13}
16	+	+	+	+	Y_{16}

TABLE 7.6

Factor Combinations for X_1X_2 and X_3X_4
and Associated Runs from Table 7.5

X_1	X_2	Runs	X_3	X_4	Runs
−	−	1, 13	−	−	1,4
+	−	6, 10	+	−	6, 7
−	+	7, 11	−	+	10, 11
+	+	4, 16	+	+	13, 16

of factors and eliminating runs, as shown in Table 7.4. The end result will be the same. When we use this method we say that we are constructing a fractional factorial design. We will illustrate this method by showing how to construct a half fraction.

7.4.1   Constructing Half Fractions

Half the number of runs in a 2^k factorial is $(1/2)(2^k) = 2^{k-1}$. This is the number of runs in a full factorial with $k - 1$ factors. The method of constructing a half fraction is then as follows:

1. Write down the **base design**—a full factorial in $k - 1$ factors.

2. Add the kth factor to the design by **confounding** it with (i.e., making it identical to) one of the columns in the calculation matrix—specifically, the column for the highest order interaction.

3. Use the original $k - 1$ columns plus the column for the kth factor to define the design.

For example, in a half fraction of a 2^4 experiment we would start with a $2^{4-1} = 2^3$ factorial and deliberately confound an added fourth factor with the three-way interaction $X_1X_2X_3$, as shown in Table 7.7. Using only the columns for the factors themselves (X_1, X_2, X_3, and X_4) gives us the design which is shown in Table 7.8. By examination, it can be seen that this design is identical to the one shown in Table 7.5 with the runs in a different order.

TABLE 7.7
Construction of a Half Fraction of a 2^4 Design

Mean	X_1	X_2	X_3	X_1X_2	X_1X_3	X_2X_3	X_4 $X_1X_2X_3$
+	−	−	−	+	+	+	−
+	+	−	−	−	−	+	+
+	−	+	−	−	+	−	+
+	+	+	−	+	−	−	−
+	−	−	+	+	−	−	+
+	+	−	+	−	+	−	−
+	−	+	+	−	−	+	−
+	+	+	+	+	+	+	+

TABLE 7.8
Resulting Design of a
Half Fraction of a 2^4
Factorial

X_1	X_2	X_3	X_4
−	−	−	−
+	−	−	+
−	+	−	+
+	+	−	−
−	−	+	+
+	−	+	−
−	+	+	−
+	+	+	+

We can understand why main effects are not confounded with each other in fractional factorial designs, because they each represent one column in the calculation matrix for a full factorial design. However, let us expand the number of columns in Table 7.8 to list the full calculation matrix in all four factors. This calculation matrix is shown in Table 7.9. In this matrix, there are columns for all possible main effects and interactions of the four factors. We can see that each main effect column is the same as (we say *confounded with*) one of the interaction columns, and other interaction columns are confounded with additional interactions. When we use only a subset of the runs in a full 2^k factorial we cannot expect to be able to estimate all effects and interactions independently. There has to be some penalty for the reduction in experiments and that penalty is confounding. Let's examine the confounding in half fractions and illustrate how we deal with them in practice.

TABLE 7.9
Full Calculation Matrix for a Half Fraction of a 2^4 Factorial Design

Mean	X_1	X_2	X_3	X_4	X_1X_2	X_1X_3	X_1X_4	X_2X_3	X_2X_4	X_3X_4	$X_1X_2X_3$	$X_1X_2X_4$	$X_1X_3X_4$	$X_2X_3X_4$	$X_1X_2X_3X_4$
+	−	−	−	−	+	+	+	+	+	+	−	−	−	−	+
+	+	−	−	+	−	−	+	+	−	−	+	−	−	+	+
+	−	+	−	+	−	+	−	−	+	−	+	−	+	−	+
+	+	+	−	−	+	−	−	−	−	+	−	−	+	+	+
+	−	−	+	+	+	−	−	−	−	+	+	+	−	−	+
+	+	−	+	−	−	+	−	−	+	−	−	+	−	+	+
+	−	+	+	−	−	−	+	+	−	−	−	+	+	−	+
+	+	+	+	+	+	+	+	+	+	+	+	+	+	+	+

7.4.2 Confounding in Half Fractions

The confounding of main effects and interactions in fractional factorial designs is direct and easy to figure out. One way is, of course, to look at the full calculation matrix in all the factors as we did in Table 7.9, and see which columns are the same. For a half fraction, they will occur in pairs. In other words, each main effect and interaction will be confounded with one other effect or interaction. But there is an easier way. To facilitate our discussion on this subject, let us define the **alias structure** or **confounding pattern** for the design as the list of factor effect names, along with each effect and interaction that is confounded with it.

Obtaining an alias structure is easiest for a half fractional factorial design, so let's start with that. In Table 7.7 the X_4 column was made identical to the $X_1X_2X_3$ column when we constructed the design. We indicate the columns are the same by the notation, $X_4 = X_1X_2X_3$, or for simplicity we will usually write only the subscripts, $4 = 123$. This equation that describes how we added the extra factor to our design is called a **generator** for the design. A half fractional factorial design has one generator. A 1/4 fractional factorial design would need two generators, a 1/8 fractional factorial would need three generators, and so on. Please remember that this generator (equation) does **NOT** mean that the effect of X_4 equals the interaction of X_1, X_2, and X_3. What it does mean is that the experimental conditions in this particular design were chosen such that the calculation column for X_4 is the same as that for $X_1X_2X_3$. The two quantities are confounded, not equal. To say it one more time: the "=" in the generator $4 = 123$ means that the columns in the calculation matrix are the same. The result is that when you calculate the effect of X_4, you are actually getting the sum of the effect of X_4 and the interaction of $X_1X_2X_3$.

For the eight run fractional factorial design shown in Table 7.8, the columns labeled X_1, X_2, X_3, and X_4 would be used to define the conditions for each of the eight experiments. When calculating the effects, however, all seven columns from Table 7.7 will be used. The effects we calculate are often called ***contrasts*** as an explicit reminder that the effects are confounded with interactions. As already stated, the effect for X_4 is confounded with the $X_1X_2X_3$ interaction. But that is not the whole story; there is more confounding taking place. To elucidate the whole confounding pattern, let us start by finding the factor which is confounded with the overall mean. To do this, multiply both sides of the generator by the left hand side. We get

$$4 \times 4 = 123 \times 4, \text{ or } \quad 4^2 = 1234, \text{ or } \quad I = 1234,$$

where the multiplication represents elementwise multiplication of the $+$ or $-$ signs in the column labeled 4 and 123, and I represents a column of $+$ signs (i.e., the sign on each data value in the sum when computing the average or mean of the data). The symbolic equation $I = 1234$, which indicates the $X_1X_2X_3X_4$ interaction is confounded with the mean, is called the ***defining relation***. It is given this distinctive name because it can be used to find the rest of the confounding pattern. To find the confounding for any factor or interaction just multiply each side of the defining relation by the symbol for that factor or interaction. Multiplying I by any factor or interaction just gives back that factor or interaction (similar to multiplying any number by $+1$ just gives back that number). And multiplying any two factors or interactions together is equivalent to replacing the two sets of subscripts by all subscripts which appear in either but not both groups. For example, to find the confounding for the X_2X_4 interaction, we multiply the defining relation by 24 and we get: $I \times 24 = 1234 \times 24$, or $24 = 12^234^2$, or $24 = 13$ (since $2^2 = I$ and $4^2 = I$). In this way, the whole confounding pattern (or alias structure) is computed from $I = 1234$ as:

$$I = 1234$$
$$1 = 234$$
$$2 = 134$$
$$3 = 124$$
$$4 = 123$$
$$12 = 34$$
$$13 = 24$$
$$14 = 23$$

By using the generator $4 = 123$ to create a half fraction of the 2^4 design, we have selected half the runs in the full factorial design in a way that no main effect is confounded with another main effect. The other half fraction (i.e., the eight runs in Table 7.4 that we eliminated) will have the same property. The generator, $4 = 123$, generates the principal fraction, and the generator, $4 = -123$ (which means let the column of signs that defines X_4 be the negative of the column of signs that defines the $X_1X_2X_3$ interaction), generates the other half fraction. In practice either fraction can be used.

We should notice that since we have conducted (or plan to conduct) eight experiments, we have eight pieces of information, and therefore we can estimate, at most, eight contrasts. Since we have 16 effects and interactions (including an overall mean) in the full 2^4 factorial, we should not expect to be able to estimate each of them cleanly when we only run half of the full design. It can be thought of as a Law of Conservation of Information! The eight contrasts that we can estimate are each the sum of two effects and/or interactions. That is the price we pay for doing fewer experiments. If we ran a 1/4 fraction of a full factorial design, each contrast that we calculate would be the sum of four effects and interactions. If we ran a 1/8 fraction of a full factorial design, each contrast that we calculate would be the sum of eight effects and interactions. And so on.

7.4.3 Simple Example of a Half Fraction

In an attempt to improve the process by decreasing variability in copper-plating thickness, a half fraction of 2^3 experiment was conducted. Table 7.10 shows the factors (identified by brainstorming and a resulting cause-and-effect diagram) and levels studied in the experiment. Since data collection was expensive, it was desirable to run only half of the 2^3 experiments required for a full factorial. Note: This is generally not an acceptable design, since it leaves no degrees of freedom to estimate experimental variability. This means that no significance testing is possible.

TABLE 7.10
Factors for Half Fraction Factorial Design

Factor	(−) Level	(+) Level
X_1 = Anode height	Up	Down
X_2 = Circuit Board Orientation	In	Out
X_3 = Anode placement	Spread	Tight

The plan for the half fraction factorial design was created by starting with base 2^2 design in Factors 1 and 2, and then assigning the added Factor 3 to the negative of the interaction column. Therefore, the

generator is: $3 = -12$. This results in the

defining relation: I $= -123$, and the

confounding pattern:

$1 = -23$
$2 = -13$
$3 = -12.$

This means that each main effect is confounded with a two-factor interaction, and that the three-factor interaction is confounded with the mean. Even though there is confounding, we still get useful information from this experiment if we can assume the interactions are negligible.

Shown in Figure 7.5 is the list of experiments and the response, which was the variance in plating thickness for replicate circuit boards coated under the same conditions. Since there are only four runs, this table is also the calculation matrix, and the calculation of the effects is shown. The effects in this case are just half the sum-of-products row.

Remember, with this data alone (only four runs) it is impossible to say which if any of the effects are significant. Also, due to the confounding, it is impossible to know if the largest effect, −4.93, is the effect of Factor 3 (Anode Placement) or is the negative of the interaction effect between Factors 1 and 2. Similar statements could be made for the other two effects. Additional experiments could have been conducted to resolve this ambiguity. However, in this case, the two-factor interactions were assumed to be negligible, and the factors were all assumed to be significant.

Interpreting the contrasts as the main effects alone, it could be seen that all effects were negative. Thus, the results indicated that using the high level of each factor (i.e., anode height = down, circuit board orientation = out, and anode placement = tight) would result

	A	B	C	D	E	F
1	**Run**	**Mean**	X_1	X_2	X_3	s^2
2	1	1	-1	-1	-1	11.63
3	2	1	1	-1	1	5.57
4	3	1	-1	1	1	3.57
5	4	1	1	1	-1	7.36
6	SUMPRODUCT	28.13	-2.27	-6.27	-9.85	
7	Effects	7.03	-1.14	-3.14	-4.93	

FIGURE 7.5: 2^{3-1} Fraction Factorial Design and Response

in a smaller response (variance of copper plating thickness). But, no experiment had been conducted at this set of factor levels. Therefore, a confirmation experiment was run at the $+$, $+$, $+$ levels of the factors resulting in a variance of 3.13, a 70% reduction in variability. Since the result of this confirmation experiment resulted in a lower response value, as had been predicted, it justified the earlier assumption that the interactions were negligible, and supported the interpretation that all main effects had negative effects. Therefore permanent changes were implemented in the process to use this $(+, +, +)$ configuration of the plating cells.

7.4.4   One-Quarter and Higher Fractional Factorials

In constructing half replicates of 2^k designs, only half the total experimental runs were made; therefore, each factor effect and interaction that could be estimated had one other interaction effect confounded with it (they are confounded in groups of two). If one-quarter of the total experimental runs were made, each factor effect and interaction would have three other interactions confounded with it (they would be confounded in groups of four), as was mentioned in the previous section. If one-eighth of the total experimental runs were made, the factors and interactions would be confounded in groups of eight, and so on—the greater the fractionation, the greater the confounding. This is logical. The interactions between the factors do not go away just because we do fewer experiments; instead, their impacts get added to the contrasts that can be estimated.

This will be illustrated by examining in detail a one-quarter replicate. To construct a one-quarter replicate, we begin with a full factorial in k-2 factors, called the **base design**. As with the half-fraction, the base design is the full factorial design which will give us the number of runs we want to make. In the case of a one-quarter replicate, we have two factors left over whose levels still must be set. These added factors are taken care of by associating (confounding) them with two of the columns of the calculation matrix. For example, in constructing a one-quarter replicate of a 2^5 factorial, we start by writing a 2^3 base design, then we associate the added factors, 4 and 5, with two interactions in the calculation matrix. Let us arbitrarily use the 12 and 13 interactions for X_4 and X_5, respectively. That is, our generators are: $4 = 12$ and $5 = 13$. This is shown in Table 7.11.

From the generators we find two of the interactions confounded with the mean, which are I = 124 = 135. The third interaction can be found by multiplication, recognizing that if 124 and 135 are both columns of "$+$" signs ($=$ I), then when we multiply them together we still get a column of "$+$" signs. Therefore, the last interaction is: I = (124)(135) = 2345, and the complete defining relation is: I = 124 = 135 = 2345.

Next, the entire confounding pattern can be obtained by multiplication as before. It is given in Table 7.12. Note that the first column corresponds to the contrasts we can estimate, which we obtain from the headings of the columns in the calculation matrix. Notice also

TABLE 7.11

Construction of a One-Quarter Fraction of a 2^5 Design

Mean	X_1	X_2	X_3	X_4 X_1X_2	X_5 X_1X_3	X_2X_3	$X_1X_2X_3$
+	−	−	−	+	+	+	−
+	+	−	−	−	−	+	+
+	−	+	−	−	+	−	+
+	+	+	−	+	−	−	−
+	−	−	+	+	−	−	+
+	+	−	+	−	+	−	−
+	−	+	+	−	−	+	−
+	+	+	+	+	+	+	+

$$\Downarrow$$

X_1	X_2	X_3	X_4	X_5
−	−	−	+	+
+	−	−	−	−
−	+	−	−	+
+	+	−	+	−
−	−	+	+	−
+	−	+	−	+
−	+	+	−	−
+	+	+	+	+

that the confounding for each contrast is a group of four as expected for a one-quarter replicate.

TABLE 7.12

Full Confounding Pattern for the 2^{5-2}
Design with Generators: 4=12 and 5=13

I	=	124	=	135	=	2345
1	=	24	=	35	=	12345
2	=	14	=	1235	=	345
3	=	1234	=	15	=	245
4	=	12	=	1345	=	235
5	=	1245	=	13	=	234
23	=	134	=	125	=	45
123	=	34	=	25	=	145

When we construct higher fractional replicates like 1/8 or 1/16, we will have the mean confounded with 7 or 15 other interactions, respectively. For the 1/8 replicate, we start by writing down the full factorial in k-3 factors and then associating the last three factors with any three interactions of the first k-3 factors (i.e., we need three generators). This, in effect, specifies three interactions that will be confounded with the mean, I. Then, the other four interactions confounded with the mean will be all of the possible pair-wise products, and the three-way product of the original three interactions in the defining relation. For the 1/16 replicate, it is necessary to specify four generators which will give us four interactions in the defining relation. Then, the remaining eleven interactions in the defining relation will be all of the possible pair-wise products, three-way products, and the four-way product of the original four.

7.4.5   Fractional Factorial Design Tables

There are usually a number of ways of associating the added p factors with interactions in a $(1/2)^p$ fractional factorial. For example, in the one-quarter fraction of a 2^5 shown above we could use the generators 4=23 and 5=123 rather than 4=12 and 5=13 that we did use. Each of the ways of choosing the generators will result in a different design and hence a different confounding pattern. Some designs will be more desirable than others since they will have fewer low order interactions confounded with main effects. Rather than leave the choice of the generators to chance, there are two approaches that can be used. One is to use statistical software that has capabilities for automatic creation and analysis of fractional factorial designs. The illustration and use of such software will be left as a laboratory exercise. In the absence of statistical software, Appendix B.1 provides simple tables of fractional factorial designs with optimal confounding patterns.

In Appendix B.1, notice that there are two 8-run fractional factorial designs, three 16-run designs, and three 32-run designs. Each of these designs represents a different fraction and confounding pattern. When using these tables you should be sure to pick the appropriate one by matching the number of factors you have with the number of factors listed at the top of the table. In this way you will get the best confounding pattern possible for the number of factors you have. The confounding patterns for the 8-, 16-, and 32-run fractional factorials given in Appendix B.1 are also listed with the tables, omitting interaction terms higher than second order.

To illustrate interpretation of the confounding patterns listed in Appendix B.1 we will refer to a design that is the same as the one-quarter replicate of a 2^5 we constructed in the last section. This design can be found in Table B.1-5, and it is reproduced in Table 7.13 for convenience. The computation table for the experiment is shown in the body of Table B.1-5, and below the table under CONFOUNDINGS, the confounding pattern is listed. In the confounding pattern only two-way interactions are shown. The experiment constructed in the last section had only five factors where Table B.1-5 can be used for up to seven factors. If we only had five factors, the columns labeled X_6 and X_7 would not be used when making a list of the experiments to be run, and interactions containing Factor 6 or Factor 7 in the confounding pattern would be deleted (such as 67 in line 1, 36 and 57 in line 2, etc.). When calculating effects the entire body of Table B.1.5 would be used for a computation table.

TABLE 7.13
Copy of Appendix Table B.1-5

Run No.	Mean	X_1	X_2	X_3	X_4	X_5	X_6	X_7	CONFOUNDINGS
1	+	−	−	−	+	+	+	−	(4=12, 5=13, 6=23, 7=123)
2	+	+	−	−	−	−	+	+	$X_1 = 1 + 24 + 35 + 67$
3	+	−	+	−	−	+	−	+	$X_2 = 2 + 14 + 36 + 57$
4	+	+	+	−	+	−	−	−	$X_3 = 3 + 15 + 26 + 47$
5	+	−	−	+	+	−	−	+	$X_4 = 4 + 12 + 37 + 56$
6	+	+	−	+	−	+	−	−	$X_5 = 5 + 13 + 27 + 46$
7	+	−	+	+	−	−	+	−	$X_6 = 6 + 17 + 23 + 45$
8	+	+	+	+	+	+	+	+	$X_7 = 7 + 16 + 25 + 34$

7.4.6 Example of Fractional Factorial in Process Improvement

A chemical process for manufacturing a new herbicide product was devised and tested in the laboratory. When the process was scaled up to run in the pilot plant, the levels of a certain impurity (by-product of the reactions) became a problem. In the laboratory there was less than 0.8% of this impurity in the final product. But, typical runs in the pilot plant had more than 2%. To make the process cost effective, and to eliminate the need for further process steps to purify the product, it was desired to reduce the level of the impurity to less than 0.75%. To find a combination of factor levels that would do this, some experiments in the pilot plant were needed. Figure 7.6 is a diagram of the pilot plant.

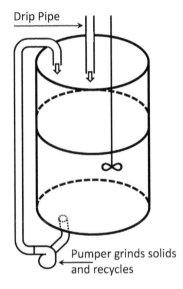

Drip Pipe

Pumper grinds solids and recycles

FIGURE 7.6: Pilot Plant Diagram

At first, trial-and-error tests were planned. However, due to limited time available in the pilot plant, it was decided to use a statistical plan to get the maximum information in a limited time. A list of the potential factors that was made in a brainstorming session is given below.

List of Potential Factors in Pilot Plant (Result of Brainstorming)
1. pH
2. Temperature
3. Addition Time
4. Stir at Completion
5. H_2O Level (Full Charge vs. Partial + Makeup)
6. Agitation Rate, RPM
7. Pumper on Loop
8. Source of Raw Material
9. 2/3 Batch or Whole
10. Close Couple vs Solid Cake
11. Drip Pipe Height
12. pH Probe Accuracy

Before the brainstorming session, the lead chemist had planned to conduct a simple 2^3 design with two replicates using the factors pH, Pumper on Loop, and RPM. But, after the brainstorming his eyes were opened by other's opinions about many other potential causes for the impurities. Since all these factors could be studied in the same number of experiments (16) that he had originally planned on running, he decided to conduct a fractional factorial experiment. Two of the factors listed above (Drip Pipe Height and pH probe accuracy) could not be varied or controlled so they were eliminated from the study, and three other factors were dropped because they were felt to be less important (source of raw material, size of batch, and close couple vs solid cake). Therefore with seven remaining factors a 2^{7-3} 16 run fractional factorial experiment was planned. The first seven columns of Appendix Table B.1-8 were used to define the runs. Table 7.14 shows these experimental conditions and the response data resulting from the experiments. The experiments were run in the random order shown in Column 2 to prevent biases from changes in levels of factors that could not be controlled and other background factors.

TABLE 7.14
Pilot Plant Experimental Results

Run	Random Order	pH X_1	Temp X_2	Time X_3	Stir X_4	H_2O X_5	RPM X_6	Pump X_7	% Impurity Y
1	5	−	−	−	−	−	−	−	2.60
2	4	+	−	−	−	+	−	+	0.89
3	14	−	+	−	−	+	+	−	1.82
4	7	+	+	−	−	−	+	+	0.49
5	15	−	−	+	−	+	+	+	1.83
6	2	+	−	+	−	−	+	−	1.26
7	1	−	+	+	−	−	−	+	2.23
8	3	+	+	+	−	+	−	−	1.98
9	8	−	−	−	+	−	+	+	0.67
10	11	+	−	−	+	+	+	−	0.07
11	12	−	+	−	+	+	−	+	0.51
12	16	+	+	−	+	−	−	−	0.72
13	10	−	−	+	+	+	−	−	1.61
14	6	+	−	+	+	−	−	+	0.08
15	9	−	+	+	+	−	+	−	0.53
16	13	+	+	+	+	+	+	+	0.16

After completing the experiments and making some initial exploratory plots, as described in Chapter 4, the next step was to analyze the data completely to determine which factors have significant effects. The effects can be calculated most easily by using a spreadsheet program (like Excel) as described in Chapter 4 (or equivalently the least squares method which is described in Chapter 8). To use either the spreadsheet or least squares method, copy the entirety of Table B.1-8 (not just the first seven columns) to use as the computation table. Equivalently, the columns of Table B.1-8 can be used as the columns of the X-matrix used in the least squares method. Remember to put a +1 for the high level (not just a +) and a −1 for the low level (not just a −) or the SUMPRODUCT function will not work, and neither will the least squares method. Figure 7.7 shows the results of these computations.

Once the effects were calculated from the data, the next step was to determine which effects are significant and interpret the results. Since there were no replicate experiments, no standard error of an effect (s_E) could be calculated, and the significant effects had to be judged using graphical methods like a Normal Probability plot of effects described in Section 2.5.3. Figure 7.8 shows a Normal plot of the effects.

This graph shows clearly that the significant effects appear to be X_1 = pH, X_4 = Stir at Completion, X_6 = H_2O Level, and X_7 = Agitation Rate (RPM). All of the effects were

Run	Mean	X_1	X_2	X_3	X_4	X_5	X_6	X_7	X_8	E_1	E_2	E_3	E_4	E_5	E_6	E_7	Y
1	1	-1	-1	-1	-1	-1	-1	-1	-1	1	1	1	1	1	1	1	2.60
2	1	1	-1	-1	-1	1	-1	1	1	-1	-1	-1	1	-1	1	1	0.89
3	1	-1	1	-1	-1	1	1	-1	1	-1	1	1	-1	-1	1	-1	1.82
4	1	1	1	-1	-1	-1	1	1	-1	1	-1	-1	-1	1	1	-1	0.49
5	1	-1	-1	1	-1	1	1	1	-1	1	-1	1	-1	-1	-1	1	1.83
6	1	1	-1	1	-1	-1	1	-1	1	-1	1	-1	-1	1	-1	1	1.26
7	1	-1	1	1	-1	-1	-1	1	1	-1	-1	1	1	1	-1	-1	2.23
8	1	1	1	1	-1	1	-1	-1	-1	1	1	-1	1	-1	-1	-1	1.98
9	1	-1	-1	-1	1	-1	1	1	1	1	1	-1	1	-1	-1	-1	0.67
10	1	1	-1	-1	1	1	1	-1	-1	-1	-1	1	1	1	-1	-1	0.07
11	1	-1	1	-1	1	1	-1	1	-1	-1	1	-1	-1	1	-1	1	0.51
12	1	1	1	-1	1	-1	-1	-1	1	1	-1	1	-1	-1	-1	1	0.72
13	1	-1	-1	1	1	1	-1	-1	1	1	-1	-1	-1	1	1	-1	1.61
14	1	1	-1	1	1	-1	-1	1	-1	-1	1	1	-1	-1	1	-1	0.08
15	1	-1	1	1	1	-1	1	-1	-1	-1	-1	-1	1	-1	1	1	0.53
16	1	1	1	1	1	1	1	1	1	1	1	1	1	1	1	1	0.16
SumProduct	17.45	-6.15	-0.57	1.91	-8.75	0.29	-3.79	-3.73	1.27	2.67	0.71	1.57	0.81	0.41	-1.09	-0.45	
Effects	1.09	-0.77	-0.07	0.24	-1.09	0.04	-0.47	-0.47	0.16	0.33	0.09	0.20	0.10	0.05	-0.14	-0.06	

FIGURE 7.7: Effects from Pilot Plant Study Calculated by Worksheet

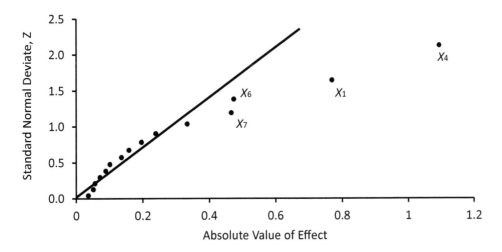

FIGURE 7.8: Half-Normal Probability Plot of Effects from Pilot Plant Study

negative indicating that the high level of each would produce the lowest impurity level.

Since none of the unassigned effects (X_8 or E_1–E_8) appear to be significant in this experiment, a simple linear prediction equation can be written to represent the results. This equation is:

$$Y(\%\text{impurities}) = 1.09 - 0.38X_1 - 0.54X_4 - 0.24X_6 - 0.23X_7.$$

It can be seen from this equation that the lowest impurities should result when all four of the significant factors are at their high level. However, the predicted value at those conditions is negative! This negative prediction, along with the fact that the response data spans over 1-1/2 orders of magnitude (i.e., 0.07 to 2.7), may indicate that the underlying relationship is nonlinear (not to mention that this also calls into question the assumption of constant variance). A better prediction equation (the calculation of the coefficients is left as an exercise) is:

$$\ln(Y) = -0.35 - 0.56X_1 - 0.73X_4$$

which results in reasonable predicted values with the high levels of $X_1 = $ pH and $X_4 = $ Stir. These results were utilized, with no further experimentation, and the process continued to produce product with acceptably low impurities. Statistical experimentation, as shown in this example, has been a valuable tool in chemical process development that tremendously reduces the experimentation needed to make processes work efficiently (compared to trial-and-error experimentation).

An interesting note about this experiment is that the 2^3 experiment initially planned by the lead chemist did not include Factor 4 (Stir). But, this factor turned out to have the largest effect and was the key to finding a combination of factor levels that resulted in low impurity levels and a cost effective process. In screening experiments, it is very wise to study as many factors listed during brainstorming as possible. The number of experiments required to study 11 factors is no more than the number required to study four factors when using a fractional factorial design. But, on the other hand, if an important factor is left out of a screening experiment the results may be inconclusive and/or incomplete.

In the example above, four or less factors appear to be significant in this 16-run screening experiment. Since continued plant operation at the conditions predicted to be best from the simple linear prediction equation (i.e., no interactions included) were successful, the results of this experiment were conclusive. However, many times after screening experiments the results will not be conclusive, due to interactions among the important factors. In these cases, additional follow-up experiments with the important factors may be required after the screening experiments.

7.4.7 Advantages of Fractional Factorial Designs

Fractional factorial designs have two advantages over the other screening designs that will be presented later in this chapter. The first advantage is called the *projection property*. After the data collection and analysis are complete, and only a few of the many factors in the design are found to be important, we may be left with a full factorial design in these key variables (which would allow us to estimate all interactions between those factors, not just their main effects). The second advantage is that confoundings between the variables (main effects and interactions) are much simpler to determine than they are for other types of screening designs.

To illustrate the first advantage, consider the eight-run design for four factors in Table B.1-6. Any three columns selected from the first four will form a full factorial design. This is illustrated in Table 7.15. The top left hand table shows that we have a full factorial if X_4 is unimportant. This was not a surprise, since the base design was a 2^3 factorial in X_1, X_2, and X_3. But the other three tables show the same result if one of the other three factors turns out to be unimportant instead. This was made more obvious by reordering the run numbers so that the same full factorial pattern can be seen for any subset of the four factors. Therefore, if we use the design in Table B.1-6 to screen four factors and at least one of the four factors turns out to be unimportant, we will have a full factorial in the key variables if we ignore the column(s) for the unimportant factor(s). If we want to study the interaction effects involving the important variables there will be no need to do follow-up experiments—we already have a full factorial design.

The second advantage of fractional factorial designs—being able to determine the confoundings quite easily—can help in deciding which follow-up experiments to perform when

TABLE 7.15
Full Factorial Designs Formed from Subsets of a Fractional Factorial

Run	X_1	X_2	X_3	X_4		Run	X_1	X_2	X_4
1	−	−	−	−		1	−	−	−
2	+	−	−	+		6	+	−	−
3	−	+	−	+		7	−	+	−
4	+	+	−	−		4	+	+	−
5	−	−	+	+		5	−	−	+
6	+	−	+	−		2	+	−	+
7	−	+	+	−		3	−	+	+
8	+	+	+	+		8	+	+	+

Run	X_1	X_3	X_4		Run	X_2	X_3	X_4
1	−	−	−		1	−	−	−
4	+	−	−		4	+	−	−
7	−	+	−		6	−	+	−
6	+	+	−		7	+	+	−
3	−	−	+		2	−	−	+
2	+	−	+		3	+	−	+
5	−	+	+		5	−	+	+
8	+	+	+		8	+	+	+

they are necessary. To be specific, let us consider an example. Suppose that in the experiment presented in the last section, the conclusion that X_1, X_4, X_6, and X_7 were important was valid, and the unassigned factor E_1 had been large (in the range 0.7 to 1.1 like Factors 1 and 4) and was considered to be significant. This unassigned effect represents a string of possible two-factor interactions. From the CONFOUNDINGS listed with Table B.1-8 we see that $E_1 = 12 + 35 + 67$. (Note: We drop 48 from the interactions in the table because there was no Factor 8 — there were only seven factors in the experiment.) With only the 16 experiments run in the fractional factorial design it is impossible to determine which of the three potential interactions this effect represents. This can be seen clearly by examining the calculation columns for these three interactions shown in top part of Table 7.16. The effect that would be computed for each of these interactions would be exactly the same using the original 16 experiments in the table. In order to de-confound these three interactions, we need to run additional experiments. If we were unable to run any additional experiments, then the best guess would be that the interaction of two important effects, X_6 and X_7, is the most likely interaction to be real.

If we have the capacity to run additional experiments, the ones shown in the bottom part of Table 7.16 are one set that will work. These runs were obtained by writing the levels for the three confounded interactions first. This was done utilizing a full factorial in X_1X_2, X_3X_5, and X_6X_7, which can be seen by examining the last three columns of Table 7.16. This pattern is used because we know that it will result in the three interactions being independently estimable. Next, any combination of levels for Factors 1–7 can be chosen for each run whose products will result in the appropriate levels for the three interactions. For example, in Run 17, any combination of levels for X_1–X_7 that will result in $X_1X_2 = -$, $X_3X_5 = -$, and $X_6X_7 = -$, will work. The combination $X_1 = -$, $X_2 = +$, $X_3 = -$, $X_4 = +$, $X_5 = +$, $X_6 = -$, $X_7 = +$ was used, but the combination $X_1 = +$, $X_2 = -$, $X_3 = +$, $X_4 = -$, $X_5 = -$, $X_6 = +$, $X_7 = -$ will work just as well. Using this logic, the levels for Factors 1–7 were completed for runs 17–24. If the experiments listed in the bottom of Table 7.16 were completed in a random order, there would be sufficient data to estimate the main

TABLE 7.16
Calculation Columns for Confounded Interactions in Pilot Plant Example

Run	pH X_1	Temp X_2	Add Time X_3	Stir X_4	H$_2$O X_5	RPM X_6	Pump X_7	X_1X_2	X_3X_5	X_6X_7
1	−	−	−	−	−	−	−	+	+	+
2	+	−	−	−	+	−	+	−	−	−
3	−	+	−	−	+	+	−	−	−	−
4	+	+	−	−	−	+	+	+	+	+
5	−	−	+	−	+	+	+	+	+	+
6	+	−	+	−	−	+	−	−	−	−
7	−	+	+	−	−	−	+	−	−	−
8	+	+	+	−	+	−	−	+	+	+
9	−	−	−	+	−	+	+	+	+	+
10	+	−	−	+	+	+	−	−	−	−
11	−	+	−	+	+	−	+	−	−	−
12	+	+	−	+	−	−	−	+	+	+
13	−	−	+	+	+	−	−	+	+	+
14	+	−	+	+	−	−	+	−	−	−
15	−	+	+	+	−	+	−	−	−	−
16	+	+	+	+	+	+	+	+	+	+

Some Additional Runs that Allow Estimation of Confounded Interactions

Run	pH X_1	Temp X_2	Add Time X_3	Stir X_4	H$_2$O X_5	RPM X_6	Pump X_7	X_1X_2	X_3X_5	X_6X_7
17	−	+	−	+	+	−	+	−	−	−
18	+	+	−	+	+	−	+	+	−	−
19	−	+	+	−	+	−	+	−	+	−
20	+	+	+	−	+	−	+	+	+	−
21	+	−	+	+	−	−	−	−	−	+
22	−	−	−	−	+	+	+	+	−	+
23	−	+	+	+	+	+	+	−	+	+
24	+	+	−	−	−	+	+	+	+	+

effects and the interactions that add together to make up the E_1 effect in the fractional factorial, but the analysis would have to be done by regression.

When we combine the two sets of experiments, the columns will no longer be orthogonal (i.e., will not form a full factorial in every pair of columns). Therefore, we cannot analyze the combined data from the two sets of experiments by the worksheet method. Instead, least squares regression analysis must be used. (Regression analysis is discussed fully in Chapter 8.) An X-matrix would be defined containing columns for $X_1 - -X_7$, X_1X_2, X_3X_5, X_6X_7, and one additional column, X_8, representing a block of experiments ($X_8 = -$ for Runs 1–16 and $X_8 = +$ for Runs 17–24). This X matrix would define a linear model with 12 coefficients (the 11 terms plus the intercept). Since there are 24 data points in the combined experiments, there would be 12 degrees of freedom left for estimating the standard error of the regression coefficients and calculating t-statistics.

7.4.8   Resolution of Fractional Factorial Designs

Since there are many ways of generating a $(1/2)^p$ fraction of a 2^k factorial design, it is useful to have a measure of how "good" a particular design (or fraction) is. One measure that is used is called **resolution**. Resolution is defined as the length of the shortest word in the defining relation. For a half fraction there is only one word in the defining relation (other than the letter I). For example, consider the half fraction of a 2^5 factorial design with the generator, 5 = 1234. The defining relation, is I = 12345. Since the shortest (only) word (12345) has five factors in it, the resolution is five. Resolution is indicated by a Roman numeral, in this case V. Had the generator for the design been 5 = 234, then the defining relation would have been I = 2345, which is a resolution IV. Or, had the generator for the design been 5 = 23, the defining relation would have been I = 235 which is resolution III. The resolution V design is most desirable, because the higher the resolution the better (at least up to resolution V).

Why is higher resolution better? Because a resolution V design has main effects and two factor interactions confounded only with three factor interactions and higher. For the example with I = 12345, we see that 5 = 1234, 1 = 2345, etc. and 12 = 345, 13 = 245, etc. For all practical purposes, all main effects and two-factor interactions are not confounded with anything, because it is very rare that a three-factor interaction exists. Also, a resolution V design has the projection property that every subset of four factors forms a full 2^4 factorial, in the sense described in Section 7.4.7. So if four or less factors are important, a full factorial in the important factors is embedded in the resolution V design. Tables B.1-9 and B.1-13 are resolution V designs in five and six factors, respectively.

A resolution IV design has main effects clear of any confounding except three-factor (or higher) interactions. But some or all of the two-factor interactions are confounded with other two-factor interactions. For our resolution IV example of 5 = 234 (or I = 2345) we have 1 = 12345, 2 = 345, etc., which are great, but 23 = 45, 24 = 35, etc. Resolution IV designs have the projection property that every subset of three factors forms a full 2^3 design. Tables B.1-6, B.1-8, and B.1-11 are resolution IV.

A resolution III design has some or all main effects confounded with two-factor interactions. For our resolution III example of 5 = 23 (or I = 235) we have 1 = 1235, 2 = 35, 3 = 25, etc. Resolution III designs have the projection property that every pair of factors forms a full 2^2 design, and they are the lowest resolution that is generally acceptable. Their use requires assuming that two-factor interactions are essentially negligible. This assumption is acceptable in screening designs because it allows us to experiment with a large number of factors with a minimal number of runs. Note that a resolution II design would have main effects confounded with other main effects which is unacceptable. So, useful resolutions are III, IV, and V. Resolution III is used mainly for screening. Tables B.1-5, B.1-7, and B.1-10 are resolution III.

7.5   Plackett–Burman Screening Designs

As an alternate way to avoid the problem of confounding the (main) effects of factors under study with each other, Plackett and Burman [1946] developed a set of tables to be used for screening designs. Their tables are resolution III designs. Table 7.17 is an example of an eight-run Plackett-Burman design. This design can be used to screen up to seven factors. This table is a subset of the 2^7 runs necessary for a full factorial in seven factors and is similar to the fractional factorial designs in that no factors are confounded in this table. Specifically, at the high$(+)$ level of each of the factors there are an equal number of experiments conducted at the high$(+)$ and low$(-)$ levels of each of the other factors. The same is true, of course, at the low level of each of the factors.

TABLE 7.17
A Plackett–Burman Design for Seven Factors

Run	X_1	X_2	X_3	X_4	X_5	X_6	X_7
1	+	+	+	−	+	−	−
2	−	+	+	+	−	+	−
3	−	−	+	+	+	−	+
4	+	−	−	+	+	+	−
5	−	+	−	−	+	+	+
6	+	−	+	−	−	+	+
7	+	+	−	+	−	−	+
8	−	−	−	−	−	−	−

To verify that this is true, Table 7.18 shows a tally of runs at all combinations of Factors 1 and 2, similar to Table 7.6. A similar table could be constructed for every other pair of factors.

TABLE 7.18
Verification That Factors Are not Confounded with Each Other

X_1	X_2	Run Numbers At These Levels
−	−	3, 8
+	−	4, 6
−	+	2, 5
+	+	1, 7

A full factorial in seven factors would take $2^7 = 128$ experiments or runs. The Plackett-Burman design allows seven factors to be examined in only eight runs. This is a **huge** reduction in the amount of experimentation, but it does not come without a penalty. There is no free lunch! The price we pay is that we cannot estimate any of the interactions. This

Plackett-Burman design is resolution III, and the interactions are confounded with the main effects of the factors. To see that this is true, let us create the column for computing the X_1X_2 interaction by multiplying the column of signs for X_1 times the column of signs for X_2 in Table 7.17 (which is done in Table 7.19). Also shown in Table 7.19 are the settings for the factor X_6, and we will see why right now.

TABLE 7.19

Confounding of Interactions in a
Plackett–Burman Design

X_1	X_2	X_6	X_1X_2	Response
+	+	−	+	Y_1
−	+	+	−	Y_2
−	−	−	+	Y_3
+	−	+	−	Y_4
−	+	+	−	Y_5
+	−	+	−	Y_6
+	+	−	+	Y_7
−	−	−	+	Y_8

It should be noticed that the resulting interaction column is exactly the opposite of the column of signs that would define the X_6 factor. In other words the difference in averages,

$$\frac{Y_2 + Y_4 + Y_5 + Y_6}{4} - \frac{Y_1 + Y_3 + Y_7 + Y_8}{4},$$

would be used to estimate either the effect of Factor 6, or the negative of the X_1X_2 interaction. Therefore, the X_6 effect is confounded with the X_1X_2 interaction. In fact, every main factor effect calculated from data derived from the design in Table 7.17 will be confounded with three 2-factor interactions (Nelson [1982]).

Thus, even though the Plackett–Burman tables were carefully constructed to avoid confounding of main effects of one factor with another, it is impossible to avoid confounding of interactions with the main effects of factors. In other words, some confounding is the price we pay for the great reduction in experimental effort. In order to use Plackett-Burman screening designs, all interactions must be assumed to be negligible. The saving grace is that this assumption is not too critical if follow-up experiments with the important factors are planned. Interactions normally only exist between important factors. Therefore, screening designs should be used to determine the important factors, and follow-up experiments can be conducted to study interactions of those important variables.

7.5.1 Tables of Plackett–Burman Designs

Although Plackett–Burman designs are available in every multiple of four runs from 4 to 100, tables are presented in Appendix B.1 for those designs which are most useful, that is, designs with 12, 20, 24, or 28 runs. Tables B.1-1, B.1-2, B.1-3, and B.1-4 show these designs. The eight-run Plackett–Burman design shown in Table 7.17 is not included because there are better eight-run fractional factorial designs which were discussed in Section 7.4. Notice that the 16-run Plackett–Burman design is also omitted for the same reason. The 12-, 20-, 24-, and 28-run Plackett–Burman designs can handle up to 11, 19, 23, or 27 factors respectively, although that many are not generally recommended. The factors are identified across the top of the page and the run number is identified along the left margin.

7.5.2   Using the Tables of Plackett–Burman Designs

To use a Plackett–Burman design, all you need to do is determine how many runs are necessary. Then you simply select the design with that many runs. The number of runs needed depends on two considerations: (a) how many factors are there to be studied, and (b) how much precision is wanted.

It seems reasonable that the more factors that you want to investigate, the more runs you need. Specifically, every run gives you one independent piece of information on a particular response (or, as we say in statistical jargon, it gives us one degree of freedom). If you are studying k factors, you want to know a minimum of k+1 pieces of information which are the k effects of the factors and the overall average response. This is most easily seen if you think in terms of the model you are fitting to your data, either explicitly or implicitly. It is the simple linear model:

$$Y = b_0 + b_1 X_1 + b_2 X_2 + b_3 X_3 + \ldots + b_k X_k$$

which has k+1 coefficients, and it requires one degree of freedom to estimate each coefficient. This model can be fit from the experimental data by regression, as described for fractional factorials in Section 7.4.7.

In general, you will not have an estimate of experimental error, σ, that you are comfortable using to determine the significance of each of the effects. Therefore, in addition to needing to estimate the coefficients in your model, you will also need to have some degrees of freedom to estimate σ from your data. From the t-table in Appendix A (Table A.3), you can see that any significance testing will suffer greatly if you have less than 4 degrees of freedom for your estimate of σ. This means that you need four more runs in your design, over and above the k+1. That is why it is recommended that no more than seven factors be studied using the 12-run Plackett–Burman design, no more than 15 be studied using the 20-run design, and no more than 23 be studied using the 28-run design.

In summary, the information we want from our data requires that:

$$n_F \geq k + 5 \qquad \text{for information.} \tag{7.1}$$

Before proceeding, we should also check to see that the design we select meets the desired precision, δ. Remember that δ is the smallest effect we do not want to overlook (as discussed in Section 4.2). The relationship derived in Chapter 4 between n_F and δ is:

$$n_F \geq (8\sigma/\delta)^2 \qquad \text{for precision.} \tag{7.2}$$

It should be kept in mind that n_F required for precision is negotiable, because δ is a compromise between cost (n_F) and benefits received from the information. That is, the sponsor may decide to settle for a bigger δ and pay less (smaller n_F). But the n_F required for information is not negotiable, and it cannot be reduced (except by reducing the number of factors to be studied).

The full procedure to determine the appropriate number of runs is to:
(a) Calculate the number of runs, n_F, required for information purposes.
(b) Calculate the number of runs, n_F, required for adequate precision.
(c) Pick the larger of the two n_F values.
(d) Round *up* to the nearest multiple of four runs.

Once the number of factors and runs has been decided upon, the appropriate table is used by assigning one factor to each column, then reading off the experiments to be run. Before actually running the experiments, it is, of course, important to randomize the run order. This can be done using any randomization technique.

Although the use of Equation 7.2 for determining the number of runs for precision was not mentioned in Section 7.4, it should be used in the same way when planning a fractional factorial design. However, there are only fractional factorial designs with 8, 16, or 32 runs, so Step (d) above would have to be modified, or include the possibility of running a Plackett–Burman design. For example, if $n_F = 20$ was calculated from Equation 7.2, you could either run a 20-run Plackett–Burman design or a 32-run fractional factorial to get the necessary precision (a 16-run fractional factorial would give less than the desired precision). The 32-run fractional factorial design would have a clearer confounding pattern, but at a large cost in terms of the number of runs.

If there are fewer factors than columns in the design (which is usually the case), the extra columns are not needed to define the list of experiments. The extra columns will be used, however, when performing the numerical computations. At that stage, the entire table from Appendix B.1 should be used as a worksheet for calculating effects. The effects calculated for unassigned columns will represent interaction effects, although it is difficult to say exactly what specific interactions they represent. (For a discussion on how to determine what combination of interactions they correspond to, see the appendix of Hunter's paper.?)

Once the number of factors and runs are determined, the Plackett–Burman tables are used for a screening design by copying the complete computation table for the design used from the appropriate table in Appendix B.1. But use only the columns for which you have factors to determine the experimental runs. For example, if you are studying 12 factors in a 20-run design, use columns X_1–X_{12} to define the experiments. Randomize the run order, then run the experiments and collect the data.

When analyzing the data, use the entire table from the appendix as the computation table or X-matrix for a regression. The relative importance of the factor effects can be obtained by examination of their magnitudes. The usual assumption in the analysis of screening designs is that only a few of the factor effects will be important. Therefore, these vital few can be identified by plotting the effects. Pareto Diagrams of the absolute values of effects, Normal Probability plots, or Half-Normal plots (as described in Chapter 2) are useful tools for identifying important effects.

Another method for identifying important effects from a screening design (either Plackett–Burman or fractional factorial) is to calculate a standard error of effects by using all the unassigned (or interaction) effects from the calculation worksheet, and then calculating t-statistics to compare to Table A.3. An equivalent way of doing this is to perform a regression analysis including only the factors used to define the experiments. By doing this, the unassigned factors will automatically be lumped together as an estimate of error. Caution must be taken when using this approach. If one of the interactions used to calculate the error term is large it may actually be important. Combining an important effect into the error will inflate s_E, or (if doing the analysis via regression) $s_{\hat{\beta}}$, and reduce the significance of the other main effects. A rule of thumb is not to include any interaction term in the error calculation if it is one of the largest effects in absolute value. We will illustrate this method with the next example.

7.5.3   An Example of Using a Plackett–Burman Design

At one step in the manufacture of multi-layered printed circuit boards, holes (drilled in the board to connect various layers) are copperplated by electrolysis. The manufacturer was concerned with getting a uniform copperplating of the correct thickness. Figure 7.9 shows a cross section of a drilled hole. A problem was being experienced in the plating cells. The copperplate was not uniform. It was thicker at top and bottom than in the middle. Also, the target thickness was not being met.

The process engineers wanted to fix the problem, but were not sure which factors were most important in determining the thickness or uniformity of the copperplate. Figure 7.10 shows a cause-and-effect diagram of factors they felt might be important. The factors listed in Table 7.20 are those they decided to include in the experiment.

A full 2^7 design would require 128 experiments which was impractical. The use of a Plackett–Burman design was a much more viable approach. For information, 12 runs were required: $n_F \geq k + 5 = 7 + 5 = 12$. This number of runs gave acceptable precision as well (the calculations are not shown), so a 12-run Plackett–Burman Design was selected.

The first seven columns from Table B.1-1 were copied into the table shown in Table 7.21 to define the 12 test conditions. The experiments were run in the random order shown in Table 7.21 and the resulting data were collected as shown on the right of the table.

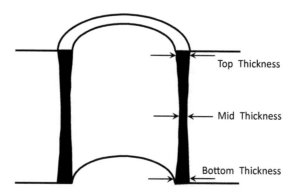

FIGURE 7.9: Cross Section of Copperplated Hole

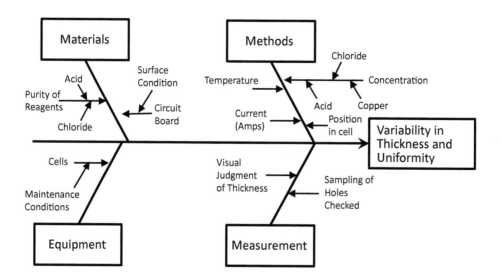

FIGURE 7.10: Cause-and-Effect Diagram of Factors Influencing Copperplating

Measurements of the copperplating thickness were made at the top, middle, and bottom of five sample drill holes on a circuit board plated under each of the conditions listed in each row of the design. The average of the five measurements at each location in the hole is presented in the table as Top, Mid, and Bot. The average of these values, $\bar{Y}$, is a summary statistic which indicates the average thickness of the copperplating. The variance, s^2, of the three averages is a measure of the uniformity of the plating thickness. Since the variance, s^2,

TABLE 7.20

Factors and Levels for Copperplating Experiment

Factors	Low Level (−)	High Level (+)
X_1 = Copper concentration	16 gm/L	19 gm/L
X_2 = Chloride concentration	65 PPM	85 PPM
X_3 = Acid (H+) concentration	198 gm/L	225 gm/L
X_4 = Temperature	72	78
X_5 = Total current	180 amp-hrs	192.5 amp-hr
X_6 = Position in cell	right	left
X_7 = Surface condition	smooth	slightly rough

TABLE 7.21

Runs and Results for Copperplating Experiment

Run Order	X_1	X_2	X_3	X_4	X_5	X_6	X_7	Plating Thickness Top	Mid	Bot	$\bar{Y}$	s^2
9	+	+	−	+	+	+	−	2.74	2.13	2.68	2.52	0.113
2	+	−	+	+	+	−	−	2.71	2.15	2.72	2.53	0.106
4	−	+	+	+	−	−	−	2.02	1.67	2.06	1.92	0.0460
5	+	+	+	−	−	−	+	1.84	1.53	1.77	1.71	0.0264
12	+	+	−	−	−	+	−	1.79	1.49	1.71	1.66	0.0241
1	+	−	−	−	+	−	+	2.42	1.78	2.39	2.20	0.130
7	−	−	−	+	−	+	+	2.05	1.80	2.10	1.98	0.0258
8	−	−	+	−	+	+	−	2.44	1.93	2.38	2.25	0.0777
10	−	+	−	+	+	−	+	2.65	2.19	2.70	2.51	0.0790
3	+	−	+	+	−	+	+	1.98	1.61	2.06	1.88	0.0576
11	−	+	+	−	+	+	+	2.40	1.70	2.30	2.13	0.143
6	−	−	−	−	−	−	−	1.80	1.43	1.75	1.66	0.0403

ranges over nearly an order of magnitude, the linear model implicit in factorial designs was thought to be inappropriate, and the $\log_e(s^2)$ was felt to be a better response to represent uniformity.

Figure 7.11 and Figure 7.12 are summaries of the calculations performed to get the factor effects. Notice that in these two figures all the columns from Table B.1-1 were copied, and the effects were calculated for each column. (As an aside, to do the calculations, the + and − levels were replaced with +1 and −1, and the SUMPRODUCT function in a spreadsheet was used.) The effects calculated for the unassigned factor columns X_8, X_9, X_{10}, and X_{11} represent interactions, although we cannot associate any one column to specific interactions as we could with the unassigned columns in the fractional factorial tables. Similar to the way the standard error of an effect was calculated using block interactions in Section 4.5.1, a standard error of the effects, s_E, was calculated using the unassigned effects X_8, X_9, X_{10}, and X_{11} in Figure 7.11. In Figure 7.12, only X_8, X_{10}, and X_{11} were used to calculate the error, since the X_9 effect was relatively large. In order to assess statistical significance, t-values were calculated by dividing each effect by the calculated s_E.

Also, Figures 7.13 and 7.14 are Normal plots of the effects. These diagrams graphically display the relative importance of the effects, and are used to help us screen out the vital few effects from the trivial or negligible ones. We can easily see that for the average thickness response, the Factors X_4 (temperature) and X_5 (total current) have the most dramatic effects, with thicker plating resulting at the higher temperature and higher total current.

	Mean	X_1	X_2	X_3	X_4	X_5	X_6	X_7	X_8	X_9	X_{10}	X_{11}	$\bar{Y}$
	1	1	1	-1	1	1	1	-1	-1	-1	1	-1	2.52
	1	1	-1	1	1	1	-1	-1	-1	1	-1	1	2.53
	1	-1	1	1	1	-1	-1	-1	1	-1	1	1	1.92
	1	1	1	1	-1	-1	-1	1	-1	1	1	-1	1.71
	1	1	1	-1	-1	-1	1	-1	1	1	-1	1	1.66
	1	1	-1	-1	-1	1	-1	1	1	-1	1	1	2.20
	1	-1	-1	-1	1	-1	1	1	-1	1	1	1	1.98
	1	-1	-1	1	-1	1	1	-1	1	1	1	-1	2.25
	1	-1	1	-1	1	1	-1	1	1	1	-1	-1	2.51
	1	1	-1	1	1	-1	1	1	1	-1	-1	-1	1.88
	1	-1	1	1	-1	1	1	1	-1	-1	-1	1	2.13
	1	-1	-1	-1	-1	-1	-1	-1	-1	-1	-1	-1	1.66
SUMPROD	24.95	0.05	-0.05	-0.11	1.73	3.33	-0.11	-0.13	-0.11	0.33	0.21	-0.11	
Effects	2.079	0.008	-0.008	-0.018	0.288	0.555	-0.018	-0.022	-0.018	0.055	0.035	-0.018	
t-values		0.24	-0.24	-0.52	8.22	15.82	-0.52	-0.62	-0.52	1.57	1.00	-0.52	

$$s_E = \sqrt{\frac{(-0.018)^2 + (0.055)^2 + (0.035)^2 + (-0.018)^2}{4}} = 0.0351$$

FIGURE 7.11: Calculation of Effects for Average Thickness, $\overline{Y}$

	Mean	X_1	X_2	X_3	X_4	X_5	X_6	X_7	X_8	X_9	X_{10}	X_{11}	$\ln(s^2)$	s^2
	1	1	1	-1	1	1	1	-1	-1	-1	1	-1	-2.18	0.113
	1	1	-1	1	1	1	-1	-1	-1	1	-1	1	-2.24	0.106
	1	-1	1	1	1	-1	-1	-1	1	-1	1	1	-3.08	0.0460
	1	1	1	1	-1	-1	-1	1	-1	1	1	-1	-3.63	0.0264
	1	1	1	-1	-1	-1	1	-1	1	1	-1	1	-3.72	0.0241
	1	1	-1	-1	-1	1	-1	1	1	-1	1	1	-2.04	0.130
	1	-1	-1	-1	1	-1	1	1	-1	1	1	1	-3.66	0.0258
	1	-1	-1	1	-1	1	1	-1	1	1	1	-1	-2.55	0.0777
	1	-1	1	-1	1	1	-1	1	1	1	-1	-1	-2.54	0.0790
	1	1	-1	1	1	-1	1	1	1	-1	-1	-1	-2.85	0.0576
	1	-1	1	1	-1	1	1	1	-1	-1	-1	1	-1.94	0.143
	1	-1	-1	-1	-1	-1	-1	-1	-1	-1	-1	-1	-3.21	0.0403
SUMPROD	0.00	0.00	0.00	0.00	0.00	0.00	0.00	0.00	0.00	0.00	0.00	0.00		
Effects	0.000	0.052	-0.091	0.174	0.093	1.111	-0.029	0.055	0.013	-0.507	-0.105	0.049		
t-values		0.78	-1.35	2.59	1.38	16.52	-0.43	0.82		-7.55				

$$s_E = \sqrt{\frac{(0.013)^2 + (-0.105)^2 + (0.049)^2}{3}} = 0.0672$$

FIGURE 7.12: Calculation of Effects for $\ln(s^2)$

For the uniformity response, ln(variance), it can be seen that the most important factor is X_5 (total current), and the lower total current tends to produce more uniform (i.e., smaller variance) copperplating. However, the unassigned effect, X_9, also appears to be relatively large as affirmed by the significantly large t-value in Figure 7.12, and also the point well off the line on the left in the Normal plot of effects in Figure 7.14. The X_9 effect

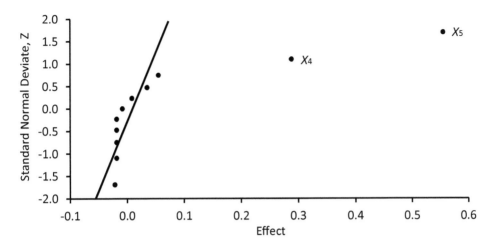

FIGURE 7.13: Normal Probability Plot of Effects for Average Thickness, $\overline{Y}$

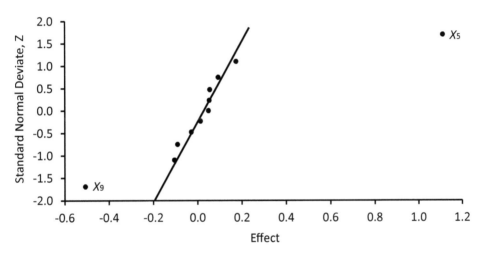

FIGURE 7.14: Normal Probability Plot of Effects for $\ln(s^2)$

demonstrates a property of Plackett–Burman designs. In a fractional factorial design, if an unassigned effect is significant, it represents a collection of several interactions, but in the case of this Plackett–Burman design, X_9 is actually confounded with of all 21 interactions. More specifically:

$$\begin{aligned}
X_9 = {} & (1/3)X_1X_2 + (1/3)X_1X_3 + (1/3)X_1X_4 + (1/3)X_1X_5 + (1/3)X_1X_6 \\
& - (1/3)X_1X_7 - (1/3)X_2X_3 - (1/3)X_2X_4 + (1/3)X_2X_5 + (1/3)X_2X_6 \\
& + (1/3)X_2X_7 + (1/3)X_3X_4 + (1/3)X_3X_5 + (1/3)X_3X_6 - (1/3)X_3X_7 \\
& + (1/3)X_4X_5 - (1/3)X_4X_6 + (1/3)X_4X_7 - (1/3)X_5X_6 + (1/3)X_5X_7 \\
& - (1/3)X_6X_7
\end{aligned}$$

This was determined by using the Alias matrix (see the appendix of Hunter's paper (Hunter [1985])).

With X_9 representing so many possible interactions, it would be very difficult to reduce the confusion without further experiments or analysis as described in Section 11.4.1 of

Chapter 11. Use of the fractional factorial designs discussed in the last section will significantly reduce the confounding of unassigned effects and may also provide the experimenter with an easy way of determining what is confounded with what.

An alternate way of obtaining the t-statistics in Figures 7.11 and 7.12 would be to use regression analysis. For example, Figure 7.15 shows the results of a regression analysis performed by MINITAB Version 16, using the seven columns of Table 7.21 as factors and the average thickness, $\overline{Y}$, as the response. The t-statistics in this figure can be seen to be the same as those shown in Figure 7.11.

```
The regression equation is
Y bar = 2.08 + 0.0042 X1  0.0042 X2  0.0092 X3 + 0.1442 X4
           + 0.2775 X5 - 0.0092 X6  0.0108 X7
```

Term	Effect	Coef	SE Coef	T	P
Constant		2.07917	0.01754	118.54	0.000
X1	0.00833	0.00417	0.01754	0.24	0.824
X2	-0.00833	-0.00417	0.01754	-0.24	0.824
X3	-0.01833	-0.00917	0.01754	-0.52	0.629
X4	0.28833	0.14417	0.01754	8.22	0.001
X5	0.55500	0.27750	0.01754	15.82	0.000
X6	-0.01833	-0.00917	0.01754	-0.52	0.629
X7	-0.02167	-0.01083	0.01754	-0.62	0.570

```
S = 0.0607591    PRESS = 0.1329
R-Sq = 98.76\%   R-Sq(pred) = 88.85\%   R-Sq(adj) = 96.59%
```

FIGURE 7.15: Regression Analysis of Average Thickness from the Plackett–Burman Design

From the results of this experiment, it could probably be assumed that no interactions are important in affecting the average thickness, and the simple linear model:

$$\overline{Y} = 2.079 + 0.144\ X_4 + 0.278\ X_5$$

derived from the significant effects could be used to identify intermediate conditions for temperature and total current that would result in an average thickness close to the target value. However, its more difficult to determine the conditions that would result in the most consistent copperplating thickness (i.e., smallest variance), since there is at least one significant interaction, incorporated in X_9, that affects the response. There are a couple of options that could be followed as the next logical step.

Number one, if time for further experimentation were available, a full factorial in the most important factors, X_3, X_4, and X_5, could be run to try and identify the important interactions affecting the ln(variance). As a second alternative, the interactions could be assumed to be less important and the low level of Factor 5, (total current = 180 amp-hours) could be chosen to minimize the variance, and the level of X_4 = temperature could be adjusted to yield the desired average thickness. Then, the copperplating cell could be run at these conditions during production to see if the resulting average thickness and uniformity were acceptable.

7.6   Other Applications of Fractional Factorials

Although Plackett–Burman designs are used strictly for screening, in some situations fractional factorial designs may be used for much more than just screening. If we are willing to assume all higher order interactions (perhaps three-way and above) are negligible, it is possible to estimate quite a few interactions from certain fractional replicate designs. This includes blocked factorials discussed in Chapter 4. Also, fractional factorial designs can be used to create screening designs for factors with more than two levels. In this section we will illustrate the use of fractional factorials in situations where it is desirable to estimate some interactions. In Section 7.7 we will show how fractional factorials can be used to create screening designs with multiple level factors.

7.6.1   Fractional Factorial Designs for Estimating Some Interactions

With regard to interactions, it is possible to construct a half replicate of a 2^6 design so that all two-way interactions are aliased with only three-way and higher interactions (see Table B.1-13 in Appendix B). This is a resolution V design as described in Section 7.4.8. Therefore, if we assume all three-way interactions are negligible (a very reasonable assumption), it is possible to use the half replicate to estimate all main effects and two-way interactions—something usually reserved for full factorial designs. The half fraction of a 2^6 design is particularly efficient (see Table B.1-9).

In general, resolution V designs have the shortest word in the defining contrast equal to five and allow estimation of all main effects and two-factor interactions. If it is desired to create a design to estimate only a few of the two-factor interactions, a resolution V design may be larger than necessary. More economical designs can be created for estimating main effects and a few specific interactions using the algorithm of Franklin and Bailey [1977], or the interaction graphs of Kackar and Tsui [1990]. In simple situations, designs can be derived for estimating main effects and a few specific interactions using the CONFOUNDINGS in the Appendix B.1 tables. The next example illustrates this.

Consider a situation where four factors are under study. To estimate the main effects and all interactions would require $2^4 = 16$ experiments. However, if some of the variables in the experiment are believed to be less important and less inclined to have interactions, a very economical fractional factorial design can be created using the tables in Appendix B.1. If Factors X_1 and X_2 were considered potentially important, and likely to interact with each other, while X_3 and X_4 are thought to be less important and less likely to interact, then the eight-run half replicate of a 2^4 presented in Table B.1-6 could be used with no penalty (except a loss in precision). From this design, all main effects, $X_1 - -X_4$, and $E_1 = 12$ (the interaction between X_1 and X_2) could be estimated if the interaction between X_3 and X_4 were assumed negligible. One practical situation where it is useful to estimate specific interactions with less experiments than required by a full factorial is in robust product design studies. That topic is beyond the scope of this handbook, but an overview can be obtained from the paper by Kackar and Shoemaker [1986].

7.6.2   Designs for Blocked Factorials in Fractional Arrangements

Another situation where the ideas of fractional factorials and confounding are very useful is in blocked designs. Recall, the blocked factorial designs presented in Section 4.5 required a complete replication of a full factorial within each block. This could result in an extensive number of experiments. If we can assume some higher order interactions are negligible, we can create much more economical blocked factorials by running only a fraction of the factorial in each block.

Figure 7.16 shows examples of how to set up various blocked designs by running fractions of a 2^k factorial as blocks. This figure refers to the tables in Appendix B.1 and B.2. In these designs some interaction terms can be estimated, and others are confounded with block effects or with the block interactions used to estimate error. When an interaction is confounded with blocks, we cannot separate its effect from the differences in blocks. If we are willing to assume the confounded interaction is negligible, there is a reduction in the number of experiments, but no loss of information. When an interaction is confounded with a block interaction used to estimate error, we should examine its effect before automatically pooling it to estimate an error term (similar to the way unassigned effects in Plackett–Burman designs were examined before creating the error term in Section 7.5.3).

The designs in Figure 7.16 should be used when the desired number of factors is larger, or the desired number of runs, or blocks size, is less than that shown in the table of blocked factorials in Chapter 4.

Consider an example of a 2^4 design in two blocks of eight taken from Johnson and Leone [1964a]. Four factors listed below were being investigated for their effect on the velocity of a projectile in a ballistics test:

X_1: propellent charge in lbs.
X_2: projectile weight in lbs
X_3: propellent web
X_4: weapon

Only eight tests could be performed in a day. Therefore it required two days to perform all 16 tests. Since extraneous factors may change between days, it was desirable to treat days as blocks to prevent possible bias. The blocked designs of Chapter 4 will not work in this situation, because they require a full 2^4 design repeated in each block. To create a blocked design with eight runs (a half fraction of the 2^4) in each block, refer to the third row of Figure 7.16. This defines a 16-run design with four factors in blocks of size eight. The figure shows us that this design is created using Table B.2-4, treating the 1234 interaction column as blocks (days). Table 7.22 shows the resulting design. All combinations of factor levels for Day 1 were run first followed by those for Day 2. Normally this would be done in a random order within each day.

In the columns on the right side of Table 7.22, we see the effects calculated. The standard error of the effects is

$$s_E = \sqrt{[(-7.125)^2 + (-0.125)^2 + (0.875)^2 + (-3.875)^2]/4} = 4.08,$$

which was calculated by pooling the block interactions that estimate error terms listed in the last column of Figure 7.16. Before calculating this standard error, one should examine the effects listed as block interactions used for error, to make sure none are unusually large. One way to do this is to make a Normal or Half-Normal plot of all the effects except those that estimate block effects listed in the second to last column of Figure 7.16. This was done in Figure 7.17.

In this figure none of the block interactions used to estimate error appear large. Only the main effects stick out from a straight line drawn through the line of points that rise from the origin. Therefore, the standard error of effects calculated above was used to calculate the t-statistics in the last column of Table 7.22 with the formula t = Effect/s_E. The degrees of freedom for these t-statistics is four, since four block interaction terms were pooled to calculate the standard error. The critical $t_{4,0.05} = 2.776$ is from Table A.3. The significant effects (whose t-statistics exceed this critical value) are asterisked in Table 7.22. It can be seen that all of the main effects appear to be significant. These are the same factors that

Factors	Appendix Table	Number of Runs	Number of Blocks	Block Size	Estimable Interactions	Interactions Defining Blocks	Interactions that Estimate Block Effects	Block Interactions Used for Error Terms
1, 2, 3	B.2-3	8	2	4	none	123	123	12, 13, 23
1, 2, 3	B.2-3	8	4	2	none	12, 13	12, 13, 23	123
1, 2, 3, 4	B.2-4	16	2	8	12, 13, 14, 23, 24, 34	1234	1234	123, 134, 124, 234
1, 2, 3, 4	B.2-4	16	4	4	none	124, 134	124, 134, 23	12, 13, 14, 24, 34, 123, 234,
1, 2, 3, 4	B.2-4	16	8	2	none	12, 23, 34	12, 13, 14, 23, 24, 34,	123, 124, 134, 234
1, 2, 3, 4, 5	B.2-5	32	2	16	12, 13, 14, 15, 23, 24, 25, 34, 35, 45	12345	12345	1234, 1235, 1245, 1345, 2345, 123, 124, 125, 134, 135, 145, 234, 235,
1, 2, 3, 4, 5	B.2-5	32	4	8	12, 13, 14, 15, 23, 24, 25, 34, 35, 45	123, 345	123, 345, 1245	124, 125, 145, 245, 235, 135, 134, 234, 1234, 1235, 1345, 2345, 12345
1, 2, 3, 4, 5	B.2-5	32	8	4	none	125, 235, 345	125, 235, 345, 13, 24, 145, 1234	12, 14, 15, 23, 25, 34, 35, 45, 123, 124, 134, 135, 234, 245, 1235, 1245, 1345, 2345,
1, 2, 3, 4, 5	B.2-5	32	16	2	none	12, 13, 34, 45	12, 13, 34, 45, 14, 15, 23, 24, 25, 35, 1234, 1235, 1245,	123, 124, 125, 134, 135, 145, 235, 234, 245, 345, 12345
1, 2, 3, 4, 5, 6	B.1-13	32	2	16	12, 13, 14, 15, 16, 23, 24, 25, 26, 34, 35, 36,	E_1	E_1	E_2, E_3, E_4, E_5, E_6, E_7, E_8, E_9, E_{10}

FIGURE 7.16: Blocked Factorial Designs

appear to be large in Figure 7.17.

All six 2-factor interactions were estimable in this experiment, but since none appear to be significant the interpretation is simple. To get the maximum projectile speed we should set the factor that has a positive effect (X_1) at its high level, and the factors that have negative effects (X_2, X_3, and X_4) at their low levels.

TABLE 7.22

2^4 Design in Two Blocks from Johnson and Leone

					DESIGN AND DATA			RESULTS		
Run	X_1	X_2	X_3	X_4	$X_1X_2X_3X_4$ Block	Velocity (coded)		Effects	t-stats	
1	−	−	−	−	Day 2	97	$1 =$	68.38	16.76	*
2	+	−	−	−	Day 1	151	$2 =$	-24.88	-6.10	*
3	−	+	−	−	Day 1	68	$3 =$	-62.38	-15.29	*
4	+	+	−	−	Day 2	150	$4 =$	-13.88	-3.40	*
5	−	−	+	−	Day 1	39	$12 =$	4.38	1.07	
6	+	−	+	−	Day 2	100	$13 =$	-5.12	1.26	
7	−	+	+	−	Day 2	15	$14 =$	6.38	1.53	
8	+	+	+	−	Day 1	66	$23 =$	-10.88	2.66	
9	−	−	−	+	Day 1	75	$24 =$	-2.88	0.70	
10	+	−	−	+	Day 2	145	$34 =$	-0.88	0.21	
11	−	+	−	+	Day 2	53	$123 =$	-7.12	error	
12	+	+	−	+	Day 1	141	$124 =$	-0.12	error	
13	−	−	+	+	Day 2	26	$134 =$	0.88	error	
14	+	−	+	+	Day 1	97	$234 =$	-3.88	error	
15	−	+	+	+	Day 1	-16	$1234 =$	2.38	Blocks	
16	+	+	+	+	Day 2	54				

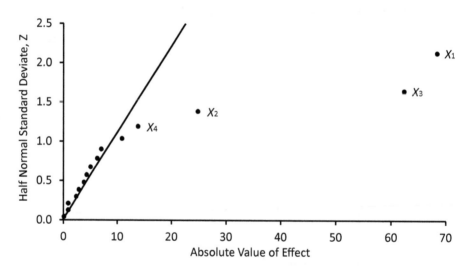

FIGURE 7.17: Half-Normal Plot of Effects from Table 7.22

7.7   Screening Designs with Multiple Level Factors

There often arise situations with qualitative factors when there are more than two levels that need to be included in the experimental study. Fortunately, it is not difficult to create screening designs for multiple level factorials using the fractional factorial tables in Appendix B and simple rules for combining and collapsing columns.

7.7.1 Combination of Factors Method (Pseudofactors)

As an example, consider an experiment with one 4-level factor and four 2-level factors. The number of experiments required for a full factorial would be $4 \times 2^4 = 64$. However, an eight-run screening design could easily be created using the two-level fractional factorial in Table B.1-5. To do this, first let the levels of factors X_1 and X_2 define the four-level factor as shown below:

Levels of X_1	Levels of X_2		Levels of 4-Level Factor
−	−	$\Longrightarrow$	1
+	−	$\Longrightarrow$	2
−	+	$\Longrightarrow$	3
+	+	$\Longrightarrow$	4

Next, assign the four 2-level factors to columns X_3, X_5, X_6, and X_7. Column X_4 is not used because it is confounded with the interaction between X_1 and X_2, as can be seen in the list of confoundings below Table B.1-5. The interaction between the two factors used to define the levels of the four-level factor is part of the four-level factor and should not be confounded with other factors. The design would appear as:

Run	A(X_1,X_2)	B(X_3)	C(X_5)	D(X_6)	E(X_7)
1	1	1	2	2	1
2	2	1	1	2	2
3	3	1	2	1	2
4	4	1	1	1	1
5	1	2	1	1	2
6	2	2	2	1	1
7	3	2	1	2	1
8	4	2	2	2	2

The analysis of a design like this would be identical to the analysis of the half fraction in Section 7.4.3. Calculate all the effects from Table B.1-5 and identify the significant factor effects using a Pareto diagram or Half-Normal plot. The effects for the two-level factors are interpreted as usual. The X_1 effect is a comparison between Levels 1 and 3 with 2 and 4 of the four-level factor. Likewise the X_2 effect is a comparison of Levels 1 and 2 with 3 and 4, while the X_1X_2 interaction (or X_4 effect) is a comparison of Levels 1 and 4 with 2 and 3.

The previous example illustrates the method of combining two-level factors to create a factor with more levels. A four-level factor can always be created by combining any two, two-level columns in a two-level fractional factorial design, as long as the interaction between the two columns is not assigned to any other two-level factor. The fractional factorial tables listed in Appendix B.1 can be used when combining two-level factors, because the CON-FOUNDINGS show us which column is confounded with the interaction of the columns we are combining, and therefore should not be assigned to another factor. Plackett–Burman designs cannot be used to create combined factors, because it is difficult to tell which columns are confounded with the two-factor interaction of interest.

The two-level factors that are combined are referred to as the pseudofactors and actually represent comparisons between different combinations of levels of the four-level factor as described previously. An eight-level factor can be created by assigning levels of the eight-level factor to the combinations of levels of three two-level pseudofactors. In this case none of the three 2-way interactions between pseudofactors nor the 3-way interaction among pseudofactors can be assigned to any other two-level factor or used to define any other four- or eight-level factors.

In general a 2^p-level factor can be created by assigning its levels to all combinations of p two-level pseudofactors. Interactions of all levels among the pseudofactors must not be used in defining any other factors.

7.7.2   Collapsing Levels (Dummy Levels)

When the number of levels of a multiple level factor is not a power of 2, the pseudofactor method cannot be used to generate the levels. Another simple method that will work is the method of collapsing levels or dummy levels. To use this method, simply assign the unneeded levels of a 2^p-level factor to previous levels. For example, suppose we want to create a screening design for a 3×2^4 factorial experiment. Start by creating the 4×2^4 experiment described above using columns X_1 and X_2 of Table B.1-5 to define the four-level factor and columns X_3, X_5, X_6, and X_7 to represent the two-level factors. Next collapse the levels of the four-level factor to three as shown below:

4-Level Factor		3-Level Factor
1	$\Longrightarrow$	1
2	$\Longrightarrow$	2
3	$\Longrightarrow$	3
4	$\Longrightarrow$	3

This would result in the following eight-run 3×2^4 design:

Run	A(X_1, X_2)	B(X_3)	C(X_5)	D(X_6)	E(X_7)
1	1	1	2	2	1
2	2	1	1	2	2
3	3	1	2	1	2
4	3	1	1	1	1
5	1	2	1	1	2
6	2	2	2	1	1
7	3	2	1	2	1
8	3	2	2	2	2

Since the number of repeats of each level is no longer equal (i.e., two 1's, two 2's, and four 3's) after collapsing levels, the standard method of analysis will not work for mixed-level screening designs created in this way. The way these designs are typically analyzed is to pool interactions to form an error term (similar to the example with a Plackett–Burman design in Section 7.5.3) and test the main effects using the GLM ANOVA that was discussed in Chapter 5.

7.7.3 L_{18} Orthogonal Array

Another tabled design which is very handy for creating screening designs with some two-level factors and other three-level factors is the L_{18} orthogonal array shown in Table B.1-15. This design table lists one 2-level factor and seven 3-level factors. Any of the three-level factors can be collapsed by the dummy level technique to create two-level factors. For example, to create an 18-run screening design for a $2^3 \times 3^5$, simply collapse the levels of Columns 2 and 3 to two-levels resulting in the design in Table 7.23.

In Table B.1-15, the interaction between Columns 1 and 2 is not confounded with any of the other columns in the design. The interactions between all other columns are partially confounded with main effects similar to the Plackett–Burman designs. Since the interaction between Columns 1 and 2 is not confounded with any other column, these two columns can be combined using the pseudofactor method to create one 6-level factor. This six-level factor can in turn be collapsed using dummy levels to create a five-level or four-level factor. Thus 18-run fractions of $6 \times 2^p \times 3^q$ or $5 \times 2^p \times 3^q$ or $4 \times 2^p \times 3^q$ type designs can be easily created using the L_{18} table.

Mixed-level designs created with the L_{18} table will not be balanced, in general, and like the designs created by the collapsing levels method, they must be analyzed using the GLM ANOVA method described in Chapter 5 by pooling interactions as an error term. One exception is when all columns are used exactly as they appear in Table B.1-15 for an 18-run fraction of a 2×3^7 design. In this case, orthogonal contrasts can be formed for each main effect and the interaction between Columns 1 and 2, as shown in Chapter 5, and a Half-Normal plot or Pareto diagram of the standardized contrasts can be used to identify important effects.

TABLE 7.23

Screening Design for $2^3 \times 3^5$ Created by Collapsing Columns 2 and 3 of Table B.1-15

Run	COLUMN							
	1	2	3	4	5	6	7	8
1	1	1	1	1	1	1	1	1
2	1	1	2	2	2	2	2	2
3	1	1	2	3	3	3	3	3
4	1	2	1	1	2	2	3	3
5	1	2	2	2	3	3	1	1
6	1	2	2	3	1	1	2	2
7	1	2	1	2	1	3	2	3
8	1	2	2	3	2	1	3	1
9	1	2	2	1	3	2	1	2
10	2	1	1	3	3	2	2	1
11	2	1	2	1	1	3	3	2
12	2	1	2	2	2	1	1	3
13	2	2	1	2	3	1	3	2
14	2	2	2	3	1	2	1	3
15	2	2	2	1	2	3	2	1
16	2	2	1	3	2	3	1	2
17	2	2	2	1	3	1	2	3
18	2	2	2	2	1	2	3	1

7.8   Summary

7.8.1   Important Terms and Concepts

The following is a list of the important terms and concepts covered in this chapter.

screening experiments
one-at-a-time experimentation
fractional factorial design
Plackett–Burman design
confounding
half fraction
base design
added factors
alias structure
confounding pattern
generator
contrast
defining relation
principle fraction
1/4 fraction
law of conservation of information
fractional factorial design tables
projection property of fractional factorial designs
regression
resolution of fractional factorial designs
Plackett–Burman design tables
n_F for information
n_F for precision
blocked factorials
screening designs for multiple level factors
pseudofactors
dummy levels
L_{18} orthogonal array

7.8.2   Important Formulas

$n_F \geq k + 5$ (for information)

$n_F \geq (8\sigma/\delta)^2$ (for information)

7.9   Exercises

1. The following data were recorded for an eight-run Plackett–Burman Screening Design (Nelson [1982]).

Run	X_1	X_2	X_3	X_4	X_5	X_6	X_7	Yield
1	+	+	+	−	+	−	−	1.1
2	−	+	+	+	−	+	−	6.3
3	−	−	+	+	+	−	+	1.2
4	+	−	−	+	+	+	−	0.8
5	−	+	−	−	+	+	+	6.0
6	+	−	+	−	−	+	+	0.9
7	+	+	−	+	−	−	+	1.1
8	−	−	−	−	−	−	−	1.4

(a) Calculate the effects for factors $X_1 - X_7$.

(b) Make a normal probability plot of the effects and a Pareto diagram of the absolute effects. Include Lenth's ME on the Pareto diagram.

(c) Which main effect is the $X_1 X_2$ interaction confounded with? $X_1 X_6$? $X_2 X_6$?

(d) What would you recommend after the analysis of this data?

2. Use Table B.1-7 and the method of pseudofactors to derive designs for the following cases.

(a) Create a 16-run fraction of a $4 \times 4 \times 2^2$ design.

(b) Create a 16-run fraction of a 8×2^3 design.

(c) Using the dummy levels method with the design you created in (a), create a 16-run $4 \times 3 \times 2^2$ design.

Chapter 8

Regression Analysis

8.1 Introduction

The goal of the vast majority of experimentation is (or ought to be) to develop a model which adequately describes the system being studied. The model can then be used for whatever the intended objective is: optimization, troubleshooting, control, etc. The model is essentially a concise summary of all the data that was taken. The model smooths out the noise (variability) in the data, and it elucidates the underlying relationships between the factors and the response(s).

As a rule, the *form* of the model is known (or specified) before any experiments are run. It could be a simple straight line, a complicated polynomial, or anything in between. It could even be a mechanistic model, although those are, in general, beyond the scope of this text. But, even though the form is known, the model has constants in it whose values are NOT known. One way to think of the goal of experimentation is that the goal is to allow the estimation of those constants (which we sometimes also call by other names like coefficients or parameters). The main thrust of this chapter describes the process we use to distill the values of those constants from our data.

We will begin this chapter with a description of the method, which is called Least Squares. We will then apply the method to a simple problem of fitting a straight line (Section 8.3). After that we will cover the more useful situation of fitting a model with several factors in it (Section 8.4). We will look at describing how well the model fits the data in Section 8.5, and how good any assumptions are that we had to make along the way (Section 8.6). It should be mentioned up front that the equations that are used for regression analysis can be expressed very succinctly in matrix form, and so matrices will be used extensively except for the very simplest of cases. If you are unfamiliar (or just rusty) on the subjects of matrix notation and matrix manipulation, there as many good and short tutorials available on-line. See, for example, stattrek.com.

8.2 Method of Least Squares

As stated in the Introduction, our goal is to find the best values for the constants in a model (i.e., the values of the constants that give the best fit to a set of data). Before we can do anything we must agree upon what we mean by "best." As the basis of our discussion, let us take an extremely simple situation, and find the best straight line through the origin and the two data points shown in Figure 8.1.

The centuries-old method of fitting the line is the "eyeball" method. This involves nothing more than putting a straightedge on the graph where it looks best by eye and then drawing the line. Although this is still a frequently used method when it comes to drawing a straight line, it has some shortcomings. First of all it is not objective; two people would not necessarily agree on the best straight line through a given set of data. We would also usually like to be able to draw some inferences about the model (like putting error limits on the coefficients and/or the model predictions) which cannot be done with eyeball lines. And last, but not least, the eyeball method can only be used with one X variable. Drawing multi-dimensional surfaces (even linear ones) by eye is virtually impossible. So we need a quantitative method.

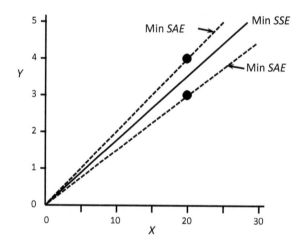

FIGURE 8.1: Possible Lines Through Two Data Points

A good fit of the line to the data means that the line should be as close to all the data points as possible. Therefore, one possible quantitative measure of goodness is the sum of the absolute values of the vertical distances from the data points to the line, or the SAE (Sum of the Absolute Errors) for short. In equation form:

$$SAE = \sum_{i=1}^{n} \left| Y_i - \hat{Y}_i \right| \qquad (8.1)$$

where Y_i is the value of the response for the ith data point, and $\hat{Y}_i$ is the predicted value of the response for the ith data point given by the line. Although this is intuitively a very reasonable measure, it is not necessarily unique. In Figure 8.1, both of the two outside lines and all lines in between them would have the same value of SAE. Therefore, any line in that interval would be judged by the SAE criterion to be equally good. This is not what we would like. We would like the line down the middle, centered between the two data points, to be best. So SAE is not such a wonderful criterion after all. And, as a practical matter, it often turns out to be difficult to find a minimum (i.e., the best model), so a better criterion is needed.

The measure that is almost universally used is the sum of the squared vertical distances from the data points to the line, or the SSE (Sum of the Squared Errors) for short. In equation form:

$$SSE = \sum_{i=1}^{n} (Y_i - \hat{Y}_i)^2. \qquad (8.2)$$

Using this criterion, there is only one best line, and it goes down the center of the data as we would expect it to. When we have more than just the two data points, it should be mentioned that minimizing the *SSE* prefers to cut down any large errors (which are squared) at the price of smaller ones. Although, as has already been indicated, there are other measures of goodness, the criterion used in this text (as well as virtually all others) will be *SSE* exclusively. When using this criterion, values for the coefficients in our model that minimize *SSE* are sought. Since we are making the sum of the squared distances as small as possible, the whole procedure is often called the **method of least squares**.

8.2.1 Estimating the Slope of a Straight Line Through the Origin

To illustrate the method for a more realistic but still simple case, consider the same straight line through the origin as discussed in relation to Figure 8.1, but applied to fitting a few more data points. The physical situation is that we want to predict the tensile strength of a standard alloy of steel as a function of its Brinell hardness number. The response, Y, is therefore the tensile strength, and the independent variable, X, is the Brinell hardness. The mathematical model is:

$$\hat{Y} = bX. \tag{8.3}$$

Samples of five steel alloys were taken and their Brinell hardness numbers were recorded in Table 8.1 . The tensile strength values were determined experimentally and also recorded in that table (along with some other calculated numbers to be used later).

TABLE 8.1
Data and Calculations for Predicting Tensile Strength of Steel from Brinell Hardness

Steel Sample	Brinell Hardness (X)	Tensile Strength (Y)	Calculation Values X^2	Y^2	XY
1	500	256	250,000	65,536	128,000
2	431	212	185,761	44,944	91,372
3	370	186	136,900	35,721	69,930
4	321	155	103,041	24,025	49,755
5	285	138	81,225	19,044	39,330
Sums:	1907	950	756,927	189,270	378,387

Now that we have data, the method of least squares can be used to estimate the unknown coefficient, b, in Equation 8.3. In other words, we would like to find the best value of b, and we now understand "best" to mean the value of b which minimizes *SSE*. If we put our model into the equation for *SSE*, we get

$$SSE = \sum_{i=1}^{n}(Y_i - \hat{Y}_i)^2 = \sum_{i=1}^{n}(Y_i - bX_i)^2 \tag{8.4}$$

Since the Y_i's and the X_i's are known quantities, the only unknown in the equation for *SSE* (Equation 8.4) is b. Therefore, we could use any minimization procedure that we choose to find the best value for b.

One possible method is "brute force." That is, we can try numerous values b, calculate the SSE for each one, plot the *SSE* values on a graph versus b, and pick the value that gives the best fit (i.e., the smallest value of *SSE*). If this were done for our data set, the

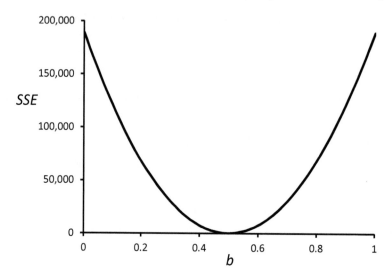

FIGURE 8.2: *SSE* as a Function of the Slope of the Line, *b*

plot shown in Figure 8.2 would be obtained from which the best value for *b* can be seen to be 0.50.

Although brute force will certainly get us to the best value of *b*, we can be much more efficient. It should be noticed that the graph of *SSE* versus *b* looks very much like a quadratic curve (a parabola), and, in fact, it is. This can be seen if we expand Equation 8.4).

$$SSE = \sum_{i=1}^{n}(Y_i^2 - 2bX_iY_i + b^2X_i^2) \tag{8.5}$$

or, upon rearranging,

$$SSE = (\sum_{i=1}^{n}X_i^2)b^2 - 2(\sum_{i=1}^{n}X_iY_i)b + \sum_{i=1}^{n}Y_i^2 \tag{8.6}$$

which is a quadratic equation in *b*, since the X_i's and Y_i's are known quantities. Using elementary calculus, the minimum is located at the point where the first derivative of *SSE* with respect to *b* is zero. If we take the first derivative we get

$$\frac{d(SSE)}{db} = 2(\sum_{i=1}^{n}X_i^2)b - 2(\sum_{i=1}^{n}X_iY_i) = 0. \tag{8.7}$$

Solving Equation 8.7) for *b*, we get

$$\hat{b} = \sum_{i=1}^{n}X_iY_i/\sum_{i=1}^{n}X_i^2 = 378,387/756,927 = 0.500. \tag{8.8}$$

This is the same answer that we got by brute force, but we got it directly. Note that the "hat" on *b* indicates that the value is not the true value (which is unknown) but rather our best guess based on the data we have and the method of least squares.

As a check that Equation 8.8) gives us a minimum (although in this case it is obvious from the graph of *SSE* versus *b*) we can check the second derivative to make sure it is positive.

$$\frac{d^2(SSE)}{db^2} = 2(\sum_{i=1}^{n} X_i^2) > 0. \qquad (8.9)$$

Since the second derivative is the sum of the X_i's squared, it will always be positive, and so this procedure will always give a minimum in *SSE*. This continues to be true for larger models also.

The data for this example is shown graphically in Figure 8.3 along with the prediction equation, $\hat{Y} = 0.500\ X$, fit by using least squares. The equation can be seen to go through the data nicely.

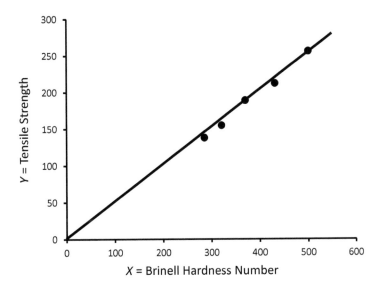

FIGURE 8.3: Brinell Hardness Data and Fitted Line Through the Origin

8.3 Linear Regression

8.3.1 Estimating the Slope and Intercept of a Straight Line

Let us now look at a situation that is just a bit more complex—the case where the regression equation is a straight line (as before) but it may have a nonzero intercept. The least squares estimates of the slope and intercept are still found in the same manner. The model is

$$\hat{Y} = a + bX \qquad (8.10)$$

and the *SSE* (the sum of the squared differences between actual data and predicted values from the equation) can be written as a quadratic function of a as well as b. This is shown in Equation 8.11 which expands to Equation 8.12, which makes it more obvious.

$$SSE = \sum_{i=1}^{n}(Y_i - \hat{Y}_i)^2 = \sum_{i=1}^{n}(Y_i - [a + bX_i])^2 \qquad (8.11)$$

$$SSE = \sum Y_i^2 - (2\sum Y_i)a - (2\sum X_i Y_i)b + na^2 + (\sum X_i^2)b^2 + (2\sum X_i)ab \qquad (8.12)$$

The values of a and b which give a minimum *SSE* can be found, as before, by taking the derivatives with respect to a and b and setting them equal to zero.

$$\frac{\partial SSE}{\partial a} = -2\sum Y_i + 2na + (2\sum X_i)b$$

$$\frac{\partial SSE}{\partial b} = -2\sum X_iY_i + (2\sum X_i)a + 2(\sum X_i^2)b. \qquad (8.13)$$

This is a pair of simultaneous linear equations in a and b whose solutions are

$$\hat{b} = \frac{\sum X_iY_i - (\sum X_i)(\sum Y_i)/n}{\sum X_i^2 - (\sum X_i)^2/n}$$

$$\hat{a} = (\sum Y_i - \hat{b}\sum X_i)/n. \qquad (8.14)$$

For the data in Table 8.1 , the calculated coefficients are

$$\hat{b} = \frac{378,387 - (1907)(950)/5}{756,927 - (1907)^2/5} = 0.5425$$

$$\hat{a} = [950 - (0.5425)(1907)]/5 = -16.92. \qquad (8.15)$$

The model (straight line) with the least squares estimates for a and b is shown in Figure 8.4 along with the data. It can be seen to be a closer fit to the data than the straight line through the origin. This is to be expected, of course, since the bigger model is more flexible.

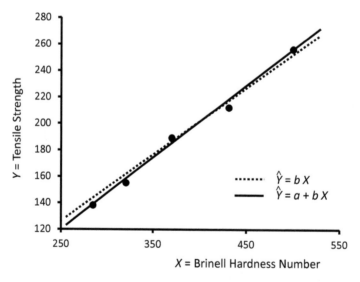

FIGURE 8.4: Brinell Hardness Data and Fitted Line, $\hat{Y} = a + bX$

The fact that the coefficients of the least squares regression line are so easily calculated has made the method very popular. Even many handheld calculators can do the calculations automatically, not to mention popular computer software packages like spreadsheets.

8.3.2 Statistical Significance of Coefficients

As early as the nineteenth century, least squares was a commonly used method of fitting regression lines to astronomical data. Karl Gauss sought to determine the conditions under which the line fitted with least squares was the best possible estimate of the true line (best in the sense of being unbiased with minimum variance). His work led to the discovery of the Normal (or Gaussian) Distribution. He also found what he was after—that the least squares estimates are best under the following conditions: (1) the errors, ϵ_i's, are Normally distributed with a mean of zero, (2) all the errors have the same variance (i.e., all data have the same precision), and (3) the errors are independent from observation to observation (i.e., knowing the value of one of them does not help you predict any others). Remember that the error, ϵ_i, is what is added to the true response for the ith observation to give the observed response.

$$Y_i = a + bX_i + \epsilon_i. \tag{8.16}$$

Figure 8.5 illustrates this situation. At any specific X value, the long run average or expected value of Y is $a + bX$, but a particular observation of Y is one random selection from the bell-shaped Normal distribution centered on the line, $Y = a + bX$. It should also be noticed that the Normal Distributions all have the same variance, σ^2. When the distribution of the

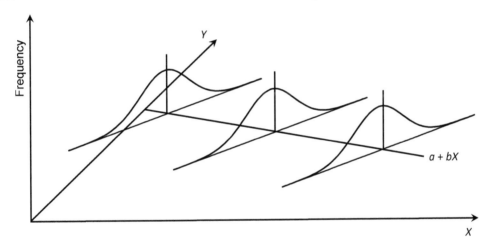

FIGURE 8.5: Assumptions Made During Regression Analysis

errors is Normal, not only do we know that the least squares estimates of a and b are best, but we are also able to use statistical theory to perform tests of significance and/or determine confidence intervals. It would be a rare situation where that would not be just as important as determining the best line.

In this section, we will focus on tests of significance. Specifically, we want to determine if the coefficients (a and b) are significantly different from zero. This is crucial if one wants to justify the assertion that X influences Y. Since b quantifies just how much Y changes when X is changed, if b could reasonably be zero, then it is plausible that X has no impact on Y.

Let us first consider the case where we have data at only two values of X: x_1 and x_2. Then, as can be seen in Figure 8.6, comparing the slope of the regression line, **b**, to zero is equivalent to comparing the mean of the Normal Distribution centered at $a + bx_1$ (denoted by μ_1) to the mean of the Normal distribution centered at $a + bx_2$ (denoted by μ_2). Recall from Chapter 2, that the signal-to-noise ratio (or two-sample t-statistic) for comparing the means of two Normal distributions is:

$$t_{n_1+n_2-2} = \frac{\bar{Y}_1 - \bar{Y}_2}{s_p\sqrt{1/n_1 + 1/n_2}}. \tag{8.17}$$

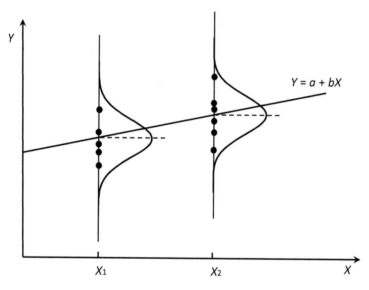

FIGURE 8.6: Comparing the Slope, b, to Zero is Equivalent to Comparing the Means of Two Distributions

If this statistic is larger than the upper tail of the tabulated student's t-distribution, it indicates that $\mu_1 - \mu_2 > 0$, or the mean of the first population is greater than the mean of the second population. If this statistic is smaller than the lower tail of the tabulated student's t-distribution it indicates that $\mu_1 - \mu_2 < 0$, or the mean of the second population is larger. In either case, if the two means are different, the slope, b, is not zero. You should notice that this is basically the same situation we are in when we analyze data from a two-level factorial design (with or without center points). In the general case where there are more than two values of X, the t-statistic for testing whether $b = 0$ is, as expected, the ratio of $\hat{b}$ to its standard deviation:

$$t_{n-2} = \hat{b}/s_{\hat{b}} \tag{8.18}$$

$$\text{where: } s_{\hat{b}} = s/\sqrt{\sum(X_i - \bar{X})^2}. \tag{8.19}$$

In this equation, s is the standard deviation of the errors in the individual responses, and $\bar{X}$ is the average of the X values ($\bar{X} = \sum X_i/n$). Although the equation for the standard deviation of $\hat{b}$ was given without any derivation (or perhaps especially because it was not derived), it should be examined to see if it makes sense. Equation 8.19) says that the variability in the estimate of the slope is proportional to the variability of the individual observations, s, which is reasonable. The only other thing that affects the precision of $\hat{b}$ is $\sqrt{\sum(X_i - \bar{X})^2}$; the bigger the sum the more precise the estimate of the slope. The sum gets larger if we have more data and/or the X values are further away from the middle. Both of these impacts make good intuitive sense—more data and more spread in the values of X seem like they would result in a better estimate of the slope.

In order to calculate the standard deviation of $\hat{b}$, we first need to know s, the standard deviation of the errors in the Y values. When we had a group of replicate data points taken at the same conditions, the formula used in Chapter 2 (as well as elsewhere) was:

$$s^2 = \sum_{i=1}^{n}(Y_i - \bar{Y})^2/(n-1). \qquad (8.20)$$

Now, however, we do not (necessarily) have any replicate data points. But, that does not mean that we have no estimate of the variability. The scatter of the data around the line is essentially the same thing, so our equation to calculate the variance becomes:

$$s^2 = \sum_{i=1}^{n}(Y_i - \hat{Y}_i)^2/(n-c) = SSE/(n-c). \qquad (8.21)$$

It should be noticed that the denominator changed from (n−1) to the more general (n−c), where c is the number of constants in the model to be estimated from the data. Before, c was one (just an average), and now c is two (a slope and an intercept).

TABLE 8.2
Worksheet for Calculating the Significance of the Linear Model
for Brinell Hardness

X_i	Y_i	$\hat{Y}_i = \hat{a} + \hat{b}X_i$	$\epsilon_i = Y_i - \hat{Y}_i$	ϵ_i^2	$(X_i - \bar{X})^2$	X_i^2
500	256	254.34	1.66	2.747	14,066.0	250,000
431	212	216.91	-4.91	24.097	2,460.2	185,761
370	189	183.82	5.18	26.881	130.0	136,900
321	155	157.23	-2.23	4.982	3,648.2	103,041
285	138	137.70	0.30	0.089	9,293.0	81,225
Sums: 1907	950			58.796	29,597.2	756,927
$\bar{X} =$ 381.4			$s^2 =$ 19.60			
			$s =$ 4.43			

For our Brinell Hardness example, the quantities needed to calculate the t-statistic are given in Table 8.2. Using them gives:

$$s_{\hat{b}} = 4.43/\sqrt{29.597.2} = 0.0257.$$

So $t_3 = 0.5425 / 0.0257 = 21.08$. As was the case for other *t*-statistics used in previous sections, a value larger than the tabled *t*-distribution with the appropriate degrees of freedom indicates that $b \neq 0$. This is far, far greater than the tabled value of 3.182, so we conclude that there is certainly (for all practical purposes) a relationship between Brinell Hardness and tensile strength.

In addition to checking the statistical significance of the slope of the line, we can also check whether the intercept, a, is non-zero. The *t*-test can be thought of as subtracting $Y = 0 + bX$ from each Y value, then performing a one-sample *t*-test to determine if the differences have zero mean. The general formula for this *t*-statistic is:

$$t_{n-2} = \hat{a}/s_{\hat{a}} \qquad (8.22)$$

where: $$s_{\hat{a}} = s\sqrt{\frac{\sum X_i^2}{n\sum(X_i - \bar{X})^2}}. \qquad (8.23)$$

For the Brinell Hardness data,

$$s_{\hat{a}} = 4.43\sqrt{756,927/(5)(29,597.2)} = 10.01.$$

Therefore, $t_3 = -16.92 / 10.01 = -1.69$. Since this is not as large as the tabled value, we conclude that the intercept could reasonably be zero. Under the philosophy, "The simpler, the better!" the intercept term in the model should be dropped, and the model of a straight line through the origin is found to be more appropriate.

Although the calculations of least squares estimates for the slope and intercept of straight lines as well as the standard deviations of those estimates can be done by hand (presumably using a calculator at minimum), computer packages now do most or all of the work for us. A sample of a typical spreadsheet regression analysis is given in Figure 8.7. The coefficients and their standard errors are calculated. The output even includes the t-statistics and the significance of the coefficients, as well as other bits of statistical information.

(a) Selected Spreadsheet Output for a Straight Line Through the Origin

Regression Statistics	
R Square	0.987
Standard Error	5.356
Observations	5

	Coefficients	Std Error	t Stat	P-value	Lower 95%	Upper 95%
Intercept	0	#N/A	#N/A	#N/A	#N/A	#N/A
X Variable 1	0.500	0.00616	81.20	1.38E-07	0.483	0.517

(b) Selected Spreadsheet Output for a Straight Line With Non-zero Intercept

Regression Statistics	
R Square	0.993
Standard Error	4.427
Observations	5

	Coefficients	Std Error	t Stat	P-value	Lower 95%	Upper 95%
Intercept	-16.9	10.01	-1.69	0.190	-49	15
X Variable 1	0.543	0.0257	21.08	2.33E-04	0.461	0.624

FIGURE 8.7: Selected Regression Output from Excel (Office for Windows 2013)

8.3.3   Confidence Intervals for Coefficients

Another very common and useful way of expressing the variability in an estimate of a coefficient in a model is with a confidence interval. It is typically used when we are sure that the parameter we are estimating cannot be zero. So we are more interested in putting reasonable error limits on our estimate than we are of testing the null hypothesis. A confidence interval is defined as an interval which has a specified probability of containing the true parameter that we are estimating. The confidence interval is based on our parameter estimate, which is our best guess of the true parameter (obtained from our data), but it recognizes the variability in the estimate.

Mathematically we can state the confidence interval as:

Probability (Lower Limit < true parameter) = confidence, or

Probability (true parameter < Upper Limit) = confidence, or

Probability (Lower Limit < true parameter < Upper Limit) = confidence.

We have three expressions, because we may want only a lower limit, only an upper limit, or both a lower and an upper limit. The lower limit and/or upper limit are calculated from our statistic (estimate of our parameter) and the probability distribution that describes the variability in that statistic. In the case of regression analysis, the appropriate probability density is the t-distribution. The concept is essentially the same as it was for our significance test—the measured value of our parameter is a random variable which is described by a t-distribution centered on the true parameter value. The true parameter value is assumed to be zero for a test of significance. Then we check to see if we could reasonably get the value we actually did get from our data (under those assumptions). If not, then the assumption must be bad, so our true parameter is not zero (or more properly, it is statistically significantly different from zero). When we calculate a confidence interval, we move the center of the distribution around until it just becomes unreasonable to get the result that we did. How far we can move the center of the distribution to either side becomes our confidence interval.

As an example, let us calculate a confidence interval for the slope of our straight line which was fit to the Brinell Hardness data when our model contains an intercept, a, and a slope, b. From our significance test, it was clear that the slope could not possibly be zero. So our attention has turned to establishing what a reasonable range of values for the slope could be. Our estimate of the slope is $\hat{b} = 0.543$. If we want a lower limit for what the slope could be, we drag the t-distribution to the left, until it starts to get unreasonable to actually observe our result. This is shown graphically in Figure 8.8(A). Before we can get a specific value for the lower limit, we need to quantify "unreasonable," which is the same as our Type I error, α, discussed in Chapter 2. If we arbitrarily pick α to be 0.05 (or 5%), the confidence level is 95% (= $1 - \alpha$). We then use the critical t-value, $t_{3,.05} = 2.353$, from Table A.3 with 3 degrees of freedom (the degrees of freedom for s) and the standard deviation of $\hat{b}$, $s_{\hat{b}} = 0.0257$, to determine our confidence limit:

Lower 95% Limit (one-sided) = $\hat{b} - t^* s_{\hat{b}} = 0.543 - 2.35(0.0257) = 0.482$.

If we wanted an upper 95% confidence limit, we would drag the t-distribution to the right until only 5% of it was to the left of our parameter estimate, as shown in Figure 8.8(B). The upper confidence limit for our example is:

Upper 95% Limit (one-sided) = $\hat{b} + t^* s_{\hat{b}} = 0.543 + 2.35(0.0257) = 0.603$.

Finally, if we wanted a two-sided 95% confidence interval, we would drag the t-distribution to the left until only $2\frac{1}{2}$% of it was to the right of our parameter estimate, and we would drag the t-distribution to the right until only $2\frac{1}{2}$% of it was to the left of our parameter estimate. This is shown in Figure 8.8(C). The confidence interval for our example is:

95% Interval (two-sided) = $\hat{b} \pm t^* s_{\hat{b}} = 0.543 \pm 3.18(0.0257) = 0.461$ to 0.624.

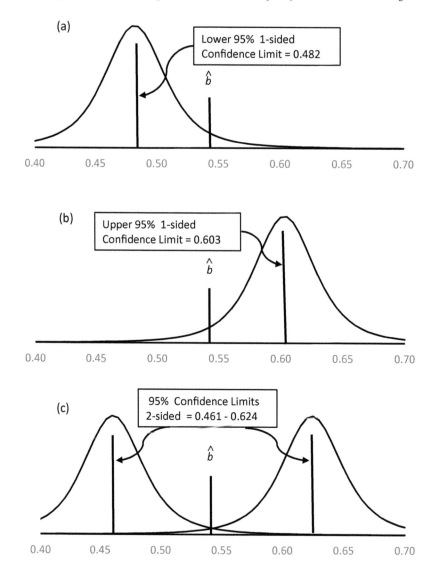

FIGURE 8.8: Confidence Limits for the Slope, *b*

As mentioned above, the confidence level used is strictly a matter of choice. If a confidence level less than 95% were used, the interval would become smaller, since the smaller interval would have less probability of containing the true parameter value. Conversely, if a confidence level greater than 95% were used, the interval would become larger, since the larger interval would have more of a chance of containing the true parameter value.

It should also be stated that confidence intervals can also be used to judge significance. If the confidence interval for the true parameter includes zero, then zero is a reasonable value of the parameter. In that event, the parameter estimate would not be statistically significant, and the null hypothesis, Ho: parameter = 0, would be accepted.

8.3.4 Precision of Predictions

In addition to obtaining the coefficients of the best fitting straight line for a set of data and checking their statistical significance, reporting the precision in predicted values obtained from the line is often important. The predicted value of Y when $X = x_o$ is the estimator of the expected value of Y when $X = x_o$, which is $a + bx_o$ as shown in Figure 8.5.

The variance of a predicted value from the equation, $\hat{Y} = \hat{a} + \hat{b}X$, is given by the formula:

$$Var(\hat{Y}|X = x_o) = \frac{\sigma^2}{n} + \frac{(x_o - \bar{X})^2 \sigma^2}{\sum (x_i - \bar{X})^2} \tag{8.24}$$

where: $Var(\hat{Y}|X = x_o)$ denotes the variance of the predicted value, $\hat{Y}$, when the independent variable $X = x_o$. As usual, n is the number of data points, and $\bar{X}$ is the average of the independent variable values. The estimate of the standard error of a predicted value is then the square root of the right hand side of Equation 8.24, with the true variance, σ^2, replaced by its estimate, s^2.

$$s_{\hat{Y}}|X = x_o = s\sqrt{\frac{1}{n} + \frac{(x_o - \bar{X})^2}{\sum (x_i - \bar{X})^2}}. \tag{8.25}$$

For the data in Table 8.2, s = 4.43, n = 5, x_o is any value of X (hopefully in the range in the data), $\bar{X} = 381.4$, and the large denominator $\sum (x_i - \bar{X})^2 = 29{,}597.2$.

The standard deviations of $\hat{Y}$ for a range of values of x_o are shown in Table 8.3. It should be noticed that the standard error of a predicted value depends upon the point, x_o, where it is calculated. This is obvious from Equation 8.25, but runs counter to many people's intuition. When x_o is equal to the average of the X values for the data, the standard error is a minimum. As the value of x_o moves toward the boundary of the data (in either direction), the standard error increases. This is shown graphically in Figure 8.9, which shows the two lines: $\hat{Y} - 2s_{\hat{Y}}$ and $\hat{Y} + 2s_{\hat{Y}}$. Although the calculations presented here may seem tedious, it should be mentioned that these standard errors are typically produced quite conveniently by regression computer programs.

TABLE 8.3

Calculation of Error Limits on a Straight Line Model

x_o	$\hat{Y}$	$s_{\hat{Y}}$	$\hat{Y} - 2s_{\hat{Y}}$	$\hat{Y} + 2s_{\hat{Y}}$
275	132.28	3.38	125.52	139.03
300	145.84	2.88	140.07	151.60
325	159.40	2.45	154.49	164.31
350	172.96	2.14	168.69	177.24
375	186.53	1.99	182.55	190.50
400	200.09	2.04	196.02	204.16
425	213.65	2.28	209.10	218.21
450	227.22	2.65	221.91	232.52
475	240.78	3.12	234.54	247.02
500	254.34	3.64	247.07	261.62
525	267.91	4.19	259.52	276.29

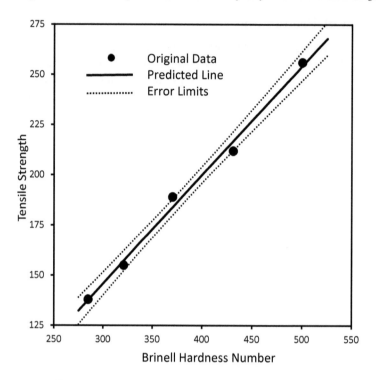

FIGURE 8.9: Error Limits on a Straight Line Model

8.4   Multiple Regression

8.4.1   Introduction

In the last section, the ideas of regression analysis were presented in terms of the simple linear model,

$$Y = b_0 + b_1 X. \tag{8.26}$$

However, the ideas presented extend very easily to the case where there are multiple independent variables, X_i's, and the equation is of the form:

$$Y = b_0 + b_1 X_1 + b_2 X_2 + \ldots + b_p X_p. \tag{8.27}$$

Once we know how to fit a ***linear*** model with any number of independent variables, it may look like we will still need to learn more in order to fit a more complex polynomial model. This is not the case. For example, if we want to fit a full factorial model in three X variables to some data, the equation we would use is:

$$Y = b_0 + b_1 X_1 + b_2 X_2 + b_3 X_3 + b_{12} X_1 X_2 + b_{13} X_1 X_3 + b_{23} X_2 X_3 + b_{123} X_1 X_2 X_3. \tag{8.28}$$

But this is really no different than the equation:

$$Y = b_0 + b_1 X_1 + b_2 X_2 + b_3 X_3 + b_4 X_4 + b_5 X_5 + b_6 X_6 + b_7 X_7 \tag{8.29}$$

where X_1, X_2, and X_3 are the three independent variables, X_4 is really $X_1 X_2$, X_5 is really $X_1 X_3$, X_6 is really $X_2 X_3$, and X_7 is really $X_1 X_2 X_3$. Therefore, in this example, X_4, X_5, X_6, and X_7 are not independently set values for any experiment, but rather they are calculated from X_1, X_2, and X_3. However, our regression analysis does not know (or care) where the values of the variables came from; it only cares that there are now seven **independent** variables (in addition to the constant). So as you can see, the term, independent, does not mean that one variable cannot be calculated from another. It does mean that two (or more) columns may not look the same, or even be a linear function of each other, so we may have to be careful that our data support the calculation of a bigger model. But, once we can fit a linear model in several factors, we will be set, in terms of methodology, to fit any factorial or polynomial model as well.

In order to simplify the formulas used in multiple regression analysis, we will use matrix terminology as mentioned in the introduction. For those unfamiliar with matrices, tutorial web sites such as stattrek.com should provide the necessary background.

8.4.2 Estimation of Coefficients

Matrix notation can be used as a very concise way of writing the formulas that are used in the various phases of multiple regression analysis. As a basis for discussion, let us continue to use the simple factorial model with three independent variables, X_1, X_2, and X_3, given in Equation 8.28, or its equivalent, Equation 8.29. Let us apply it, for example, to the data in Figure 3.11, which were from a 2^3 factorial design replicated a second time for a total of 16 experiments.

First, let us recognize that our responses, Y_i's, from our experiments, taken collectively can be thought of as an n×1 matrix (or column vector), $\mathbf{Y}$, where n is the number of runs (n=16 in this example). And, likewise, the settings of our independent variables are seen to be an n×k matrix, where k is the number of independent variables (k=3 in this example). If that matrix is expanded to include a column for every coefficient in our model, then we end up with an n×(p+1) matrix which we will denote by $\mathbf{X}$. In this case, p, the number of coefficients in the model exclusive of the constant term, is seven, so the (p+1)×1 matrix of coefficients in our model, $\mathbf{B}$, contains eight coefficients in all. The $\mathbf{X}$ and $\mathbf{Y}$ matrices are given below, generically in terms of the number of data points, but for our specific number of constants. To reiterate, we had to add columns for the three two-factor interaction terms and the single three-factor interaction term and also a column of 1's (because of the b_o term) to form the total $\mathbf{X}$ matrix. (Remember that we are defining X_4, X_5, X_6, and X_7 to be X_1X_2, X_1X_3, X_2X_3, and $X_1X_2X_3$, respectively.) Let us further define two more matrices: the (p+1)×1 matrix, $\mathbf{B}$, of coefficients in our model, and the n×1 matrix of predictions from our model, $\hat{\mathbf{Y}}$.

$$\mathbf{Y} = \begin{bmatrix} y_1 \\ y_2 \\ y_3 \\ \vdots \\ y_n \end{bmatrix} \quad \mathbf{X} = \begin{bmatrix} 1 & x_{11} & x_{12} & x_{13} & x_{14} & x_{15} & x_{16} & x_{17} \\ 1 & x_{21} & x_{22} & x_{23} & x_{24} & x_{25} & x_{26} & x_{27} \\ 1 & x_{31} & x_{32} & x_{33} & x_{34} & x_{35} & x_{36} & x_{37} \\ \vdots & \vdots & \vdots & \vdots & \vdots & \vdots & \vdots & \vdots \\ 1 & x_{n1} & x_{n2} & x_{n3} & x_{n4} & x_{n5} & x_{n6} & x_{n7} \end{bmatrix} \quad \mathbf{B} = \begin{bmatrix} b_0 \\ b_1 \\ b_2 \\ b_3 \\ b_4 \\ b_5 \\ b_6 \\ b_7 \end{bmatrix} \quad \hat{\mathbf{Y}} = \begin{bmatrix} \hat{y}_1 \\ \hat{y}_2 \\ \hat{y}_3 \\ \vdots \\ \hat{y}_n \end{bmatrix}.$$

We are now in a position to determine the least squares estimates for our coefficients, $\mathbf{B}$. The measure we are minimizing is still *SSE*, the sum of the squared errors. The error for experiment i, ϵ_i, is (as always) the difference between the actual response, y_i, and the response predicted by our equation, $\hat{y}_i$. The collection of all the errors is an n×1 matrix, $\boldsymbol{\epsilon}$, which is:

$$\epsilon = \mathbf{Y} - \hat{\mathbf{Y}}. \tag{8.30}$$

The sum of the squared errors can be computed very simply once we have $\hat{\mathbf{Y}}$:

$$SSE = \epsilon^T \epsilon = (\mathbf{Y} - \hat{\mathbf{Y}})^T (\mathbf{Y} - \hat{\mathbf{Y}}) \tag{8.31}$$

where the superscript T on a matrix indicates the transpose of that matrix. The $\hat{\mathbf{Y}}$ in Equation 8.31 can be expressed in terms of $\mathbf{B}$ as:

$$\hat{\mathbf{Y}} = \mathbf{X}\,\mathbf{B} \tag{8.32}$$

so that our equation for *SSE* becomes:

$$SSE = (\mathbf{Y} - \mathbf{X}\,\mathbf{B})^T (\mathbf{Y} - \mathbf{X}\,\mathbf{B}) = \mathbf{Y}^T\mathbf{Y} - 2\mathbf{B}^T\mathbf{X}^T\mathbf{Y} + \mathbf{B}^T\mathbf{X}^T\mathbf{X}\mathbf{B}. \tag{8.33}$$

To find the minimum in *SSE* we proceed, just as we did for the simple linear model, by taking the derivative of *SSE* with respect to each element of $\mathbf{B}$ and setting the derivatives equal to zero. We get a set of simultaneous linear equations which can be written again in matrix form as:

$$\frac{d\,SSE}{d\,\mathbf{B}} = -2\mathbf{X}^T\mathbf{Y} + 2\mathbf{X}^T\mathbf{X}\mathbf{B} = 0 \tag{8.34}$$

or

$$\mathbf{X}^T\mathbf{X}\hat{\mathbf{B}} = \mathbf{X}^T\mathbf{Y} \tag{8.35}$$

The hat on $\mathbf{B}$ indicates it is the best (least squares) estimate of the values of the coefficients in our model. The equations implied by Equation 8.35 are called the ***least squares normal equations***. Since $(\mathbf{X}^T\mathbf{X})$ is a full rank square matrix, it has an inverse defined. Therefore, the solution of Equation 8.35 can be obtained by multiplying both sides of the equation by $(\mathbf{X}^T\mathbf{X})^{-1}$ to get

$$\hat{\mathbf{B}} = (\mathbf{X}^T\mathbf{X})^{-1}\,\mathbf{X}^T\mathbf{Y}. \tag{8.36}$$

Equation 8.36 is a simple but very powerful matrix equation which defines the least squares estimates for all the regression coefficients simultaneously. To see how it works, we will calculate the coefficients in the factorial model for the fly ash example in Chapter 3. The data are given in Figure 3.11. Using the coded values of the X's, we get the following $\mathbf{X}$ and $\mathbf{Y}$ matrices:

$$
\begin{array}{ccccccccc}
\text{Int.} & X_1 & X_2 & X_3 & X_1X_2 & X_1X_3 & X_2X_3 & X_1X_2X_3 & \text{Strength}
\end{array}
$$

$$
\mathbf{X} =
\begin{bmatrix}
1 & -1 & -1 & -1 & 1 & 1 & 1 & -1 \\
1 & -1 & -1 & -1 & 1 & 1 & 1 & -1 \\
1 & 1 & -1 & -1 & -1 & -1 & 1 & 1 \\
1 & 1 & -1 & -1 & -1 & -1 & 1 & 1 \\
1 & -1 & 1 & -1 & -1 & 1 & -1 & 1 \\
1 & -1 & 1 & -1 & -1 & 1 & -1 & 1 \\
1 & 1 & 1 & -1 & 1 & -1 & -1 & -1 \\
1 & 1 & 1 & -1 & 1 & -1 & -1 & -1 \\
1 & -1 & -1 & 1 & 1 & -1 & -1 & 1 \\
1 & -1 & -1 & 1 & 1 & -1 & -1 & 1 \\
1 & 1 & -1 & 1 & -1 & 1 & -1 & -1 \\
1 & 1 & -1 & 1 & -1 & 1 & -1 & -1 \\
1 & -1 & 1 & 1 & -1 & -1 & 1 & -1 \\
1 & -1 & 1 & 1 & -1 & -1 & 1 & -1 \\
1 & 1 & 1 & 1 & 1 & 1 & 1 & 1 \\
1 & 1 & 1 & 1 & 1 & 1 & 1 & 1
\end{bmatrix}
\qquad
\mathbf{Y} =
\begin{bmatrix}
43.4 \\ 42.8 \\ 50.1 \\ 49.5 \\ 27.3 \\ 29.1 \\ 47.0 \\ 48.4 \\ 39.4 \\ 38.4 \\ 43.8 \\ 45.0 \\ 25.0 \\ 24.2 \\ 40.7 \\ 42.9
\end{bmatrix}.
$$

The least squares normal equation (Equation 8.35) for this example is

$$
\underset{\mathbf{X}^T\mathbf{X}}{
\begin{bmatrix}
16 & 0 & 0 & 0 & 0 & 0 & 0 & 0 \\
0 & 16 & 0 & 0 & 0 & 0 & 0 & 0 \\
0 & 0 & 16 & 0 & 0 & 0 & 0 & 0 \\
0 & 0 & 0 & 16 & 0 & 0 & 0 & 0 \\
0 & 0 & 0 & 0 & 16 & 0 & 0 & 0 \\
0 & 0 & 0 & 0 & 0 & 16 & 0 & 0 \\
0 & 0 & 0 & 0 & 0 & 0 & 16 & 0 \\
0 & 0 & 0 & 0 & 0 & 0 & 0 & 16
\end{bmatrix}}
\underset{\hat{\mathbf{B}}}{
\begin{bmatrix}
\hat{b}_0 \\ \hat{b}_1 \\ \hat{b}_2 \\ \hat{b}_3 \\ \hat{b}_4 \\ \hat{b}_5 \\ \hat{b}_6 \\ \hat{b}_7
\end{bmatrix}}
=
\underset{\mathbf{X}^T\mathbf{Y}}{
\begin{bmatrix}
318.5 \\ 48.9 \\ -33.9 \\ -19.1 \\ 24.5 \\ -3.5 \\ 0.1 \\ -1.1
\end{bmatrix}}
$$

and the solution (Equation 8.36) is:

$$
\underset{\hat{\mathbf{B}}}{
\begin{bmatrix}
\hat{b}_0 \\ \hat{b}_1 \\ \hat{b}_2 \\ \hat{b}_3 \\ \hat{b}_4 \\ \hat{b}_5 \\ \hat{b}_6 \\ \hat{b}_7
\end{bmatrix}}
=
\underset{(\mathbf{X}^T\mathbf{X})^{-1}}{
\begin{bmatrix}
\frac{1}{16} & 0 & 0 & 0 & 0 & 0 & 0 & 0 \\
0 & \frac{1}{16} & 0 & 0 & 0 & 0 & 0 & 0 \\
0 & 0 & \frac{1}{16} & 0 & 0 & 0 & 0 & 0 \\
0 & 0 & 0 & \frac{1}{16} & 0 & 0 & 0 & 0 \\
0 & 0 & 0 & 0 & \frac{1}{16} & 0 & 0 & 0 \\
0 & 0 & 0 & 0 & 0 & \frac{1}{16} & 0 & 0 \\
0 & 0 & 0 & 0 & 0 & 0 & \frac{1}{16} & 0 \\
0 & 0 & 0 & 0 & 0 & 0 & 0 & \frac{1}{16}
\end{bmatrix}}
\underset{\mathbf{X}^T\mathbf{Y}}{
\begin{bmatrix}
318.5 \\ 48.9 \\ -33.9 \\ -19.1 \\ 24.5 \\ -3.5 \\ 0.1 \\ -1.1
\end{bmatrix}}
=
\begin{bmatrix}
39.81 \\ 6.11 \\ -4.24 \\ -2.39 \\ 3.06 \\ -0.44 \\ 0.01 \\ -0.14
\end{bmatrix}.
$$

These values can be compared to the mean, effects, and interactions given in Figure 3.14, keeping in mind that coefficients are one-half of effects (except for the mean). Or they can be compared to the coefficients in the sample output from MINITAB^{TM}in Figure 3.18 directly. Naturally, they agree. This shows that the worksheet method used for calculating the factorial effects shown in Chapter 3 was simply a step-by-step method for obtaining the least squares estimates.

8.4.3   Statistical Significance of Coefficients

If we group the $\mathbf{X}$ matrices in Equation 8.36, the least squares coefficients in our model can be seen to be calculated as a sum of the y_i's, with each y_i being multiplied by a constant.

$$\hat{\mathbf{B}} = [(\mathbf{X}^T\mathbf{X})^{-1}\,\mathbf{X}^T]\mathbf{Y} \equiv \mathbf{CY}. \tag{8.37}$$

Since the values of the X variables are assumed to be known without error, the matrix, $\mathbf{C}$, is a matrix of constants that can be calculated exactly. Therefore, any variability in $\hat{\mathbf{B}}$ comes from the variability in the y_i's. When this sort of relationship exists between two matrices (as between $\hat{\mathbf{B}}$ and $\mathbf{Y}$), then it has been shown by others that the variance of the calculated matrix is given by:

$$Var(\hat{\mathbf{B}}) = \mathbf{C}\,Var(\mathbf{Y})\,\mathbf{C}^T$$
$$\text{or } Var(\hat{\mathbf{B}}) = [(\mathbf{X}^T\mathbf{X})^{-1}\,\mathbf{X}^T]Var(\mathbf{Y})[(\mathbf{X}^T\mathbf{X})^{-1}\,\mathbf{X}^T]^T. \tag{8.38}$$

If we further assume, as we did in Section 8.3.2, that the errors in $\mathbf{Y}$ are independently distributed like a Normal distribution with a mean of zero and all have the same variance, σ^2, then the variance of $\mathbf{Y}$ can be written as $\mathbf{I}_n\sigma^2$. $\mathbf{I}_n$ is the n×n identity matrix, which has ones down the diagonal and has zeros everywhere else. Putting this result into Equation 8.38 we get:

$$Var(\hat{\mathbf{B}}) = [(\mathbf{X}^T\mathbf{X})^{-1}\,\mathbf{X}^T]\mathbf{I}_n\sigma^2[(\mathbf{X}^T\mathbf{X})^{-1}\,\mathbf{X}^T]^T$$
$$\text{or } Var(\hat{\mathbf{B}}) = (\mathbf{X}^T\mathbf{X})^{-1}\,\mathbf{X}^T\mathbf{I}_n\,[\mathbf{X}^T]^T[(\mathbf{X}^T\mathbf{X})^{-1}]^T\sigma^2. \tag{8.39}$$

Equation 8.39 can be simplified greatly if we recognize the following: (1) multiplication by the identity matrix does not change anything (much like multiplying an algebraic expression by "1"), and so it can be ignored, (2) the transpose of the transpose of a matrix is the original matrix, and (3) the transpose of a square symmetric matrix (like $\mathbf{X}^T\mathbf{X}$ or $(\mathbf{X}^T\mathbf{X})^{-1}$) is equal to the matrix itself. Using all these rules we get:

$$Var(\hat{\mathbf{B}}) = (\mathbf{X}^T\mathbf{X})^{-1}\,\mathbf{X}^T\mathbf{X}(\mathbf{X}^T\mathbf{X})^{-1}\sigma^2. \tag{8.40}$$

Finally, since multiplying a matrix like $\mathbf{X}^T\mathbf{X}$ by its inverse results in an Identity matrix which can be ignored, Equation 8.40 becomes:

$$Var(\hat{\mathbf{B}}) = (\mathbf{X}^T\mathbf{X})^{-1}\,\sigma^2. \tag{8.41}$$

This equation, like Equation 8.36, is a very succinct yet very powerful formula. It gives the variability in our coefficients for any experimental design, $\mathbf{X}$. The result is a square symmetric matrix whose diagonal terms give the variances of the coefficients, and the off-diagonal terms give the covariances of pairs of the coefficients. It is the diagonal terms we will use when we test the statistical significance of our coefficients.

There is one hitch to using Equation 8.41—we rarely know σ^2. That does not stop us, however. We can replace it with its estimate, s^2, and proceed. The only change is that we

TABLE 8.4
Calculation of s^2 for Fly Ash Example

Y	$\hat{Y}=X\hat{B}$	$\epsilon = Y-\hat{Y}$	ϵ^2	
43.4	43.1	0.3	0.09	$s^2 = SSE/[n\text{-}c]$
42.8	43.1	-0.3	0.09	$s^2 = SSE/[n\text{-}(p+1)]$
50.1	49.8	0.3	0.09	$s^2 = 6.92/8$
49.5	49.8	-0.3	0.09	$s^2 = 0.865$
27.3	28.2	-0.9	0.81	$s = 0.930$
29.1	28.2	0.9	0.81	
47.0	47.7	-0.7	0.49	
48.4	47.7	0.7	0.49	
39.4	38.9	0.5	0.25	
38.4	38.9	-0.5	0.25	
43.8	44.4	-0.6	0.36	
45.0	44.4	0.6	0.36	
25.0	24.6	0.4	0.16	
24.2	24.6	-0.4	0.16	
40.7	41.8	-1.1	1.21	
42.9	41.8	1.1	1.21	
		SSE =	6.92	

must then use the t-distribution (with the degrees of freedom in s^2) to test for statistical significance, rather than the Normal distribution.

As a demonstration of the use of Equation 8.41, let us continue with the fly ash example. The estimate of s^2 is determined from *SSE*, which we can calculate since we have already found $\hat{B}$. The calculation of *SSE* is detailed in Table 8.4. Once we have *SSE*, we use Equation 8.21 to evaluate s^2. In this example, the number of constants in our model that we are evaluating from our data, c, is 8 (=p+1) including the intercept. Therefore our degrees of freedom (= n−c) is 16−8, or 8. So our estimate of s^2 is $SSE/(n-c) = 6.92/(16-8) = 0.930$, and our estimate of s is $\sqrt{0.865} = 0.930$. This value of s can be compared to the value obtained by using the pooled standard deviation (which was computed in Section 3.8.2). It is seen to be identical to that value and it has the same degrees of freedom. The reason they agree is that the full factorial model predicts the corner point average at each corner. The pooled standard deviation also computed deviations from the corner point averages. So they were both, in fact, computed in exactly the same way.

Since we already know $(X^T X)^{-1}$ (which was calculated when we estimated the coefficients in our model—see the previous section), we can now use Equation 8.41 to estimate $Var(\hat{B})$:

$$Var(\hat{B}) = \begin{bmatrix} 1/16 & 0 & 0 & 0 & 0 & 0 & 0 & 0 \\ 0 & 1/16 & 0 & 0 & 0 & 0 & 0 & 0 \\ 0 & 0 & 1/16 & 0 & 0 & 0 & 0 & 0 \\ 0 & 0 & 0 & 1/16 & 0 & 0 & 0 & 0 \\ 0 & 0 & 0 & 0 & 1/16 & 0 & 0 & 0 \\ 0 & 0 & 0 & 0 & 0 & 1/16 & 0 & 0 \\ 0 & 0 & 0 & 0 & 0 & 0 & 1/16 & 0 \\ 0 & 0 & 0 & 0 & 0 & 0 & 0 & 1/16 \end{bmatrix} \times 0.865.$$

As mentioned earlier in this section, the diagonal terms in this matrix give us the variances of each of the coefficients in our model. Usually, the diagonal elements in the matrix will be different, but in the case of a 2^k-factorial design (with no center points), they are all they same, namely the pooled variance divided by the number of factorial points. So we have found, by using Equation 8.41, that the variance of each of the coefficients in the factorial model is:

$$Var(\hat{b}_i) = s_p{}^2/n_F. \tag{8.42}$$

Since we actually want the standard deviation, we must take the square root of that:

$$\text{Standard Deviation}(\hat{b}_i), s_{\hat{b}_i} = s_p/\sqrt{n_F}. \tag{8.43}$$

By comparing Equation 8.43 to Equation 3.2 (which is, $s_E = 2s_P/\sqrt{n_F}$), you can see that the standard deviation of a coefficient is half of the standard deviation of an effect. This is as expected, since the coefficient is half of an effect. So the standard deviation of each $\hat{b}_i$ in this example is $0.930/\sqrt{16} = 0.2325$. This can also be compared to the standard deviation of the coefficients column ("SE Coef") in Figure 3.18, and again it is the same as expected.

Once we have all of the standard deviations of the coefficients, all that remains is to calculate the t-statistics,

$$t_i = \hat{b}_i/s_{\hat{b}_i}, \tag{8.44}$$

and compare them to the critical (tabled) value with the appropriate degrees of freedom (8 for this example). The t-statistics for our example were calculated using Equation 8.44. They are the same in the sample output from MINITAB™ in Chapter 3, Figure 3.18, and so they are not shown again. The effects of X_1, X_2, X_3 and the X_1X_2-interaction were found to be statistically significant.

8.4.4   Confidence Intervals for Coefficients

Confidence intervals were discussed in Section 8.3.3 for the coefficients in a straight linc model. The confidence intervals for the coefficients in a multiple regression model are exactly the same. Once we have the coefficients in our model, $\hat{b}_i$ (from Section 8.4.2), and the standard deviations of those coefficients, $s_{\hat{b}_i}$ (from Section 8.4.3), we can express the precision of any of the coefficients as a confidence interval:

Mathematically we can state the confidence interval as:

Probability (Lower Limit < true parameter) = confidence, or
Probability (true parameter < Upper Limit) = confidence, or
Probability (Lower Limit < true parameter < Upper Limit) = confidence.

We have three expressions, because we may want only a lower limit, only an upper limit, or both a lower and an upper limit. The lower limit and/or upper limit are calculated from our statistic (estimate of our parameter) and the probability distribution that describes the variability in that statistic. In the case of regression analysis, the appropriate probability density is the t-distribution. The concept is essentially the same as it was for our significance test—the measured value of our parameter is a random variable which is described by a t-distribution centered on the true parameter value. The true parameter value is assumed to be zero for a test of significance. Then we check to see if we could reasonably get the value we actually did get from our data (under those assumptions). If not, then the assumption must be bad, so our true parameter is not zero (or more properly, it is statistically significantly different from zero). When we calculate a confidence interval, we move the center of the distribution around until it just becomes unreasonable to get the result that we did. How far we can move the center of the distribution to either side becomes our confidence interval.

For example, let us use two-sided confidence intervals to express the precision of the coefficients in the model for our fly ash example. The standard deviations of all the coefficients are the same, $s_{\hat{b}_i} = 0.2325$ with 8 degrees of freedom as determined in the previous section. The critical t-value for 95% confidence is $t_{8,.05} = 2.306$ (two-sided). So the confidence interval for each of the coefficients is:

$$95\% \text{ Confidence Interval} = \hat{b} \pm (2.306)(0.2325) = \hat{b} \pm 0.536.$$

8.4.5 Precision of Predictions

It is often important to report the precision of the predictions calculated using our model. Our predictions at any sets of conditions (we will denote the conditions by $\mathbf{X}_o$) are, of course, calculated using our least squares estimates of $\mathbf{B}$ via the Equation 8.32:

$$\hat{\mathbf{Y}} = \mathbf{X}_o \, \hat{\mathbf{B}}. \tag{8.45}$$

This equation is totally analogous to Equation 8.37 which said that our estimate of $\mathbf{B}$ is calculated as the product of two matrices—one that is precisely known, and one that has variability. Therefore, we can use the relationship given in Equation 8.38 (the first line) to calculate the variance of $\hat{\mathbf{Y}}$.

$$Var(\hat{\mathbf{Y}}) = \mathbf{X}_o \, Var(\hat{\mathbf{B}}) \, \mathbf{X}_o^T = \mathbf{X}_o \, [(\mathbf{X}^T\mathbf{X})^{-1}\sigma^2]\mathbf{X}_o^T \tag{8.46}$$

or

$$Var(\hat{\mathbf{Y}}) = \mathbf{X}_o \, (\mathbf{X}^T\mathbf{X})^{-1}\mathbf{X}_o^T\sigma^2. \tag{8.47}$$

Remember that $\mathbf{X}$ is the n sets of conditions at which experiments were run, and thus it does not change. On the other hand, $\mathbf{X}_o$ is the collection of points at which we would like to calculate the variance of $\hat{\mathbf{Y}}$. That can be as little as only a single point or a very large number of points, and the matrix changes accordingly.

Continuing with our fly ash example, let us say we are interested in the line of predicted 14-Day Compressive Strength, $\hat{\mathbf{Y}}$ versus Water/Cement Ratio (X_2), when the fly ash source is B ($X_1 = +1$) and the set up temperature is 30°C ($X_3 = +1$). We can calculate predictions (the line) using Equation 8.45 and the variability in those predictions using Equation 8.47. The results of those calculations are shown numerically in Table 8.5 and graphically in Figure 8.10.

TABLE 8.5
Calculation of Error Limits for Fly Ash Example

Water/Cement Ratio	X_2	$s_{\hat{Y}}$	$\hat{Y}$	$\hat{Y} - 2s_{\hat{Y}}$	$\hat{Y} + 2s_{\hat{Y}}$
0.35	-1	0.52	44.71	43.67	45.75
0.4	-0.5	0.43	44.13	43.26	44.99
0.45	0	0.40	43.54	42.73	44.34
0.5	0.5	0.43	42.95	42.08	43.82
0.55	1	0.52	42.36	41.32	43.40

It should be noted that the model used (in Table 8.5 and Figure 8.10) had all the non-significant terms deleted. It is definitely not a good idea to keep all the non-significant terms in the model, because the variability in the predictions increases, and there is no offsetting

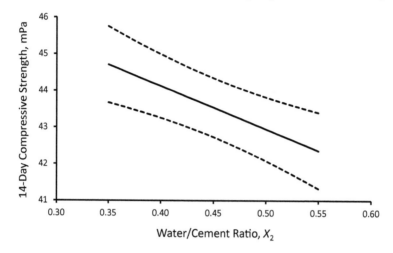

FIGURE 8.10: Error Limits for Fly Ash Model When $X_1 = +1$ and $X_3 = +1$
Reduced Model: $\hat{Y} = 39.81 + 6.11X_1 - 4.24X_2 - 2.39X_3 + 3.06X_1X_2$

benefit. This concept of trimming insignificant terms from a model is discussed in much more detail in Chapter 10.

To make sure that the use of Equation 8.47 is clear, let us calculate the variability of $\hat{Y}$ using fly ash source B, a water/cement ratio of 0.55, and a set-up temperature of 30°. Remembering our coding:

$$X_1 = [(\text{fraction of fly ash source B}) - 0.5]/0.5 = (1.0-0.5)/0.5 = +1$$
$$X_2 = [(\text{water/cement ratio}) - 0.45]/0.10 = (0.55-0.45)/0.10 = +1$$
and $X_3 = [(\text{set-up temperature}) - 20\ ^\circ\text{C}]/10\ ^\circ\text{C} = (30-20)/10 = +1$.

The $\mathbf{X}_o$ matrix (vector in this case) for the reduced model, including a "1" for the constant term, then becomes:

$$\begin{array}{cccccc} & \text{Int.} & X_1 & X_2 & X_3 & X_1X_2 \\ \mathbf{X}_o = [& 1 & 1 & 1 & 1 & 1 \end{array}].$$

When we use this $\mathbf{X}_o$ matrix in Equation 8.47, and substitute s^2 for σ^2, we get:

$$\text{Var}(\hat{\mathbf{Y}}) = \begin{bmatrix} 1 & 1 & 1 & 1 & 1 \end{bmatrix} \begin{bmatrix} 1/16 & 0 & 0 & 0 & 0 \\ 0 & 1/16 & 0 & 0 & 0 \\ 0 & 0 & 1/16 & 0 & 0 \\ 0 & 0 & 0 & 1/16 & 0 \\ 0 & 0 & 0 & 0 & 1/16 \end{bmatrix} \begin{bmatrix} 1 \\ 1 \\ 1 \\ 1 \\ 1 \end{bmatrix} (0.865)$$

or $\text{Var}(\hat{\mathbf{Y}}) = (5/16)(0.865) = 0.270$.

Notice that the $(\mathbf{X}^T\mathbf{X})^{-1}$ matrix also corresponds to the reduced model, and therefore only has five rows and five columns. Notice also that the variance of the predicted response for the factorial model at the corners of the design is given by:

$$Var(\hat{\mathbf{Y}}) = c\sigma^2/n \tag{8.48}$$

where c is the number of coefficients in the prediction equation (i.e. model). Thus, we can see why we want to eliminate any unnecessary coefficients from our model. Although Equation 8.48 is valid only at the corners of the design, it is often used to give a "quick and dirty" approximation of the variance at other conditions as well. Likewise, even though it is strictly valid only for a factorial design, the concept that the variance increases with the number of coefficients in the model, and decreases if we take more data, remains quite true and very important.

The error limits on the equation values are typically reported as plus or minus twice the standard deviation of $\hat{\mathbf{Y}}$, which are approximate 95% confidence limits on the curve. For these values of X_1, X_2, and X_3 the error limits are $\pm 2\sqrt{0.270} = \pm 0.52$. The other error limits in Table 8.5 were calculated in a similar fashion, changing $\mathbf{X}_o$ appropriately.

8.5 Quantifying Model Closeness ($\mathbf{R}^2$)

Once a least squares model, $\hat{Y}$, has been fit to a set of data, generally the first question of interest is: "How well does the equation fit?" Of course, the minimized sum of squared errors,

$$SSE = \sum (Y_i - \hat{Y})^2, \tag{8.49}$$

is a direct measure of how well the model fits. However, it is highly dependent on the scale of the data, Y_i (not to mention the number of data points). For example, if we are modeling the gross national product, the sum of squares will be very large regardless of how good our model is. On the other hand, if we are modeling the concentration of ozone in air, we will get a very small sum of squares regardless of how poor our model is. Therefore, comparing the sums of squares for two different problems is definitely a situation where we would be comparing apples and oranges. What is needed is some sort of normalization. The normalization that is most commonly used is accomplished by comparing SSE to the total variability in the data, SST (which stands for sum of squares, total). SST is defined as the sum of the squared deviations of the Y data from their average. In equation form,

$$SST = \sum (Y_i - \bar{Y})^2. \tag{8.50}$$

The ratio of SSE to SST represents the fraction of the total variability in the data that is not explained by the model. We could use this statistic to compare the models for two different situations, because the ratio is unit-less (i.e., it does not depend on the scale of the data, and it essentially makes no difference how much data was used in the analysis).

People actually use the complementary statistic—the fraction of the total variability in the data that IS explained by the model— which is given the name, R^2.

$$R^2 = 1 - SSE/SST = (SST - SSE)/SST. \tag{8.51}$$

R^2 is generally greater than 0, and it cannot exceed 1. The bigger it is, the better it is. An R^2 value of 1 means that SSE is zero, and therefore the model fits the data perfectly. At the other end of the scale, an R^2 value of zero means that $SSE = SST$, or the model is no better than just using the average of the data. As long as the model has a constant term, it is not possible for R^2 to be less than zero.

The R^2 value for the fly ash example is given in Table 8.6, and we see that the value of R^2 is 0.991. The value of R^2 for that example is extremely high (close to 1.0). This means that there is only a small bit of variation in the data (namely 0.9% of it) that is not explained by the model.

Figure 8.11 (a) shows graphically how the data and the model compare for the polypropylene pyrolysis example by plotting the data, Y (on the y-axis), vs. the model predictions, $\hat{Y}$

TABLE 8.6

Detailed Calculation of R^2 for Fly Ash Example

Y_i	$\bar{Y}$	$Y_i\text{-}\bar{Y}$	$(Y_i\text{-}\bar{Y})^2$	$\hat{Y}_i$	$Y_i\text{-}\hat{Y}_i$	$(Y_i\text{-}\hat{Y}_i)^2$
43.4	39.81	3.59	12.9	43.39	0.01	0.00
50.1	39.81	10.29	105.8	49.49	0.61	0.38
27.3	39.81	-12.51	156.6	28.79	-1.49	2.21
47.0	39.81	7.19	51.7	47.14	-0.14	0.02
39.4	39.81	-0.41	0.2	38.61	0.79	0.62
43.8	39.81	3.99	15.9	44.71	-0.91	0.83
25.0	39.81	-14.81	219.4	24.01	0.99	0.98
40.7	39.81	0.89	0.8	42.36	-1.66	2.76
42.8	39.81	2.99	8.9	43.39	-0.59	0.35
49.5	39.81	9.69	93.8	49.49	0.01	0.00
29.1	39.81	-10.71	114.8	28.79	0.31	0.10
48.4	39.81	8.59	73.7	47.14	1.26	1.59
38.4	39.81	-1.41	2.0	38.61	-0.21	0.05
45.0	39.81	5.19	26.9	44.71	0.29	0.08
24.2	39.81	-15.61	243.8	24.01	0.19	0.04
42.9	39.81	3.09	9.5	42.36	0.54	0.29
		$SST =$	1136.7		$SSE =$	10.29

$$R^2 = (SST\text{-}SSE)/SST = (1136.7 - 10.29)/1136.7 = 0.991$$

(on the x-axis). The 45 degree line on the graph represents the perfect fit (where $Y = \hat{Y}$). As expected with a high R^2 value, the points are quite close to the 45 degree line. The three other graphs in Figure 8.11 show the kind of scatter that would be seen for other values of R^2. As R^2 decreases from 0.99 to 0.50, the scatter around the 45 degree line gets more and more pronounced.

Although R^2 is a useful indicator of how closely the regression equation matches the data, it should not be used as the only criterion for judging models. The reason is that R^2 is not a definitive measure of how good a model is. Specifically, a high R^2 does not necessarily mean that the model is good, nor does a low R^2 necessarily mean the model is poor. For example, if we have the simple Y vs. X data shown in Figure 8.12 (a), a straight line model would have a high R^2 value ($R^2 = 0.95$). But it is clear from the plot that a straight line is a poor model and a quadratic is needed to describe the data. On the other hand, if we have the Y vs. X data shown in Figure 8.12 (b), the data definitely show a trend of increasing Y with increasing X. A straight line is statistically significant and all that is warranted. Therefore the model is good even though the R^2 value is low ($R^2 = 0.30$). The low R^2 value is an indication that there is high random variability in the data, not that the model is poor.

In the case where there are multiple independent X variables, checking for systematic versus random deviations is more difficult, and other procedures are required. These procedures are discussed in the next two sections in this chapter.

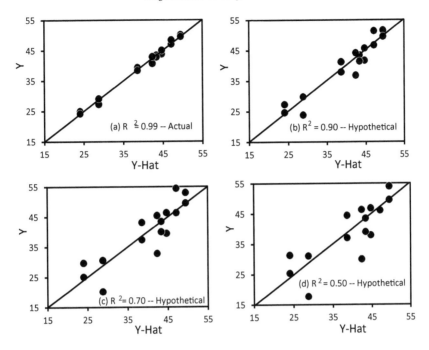

FIGURE 8.11: Data Values vs. Model Predictions for the Fly Ash Example

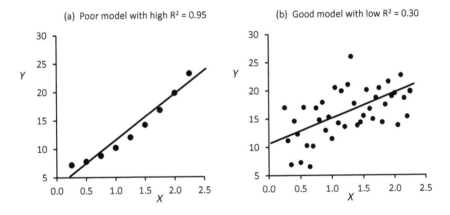

FIGURE 8.12: Models with Deceptive R^2 Values

8.6   Checking Model Assumptions (Residual Plots)

We typically invest quite a bit of time and money in our modeling efforts in order to come up with a model we can use to make our system better. But before we use it, we need to be careful that it does not have any major shortcomings. This amounts to checking our assumptions:

1. The errors are Normally distributed.

2. The errors have the same variance.

3. The errors are independent (or the experiment was randomized).

4. The errors have a mean of zero (i.e., the model is adequate).

If one or more of our assumptions turns out to be incorrect, we must reanalyze our data taking this into account. We could then end up with a different (more correct) model. The new model is the one on which we should be basing our conclusions.

Notice that all four of our assumptions deal directly with the errors in our data. Since our estimates of the errors are the residuals, $\epsilon_i = Y_i - \hat{Y}_i$, it makes sense that the residuals need to be examined closely. The residuals for our fly ash example are given again in Table 8.7 (along with the calculations for Normal plots).

TABLE 8.7
Residuals for the Fly Ash Example and Detailed Calculations
Required for the Full-Normal and Half-Normal Plots

Run	Y_i	$\hat{Y}_i$	For Full-Normal Plot				For Half-Normal Plot			
			ϵ_i	$\epsilon_{(i)}$	$\frac{(i)-0.5}{n}$	z	$|\epsilon_i|$	$|\epsilon_{(i)}|$	$0.5 + \frac{(i)-0.5}{2n}$	z
1	43.4	43.39	0.01	-1.66	0.031	-1.86	0.012	0.012	0.516	0.04
2	50.1	49.49	0.61	-1.49	0.094	-1.32	0.612	0.012	0.547	0.12
3	27.3	28.79	-1.49	-0.91	0.156	-1.01	1.488	0.138	0.578	0.20
4	47.0	47.14	-0.14	-0.59	0.219	-0.78	0.138	0.187	0.609	0.28
5	39.4	38.61	0.79	-0.21	0.281	-0.58	0.787	0.213	0.641	0.36
6	43.8	44.71	-0.91	-0.14	0.344	-0.40	0.913	0.288	0.672	0.45
7	25.0	24.01	0.99	0.01	0.406	-0.24	0.987	0.312	0.703	0.53
8	40.7	42.36	-1.66	0.01	0.469	-0.08	1.663	0.537	0.734	0.63
9	42.8	43.39	-0.59	0.19	0.531	0.08	0.588	0.588	0.766	0.72
10	49.5	49.49	0.01	0.29	0.594	0.24	0.012	0.612	0.797	0.83
11	29.1	28.79	0.31	0.31	0.656	0.40	0.312	0.787	0.828	0.95
12	48.4	47.14	1.26	0.54	0.719	0.58	1.262	0.913	0.859	1.08
13	38.4	38.61	-0.21	0.61	0.781	0.78	0.213	0.987	0.891	1.23
14	45.0	44.71	0.29	0.79	0.844	1.01	0.288	1.262	0.922	1.42
15	24.2	24.01	0.19	0.99	0.906	1.32	0.187	1.488	0.953	1.68
16	42.9	42.36	0.54	1.26	0.969	1.86	0.537	1.663	0.984	2.15

8.6.1   Checking for Normal Distribution of Errors

The first assumption we made was that the errors in our data are Normally distributed. Therefore, if we plot the residuals, they should look like a random sample from that bell-shaped distribution. The obvious plot to make would be a dot diagram, in which the values are plotted as dots on the x-axis. This was done for the fly ash example in Figure 8.13 (a). The plot seems to look much like a Normal Distribution, but it is actually difficult to tell.

A more sensitive tool is the Normal plot (or Half-Normal plot) discussed in Chapter 2. The information needed to construct a Full-Normal (or Half-Normal) plot is also given in Table 8.7 as a review of the procedure. The plots are shown in Figures 8.13 (b) and (c). Remember that if the points come from a Normal Distribution, the plots will be straight lines. Again, the plots look good, indicating that our assumption is fine. In general, if an outlier is detected, the point should be removed and the analysis should be done again. This was not needed for this example.

A word of **CAUTION**: dropping a data point should not be done lightly. A reason for the bad result should be diligently sought in order to more fully justify deleting the point

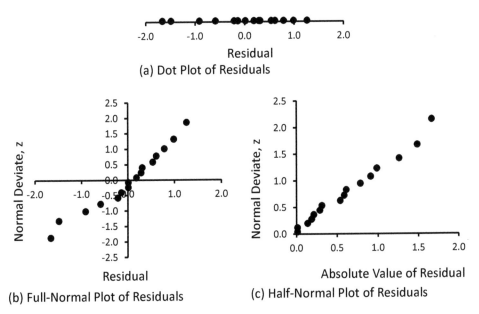

FIGURE 8.13: Plots of Residuals to Check Normality of Errors for the Fly Ash Example

and in order to help prevent more bad data in the future. Also, the experiment should be repeated if at all possible so that the integrity of the statistical design is maintained.

8.6.2 Checking for Constant Variance

An important assumption in regression analysis is that each of the data points has the same precision, and therefore each of the data points should be given equal weight. However, it is possible that some of the data have greater precision than others (i.e., the errors in those more precise data have a smaller variance than the others). The most common case of this is the situation in which the variance gets bigger as the response gets bigger. Therefore, to check the assumption, the residuals should be plotted against $\hat{Y}$ as shown in Figure 8.14 for the fly ash example. In this example it may appear as if the scatter is a bit bigger at higher values of $\hat{Y}$, but there are more data at higher values, and so there would naturally be more scatter. Thus, the assumption of constant variance seems to be valid. But the other plots of residuals shown in Section 8.6.4 should also be checked for evidence of non-constant variance, as well as for trends (which is the main goal of that section). They too seem to support the assumption of constant variance, so we can move on to checking the other assumptions for the fly ash example.

This test is especially important if the response data span an order of magnitude or more (which was not the case for the fly ash example). If the assumption of constant variance is seen to be poor, two main remedies are available. The first would be to transform the responses to new ones like taking the logarithm if the responses had a constant percentage error. The reader may recall that is what was done for the stack gas treatment example in Section 3.9. A second solution would be to use a weighted least squares analysis, in which each data point would be given a weight inversely proportional to its variance. A discussion of these procedures is beyond the scope of this text, and the reader is referred to a book on regression analysis (e.g., Draper and Smith [1966]).

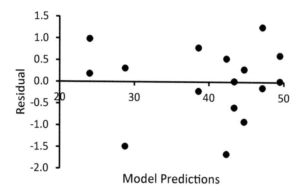

FIGURE 8.14: Plots of Residuals vs. Model Predictions for the Fly Ash Example

8.6.3   Checking for Independence of Errors

Although there are some checks for independence of errors (such as the calculation of an auto-correlation coefficient), correcting for correlations and biases in data is difficult at best, and impossible at worst. The best solution to this problem is to nip it in the bud by the diligent use of ***randomization***. The randomization of run order (within blocks) is critical to any experimental program, and virtually guarantees valid inferences from the final model.

8.6.4   Checking for a Mean of Zero for the Errors

If our model is adequate, the values predicted by it will be the true responses (within experimental error). It follows that, if our model predicts the true responses without any biases, the residuals will be scattered around a mean of zero. In other words, the assumption that the errors have a mean of zero is the same as the assumption that we have an adequate model. To check this assumption, we look for trends in the residuals. No trends should exist if our model is adequate. The residuals should have no information contained in them other than how big the random variability is for our system. The other information in our data was presumably all accounted for (and hence removed) by our model. Any non-randomness is information, and our model could be expanded to take the specific trend into account, thereby making it better (more precise).

The most likely modeling problems are failure to take an independent variable into account correctly, or to overlook an important factor. Therefore, the residuals should be plotted against each independent variable in the experimental design, and they should be plotted against any suspected factors that were not part of the design but inadvertently changed from one experiment to the next. They should also be plotted against run order (to look for a time trend), block, and anything else that might show a trend. The suggested plots were made for the fly ash example, and the plots are shown in Figure 8.15. None of the plots show any trends, and so the model appears to be adequate in this regard.

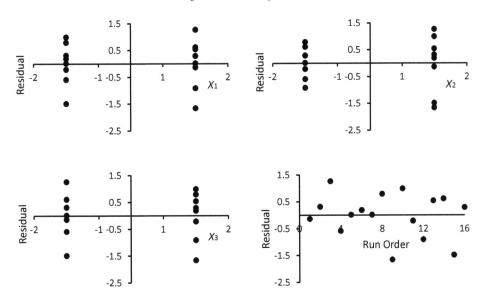

FIGURE 8.15: Plots of Residuals vs. Independent Variables and Run Order
to Check Model Adequacy for Fly Ash Example

8.7 Data Transformation for Linearity

In this chapter and previous ones, we have been fitting linear models to data. But, if a two-level factorial design (without center points) is used to collect data, there is no statistical test we can perform to determine if the linear model is adequate. With only two levels of a factor, one at $X = -$, and the other at $X = +$, a straight line joining the mean response at each level would seem to be the best we can do. However, there are clues in the data and residual plots that can make us aware of nonlinearities in the underlying relationship.

When the response data spans two or more orders of magnitude (i.e., 1, 10, 100) it is a clue that there is a nonlinear relationship between the response and factors. Another related clue that usually occurs in these situations is a tendency for the variance to increase as the response value increases. For example, if there is an exponentially increasing relationship between the response, Y, and the factor, X, then at the same time we usually see that there is more variability in the large response values than there is in the small response values.

In a situation like this the response data will normally range over orders of magnitude, and a plot of the residuals versus the predicted values will show a megaphone shape with a wider spread in the residuals at the right than on the left of the graph. The accuracy of a model fit by least squares can be improved in situations like this by transforming the response data prior to fitting the model. For example, if the true relationship is $Y = ae^{bX}$, then there is a linear relationship $\ln(Y) = b_o + bX$, where $b_o = \ln(a)$.

Beecher, Gomm, and Horn conducted a 2^3 factorial experiment to study the effect of X_1 = Laser Current, X_2 = Monochronometer Exit Slit Width, and X_3 = Photomultiplier Voltage upon the Detector Response, Y (in voltage), to determine the sensitivity of a fluorescence instrument used in spectroscopic determination and identification of chemical compounds. Table 8.8 shows the factor levels and actual response data. We can see that the response values range over more than an order of magnitude from 0.7 to 10.4. The least squares linear model fit to the data was:

$$Y = 3.754 + 3.142X_1 + 2.308X_2 + 3.242X_3$$
$$+ 0.942X_1X_2 + 1.342X_1X_3 + 0.975X_2X_3 + 0.408X_1X_2X_3$$

TABLE 8.8
Data from Fluorescence Spectroscopic Experiment

Run	X_1 $-=15A$ $+=20A$	X_2 $-=25\mu m$ $+=50\mu m$	X_3 $-=1000V$ $+=1100V$	Y Detector Response, Volts
1	$-$	$-$	$-$	0.9, 0.9, 0.7
2	$+$	$-$	$-$	2.1, 2.1, 2.1
3	$-$	$+$	$-$	1.7, 1.5, 1.7
4	$+$	$+$	$-$	3.9, 4.1, 3.9
5	$-$	$-$	$+$	2.1, 2.3, 2.1
6	$+$	$-$	$+$	5.3, 5.3, 5.3
7	$-$	$+$	$+$	4.3, 3.9, 4.1
8	$+$	$+$	$+$	10.4 10.1, 9.3

where X_i are the coded factor levels. All coefficients in the model were significant at the 99% confidence level, but the plot of residuals versus predicted values and the Normal Probability plot of residuals (not shown) both exhibited the characteristic pattern of a non-linearity and non-constant variance. Specifically, the plot of residuals versus predicted values showed the megaphone pattern with wider spread at the right than on the left. The Normal plot of residuals exhibited a non-Normal pattern characteristic of distributions with long tails. This would result when fitting a linear model to an exponential relationship, because some residuals would come from the narrow distribution at the left while others would come from the wide distribution at the right.

The model can be substantially improved if we take a natural log transformation of the response data. Doing this, the natural logs of the response data were computed, and then the least squares method was used fit the model. The results were:

$$\ln(Y) = 1.071 + 0.900X_1 + 0.645X_2 + 0.931X_3.$$

The interaction terms are not included in this model this time, because they were not significant. On the original scale of the data, interactions were significant in the model because they tried to account for the non-linearity. When a transformation is used that linearizes the relationship, the interactions are no longer important in the model. In addition to the removal of interactions from the model, the transformation also changed the plot of residuals versus predicted values. It no longer took on the characteristic megaphone shape shown. Also the Normal plot of residuals appeared more linear.

As shown in this example, non-constant variance, non-Normality of residuals, and non-linearity of the model generally go hand in hand. A simple transformation, like the logarithm, can often linearize the model and provide a better fit by justifying the assumptions that make the least squares estimates best.

To handle a wider spectrum of problems, Box and Cox proposed a whole family of transformations that can be used to linearize a model and simultaneously make the variance of residuals constant. Their transformations are of the form Y^λ, where in the special case of $\lambda = 0$ we use $\ln(Y)$. Table 8.9 shows some typical values of λ that are useful.

TABLE 8.9
Typical Box–Cox Transformations

λ	Transformation
2	Y^2
1.5	$Y^{1.5}$
1	Y (Original Scale, No Transformation)
0.5	$Y^{0.5} = \sqrt{Y}$
0	$\ln(Y)$
-0.5	$Y^{-0.5} = 1/\sqrt{Y}$
-1	$Y^{-1} = 1/Y$

To choose an appropriate transformation from the family, first fit the linear model in the original scale of the data and make a plot of the residuals versus predicted values. If the range of the residuals increases as a function of the predicted values (like a megaphone pattern) choose a value of $\lambda < 1$. On the other hand, if the range of the residuals tends to decrease as a function of the predicted value, then choose a value of $\lambda > 1$. After applying the transformation and refitting the data, again look at the plot of residuals versus predicted values and the Normal plot of residuals to see if the problem has been solved. If not choose a different value of λ.

A more formal way of choosing the appropriate value of λ is to try several values, and examine the R^2 statistic for each fitted equation. For example, if we had tried the values $\lambda = -0.50, -0.25, 0.00, 0.25, 0.50,$ and 1.00 for the data in Table 8.8, the resulting R^2 would show a maximum occurs at a value of $\lambda = 0.5$, which means that $\sqrt{Y}$ gives the best fit to this data. The square root transformation is even better than the natural log shown earlier. The plot of residuals versus predicted values and Normal plot of residuals also looks best when this square root transformation is used.

8.8 Summary

8.8.1 Important Equations

Estimation of Coefficients, $\hat{\mathbf{B}}$ (Equation 8.36)

$$\hat{\mathbf{B}} = (\mathbf{X}^T\mathbf{X})^{-1}\,\mathbf{X}^T\mathbf{Y}$$

Variability of Coefficients, $\hat{\mathbf{B}}$ (Equation 8.41)

$$Var(\hat{\mathbf{B}}) = (\mathbf{X}^T\mathbf{X})^{-1}\,\sigma^2$$

Confidence Interval for a Coefficient (Section 8.4.4)

$$100(1-\alpha)\% \text{ Confidence Interval (two-sided)} = \hat{b} \pm t^*_{\alpha/2}\,s_{\hat{b}}$$

Variability in Model Predictions, $\hat{\mathbf{Y}}$ (Equations 8.47 and 8.48)

$$Var(\hat{\mathbf{Y}}) = \mathbf{X}_0(\mathbf{X}^T\mathbf{X})^{-1}\mathbf{X}_0^T\sigma^2$$

$$Var(\hat{\mathbf{Y}}) \approx c\sigma^2/n$$

8.8.2   Important Terms and Concepts

The following is a list of the important terms and concepts covered in this chapter.

method of least squares
linear regression
multiple regression
estimation of coefficients
statistical significance of coefficients
confidence interval for coefficients
precision of predictions
R^2
residual plots
outlier
power transformation

8.9   Exercises

1. A study of the relationship between rough weight (X) and finished weight (Y) of castings was made. A sample of 12 castings was examined and the data presented below:

X Rough Weight	Y Finished Weight
3.715	3.055
3.685	3.020
3.680	3.050
3.665	3.015
3.660	3.010
3.655	3.015
3.645	3.005
3.630	3.010
3.625	2.990
3.620	3.010
3.610	3.005
3.595	2.985

(a) Make a scatter plot to see the relationship between X and Y.

(b) Use the method of least squares to calculate the coefficients in the simple linear regression model $Y = a + bX$.

(c) Calculate the standard errors of the estimated coefficients and determine if they are significant at the 95% confidence level.

2. An equation is to be developed from which we can predict the gasoline mileage (city driving) of an SUV automobile based on its weight and the temperature at the time of operation. The model being estimated is: $Y = b_0 + b_1 X_1 + b_2 X_2$. The following data are available:

Car Number	X_1 Weight tons	X_2 Temperature °F	Y Miles per Gallon
1	1.35	90	17.9
2	1.80	40	16.8
5	1.30	35	18.8
6	2.05	45	15.5
7	1.60	50	17.5
8	1.80	60	16.4
9	1.85	65	15.9
10	1.40	30	18.3

(a) Estimate the coefficients in the model using least squares. Use Equation 8.36. Note that all spreadsheet programs have functions for matrix multiplication, matrix transposition, and matrix inversion. Verify your answers using the regression function in a spreadsheet program.

(b) Estimate the standard deviation of the errors, s, by calculating the predicted value (using the coefficients from Part (a)) at each data point, and then calculating the standard deviation of the residuals (the differences between the data points and the predictions). How many degrees of freedom does the estimate, s, have? Verify your answers using the spreadsheet regression function output from Part (a).

(c) Calculate the statistical significance of the coefficients, b_1 and b_2.

(d) Calculate the predicted mileage (mpg) at $X_1 = 1.8$ tons and $X_2 = 70$ °F. Also calculate the error limits on that prediction. The error limits are usually just reported as $\hat{y} \pm 2s_{\hat{Y}}$ for simplicity.

(e) Calculate R^2 for the model. Verify your answers using the spreadsheet regression function output from Part (a).

(f) Check your model assumptions by using plots of residuals. These should include a half-Normal plot and plots versus X_1, X_2, and $\hat{Y}$. Make sure you comment on what you learn/verify from each graph.

Chapter 9

Response Surface Designs

9.1 Response Surface Concepts and Methods

Response surface designs are normally used at the last stage of experimentation—if they are required at all. The important factors have already been determined by screening experiments, experience, or sound theory. The goal at this stage of experimentation is to describe in detail the relationship between the factors and a response. It is now known (or strongly anticipated) that a simple linear model – even with interactions – is not good enough to adequately represent that relationship. If we look at Figure 9.1, we see that we want to move far to the right on our Knowledge axis.

	Present ⇓		**Goal** ⇓	
0%		**Knowledge**		**100%**
Objective: Preliminary Exploration	Screening Factors	Effect Estimation	Optimization	Mechanistic Modeling
No. of Factors	5 - 20	3 - 6	2 - 4	1 - 5
Model: Variance Components	Linear	Linear + Interactions	Linear + Interactions + Quadratics	Mechanistic Model
Purpose: Identify Sources of Variability	Identify Important Factors	Estimate Factor Effects + Interactions	Fit Empirical Model Interpolate	Estimate Parameters of Theory Extrapolate
Designs: Nested Designs	Fractional Factorials	Two-level Factorials	RSM	Special Purpose

FIGURE 9.1: Objective of Response Surface Methods

As a practical matter, since we want to know a lot about our system, it will require us to take quite a bit of data. (There is NO FREE LUNCH!) This is usually prohibitive with more than six independent continuous factors. Even five or six factors require more experiments (as we will soon see) than can usually be justified. In all cases, efficiency is very important so as to minimize the number of experiments that we must run. The experimental designs discussed in this chapter were derived with this practical objective in mind. They also have other important characteristics which we will discuss.

The term "Response Surface Methodology" (or RSM for short) refers to the complete package of statistical design and analysis tools which generally are used for the following three steps:

(1) design and collection of experimental data which allow fitting a general quadratic equation for smoothing and prediction, (2) regression analysis to select the best equation for description of the data, and

(3) examination of the fitted surface via contour plots and other graphical and numerical tools.

In this chapter and the next, all three of the above steps will be described.

9.2   Empirical Quadratic Model

Up to this point, we have been discussing designs that allow only the estimation of linear effects of the X_i's on the response, Y. In graphical form these relationships would plot as straight lines, as shown for two X's in Figure 9.2(a). Even if interactions are estimated, they only allow the slope of the plot (of Y versus one of the independent variables, say X_1) to change depending on the value of the other independent variable(s), as shown in Figure 9.2(b). But it is important to remember that, even with an interaction, the plot of Y versus X_1 is always a straight line. Clearly, that simple representation of the true situation is not always adequate. When we have already run some experiments and found curvature to be present (presumably using a factorial design plus center points), or when we *a priori* expect a plot of Y versus one of the X_i's to be curved, then we must run enough experiments to fit a more complicated model that allows for this curvature.

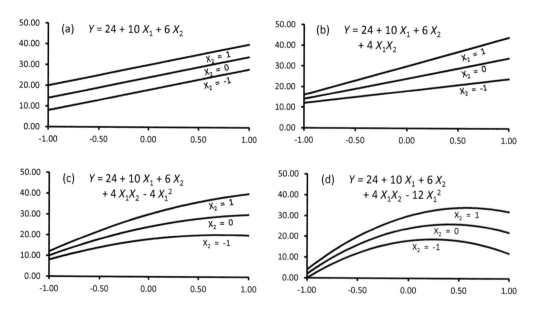

FIGURE 9.2: Impact of Terms Added to Polynomial Equation

An example with a small amount of curvature is shown in Figure 9.2(c). The optimum is still at or near the boundary, but a curved line is needed to predict Y reasonably well over the whole region of X_1. Therefore, a linear model (with interactions) would still have been adequate if the goal was simply to maximize Y, but not if we were interested in knowing Y for all values of the X's. A more severely curved response is shown in Figure 9.2(d). In this case, a straight line model (even with interactions) is not adequate for any purpose. Not only would the response be poorly predicted over the whole X_1 region, but the optimum would have been badly missed. In this case, a full quadratic model is needed to do anything. The general quadratic model for k independent variables is:

$$\hat{Y} = b_0 + \sum_{i=1}^{k} b_i X_i + \sum_{i=1}^{k} b_{ii} X_i^2 + \sum_{i=1}^{k-1} \sum_{j=i+1}^{k} b_{ij} X_i X_j. \tag{9.1}$$

For one independent variable, this equation becomes

$$\hat{Y} = b_0 + b_1 X_1 + b_{11} X_1^2, \tag{9.2}$$

while for two independent variables, the equation is

$$\hat{Y} = b_0 + b_1 X_1 + b_2 X_2 + b_{11} X_1^2 + b_{22} X_2^2 + b_{12} X_1 X_2. \tag{9.3}$$

This is the simplest model which will still allow curvature in a graph of $\hat{Y}$ versus one of the independent variables, X_i. In fact, it may appear that it is much too simple to be able to describe any realistic situation with some degree of accuracy, but that is not true. First of all, the equation is quite flexible, and with the appropriate coefficients, it can describe a wide variety of different surfaces (see Figures 9.3 and 9.4). These surfaces include hilltops, valleys, ridges, rising ridges, and falling ridges. These are the most commonly seen surfaces in practice. But the equation will even describe more unusual surfaces like saddle points if necessary. So it is very useful indeed.

As further justification of the utility of the simple quadratic equation, consider the following argument. Any situation, if we know enough about it, could be described by a mathematical model. The quadratic equation is somewhat equivalent to expanding the true model in a Taylor series about the center of the region and then dropping all terms of order greater than two. The difference is that the Taylor series fits exactly at the center of the region and gets progressively worse as we move away from the center, but the quadratic equation (which we obtained via a least squares fit to data over the whole region) spreads the error in fitting to all points. That is, it gives up some accuracy in the center if necessary to do a better job over the rest of the region.

If the region of interest is of modest size, then the polynomial (quadratic) fit is quite good. For example, let us say that we are interested in the chemical reaction:

$$\text{Reactant} \overset{k_1}{\Longrightarrow} \text{Product} \overset{k_2}{\Longrightarrow} \text{Decomposition Products}$$

and the reactions follow simple, first-order kinetics. Then the concentration of product, [P], can be modeled as a function of time by the highly non-linear model:

$$[P] = [R]_0 \frac{k_1}{k_1 - k_2} \{\exp(-k_1 t) - \exp(-k_2 t)\}.$$

If k_1 and k_2 can be given as functions of temperature by the Arrhenius expressions:

$$k_1 = 0.5 \exp\left[-10{,}000 \left(1/T - 1/400\right)\right] \text{ and}$$

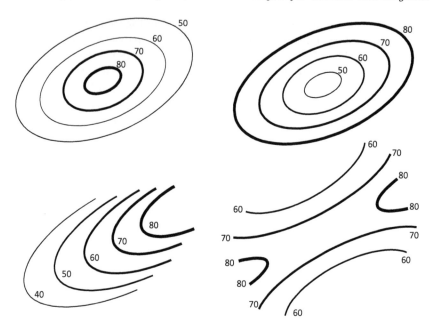

FIGURE 9.3: Some Types of Two-Dimensional Surfaces Described by a Quadratic Equation

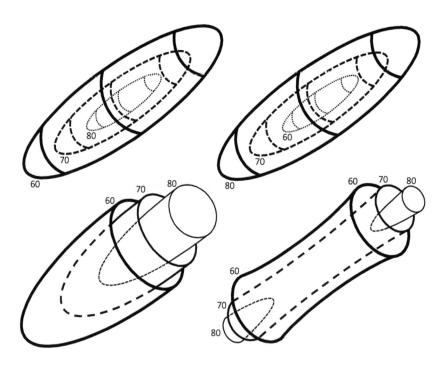

FIGURE 9.4: Some Types of Three-Dimensional Surfaces Described by a Quadratic Equation

$$k_2 = 0.2 \exp\left[-12{,}500\left(1/T - 1/400\right)\right],$$

then contours of yield of product as a function of time and temperature can be predicted as shown in Figure 9.5. Over the region of time from 5 to 15 hours and temperature from 380 to 400 °K, the quadratic fit, as shown in Figure 9.6, is seen to be quite good. It is clearly not perfect, but the quadratic equation certainly describes the main features of the response contour plot.

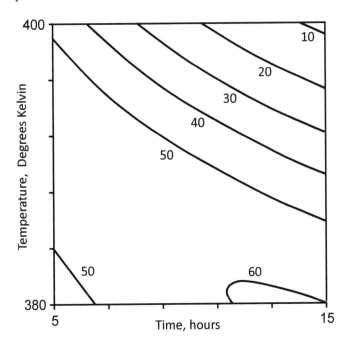

FIGURE 9.5: Actual Yield Contours

Before we end our discussion about how good and useful quadratic equations are, two words of caution are in order. First of all, a quadratic equation is usually not able to describe adequately what is going on if the region of interest is too large such as that shown in Figure 9.7. If it were necessary to predict [P] over such a large span of temperature and time, then a more complicated model, perhaps a theoretical one with mechanistic interpretation, would have to be used. But mechanistic models shall not be discussed at any length in this text.

Secondly, it must also be pointed out that the polynomial model, while being a good mathematical "French curve" for smoothing data in the experimental region, is atrocious for extrapolation of any sizable extent away from where the data were taken. As can be seen from Figure 9.8, the predictions are quite likely to be utter nonsense. Thus, it must be remembered that the quadratic equation should not be used outside of the experimental region, even though it is mathematically easy to do so.

9.3   Design Considerations

So now that we have decided that we will use a quadratic equation to describe the relationship between our response, Y, and the independent variables, the X_i's, we must next decide what experiments we will run in order to accomplish that goal. As in previous chapters, we call that collection of experiments an experimental design. Some of the important

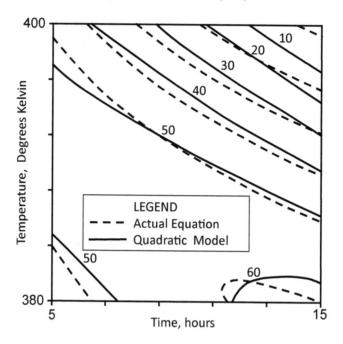

FIGURE 9.6: Quadratic Model Contours

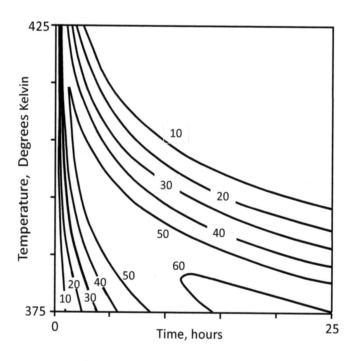

FIGURE 9.7: Actual Yield Contours

properties that we would like our "second order" experimental design to have are (in order of priority):

- Allow the coefficients in the quadratic model, Equation 9.1, to be estimated.

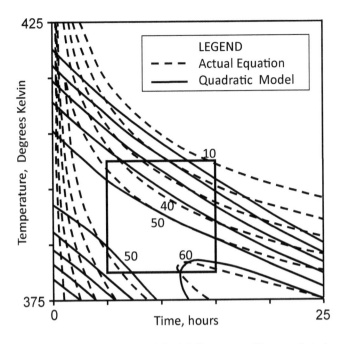

FIGURE 9.8: Quadratic Model Contours, Extrapolated

- Have a small number of runs.

- Allow sequential buildup (i.e., first order design + some more points = second order design).

- Allow detection of lack of fit.

- Permit blocking.

The first of these criteria—being able to estimate all the terms in a full quadratic model—is of course critical. If the experiments don't permit estimating the full quadratic equation, we do not have a second order design. In order to satisfy this criterion, we must have at least three levels of each X_i variable. The simplest design that meets this requirement is the 3^k factorial design. In this design, all possible combinations of the three levels $(-, 0, +)$ for each independent variable are run as our experiments.

However, for larger values of k, this design does not meet our second objective—to have a small number of runs—as can be seen from Table 9.1, below. The number of runs that is a theoretical minimum would be one run for every coefficient to be estimated. However, if this were all the experiments that were run, the equation would go right through all of the data points. There would be no smoothing taking place, there would be no estimate of error, and there would be no opportunity to test for lack of fit. Thus, even with efficient designs, the number of experiments performed is usually between 1.5 to 2.0 times the number required for estimation of the coefficients alone (unless more are necessary for the desired precision of estimation). But even by this more realistic standard, 3^k designs for more than three factors are seen to be very wasteful.

TABLE 9.1

Number of Runs for a 3^k Design

Number of Factors k	Number of Runs in 3^k Design, N	Number of Coefficients in Quadratic Equation
2	9	6
3	27	10
4	81	15
5	243	21
6	729	28
7	2,187	36

9.4   Central Composite Designs

A class of designs which is more frugal with experiments is called **central composite designs**. Central composite designs build upon the two-level factorial design that was discussed already in Chapter 3. Remember that the model used to fit the data from a 2^k-factorial design was of the form:

$$\hat{Y} = b_0 + \sum_{i=1}^{k} b_i X_i + \sum_{i=1}^{k-1} \sum_{j=i+1}^{k} b_{ij} X_i X_j \tag{9.4}$$

with interaction terms of higher order usually neglected. Equation 9.4 reminds us that the 2^k-factorial design allows the estimation of all main effects and two-factor interactions. The only terms missing to give us a full quadratic equation are the squared terms in each X_i. In order to permit the estimation of these terms, the central composite design adds a set of axial points (called **star points**) and some (more) center points. The axial points combined with the center points are essentially a set of one-at-a-time experiments, with three levels of each of the independent variables, denoted by $-\alpha$, 0, and α (where α is the distance from the origin to the axial points in coded units). With the three levels of each X_i, the quadratic coefficients can be obtained. A central composite design for two factors is shown in Figure 9.9 and one for three factors is shown in Figure 9.10.

It should be noticed from the figures that α is not 1.0, so that the star points extend out beyond the cube of the factorial points. The only things to decide are how large α should be and how many replicated center points should be included in the design. To answer these questions, statisticians have brought in two additional criteria: rotatability and uniform precision. Rotatability implies that the accuracy of predictions from the quadratic equation only depends on how far away from the origin the point is, not the direction. This criterion fixes α. The other criterion, uniform precision, means that the variance of predictions should be as small in the middle of the design as it is around the periphery. This fixes the number of center points. The detailed calculations were done by others (3) and will not be discussed here. However, the recommended values are given in Table 9.2 and Appendix B.3.

It can be seen from Figure 9.9 that for two factors, rotatability dictates that $\alpha = \sqrt{2}$, which puts all the points (except for the center points) on a circle of radius, α. The value of α for all the designs was actually calculated using the formula: $\alpha = \sqrt[4]{(\text{number of factorial points})}$. For three factors, rotatability gives $\alpha = 1.68 \approx \sqrt{3}$, which puts all the points (besides the center points) close to the surface of a sphere of radius α. This can be generalized to four factors as well (with $\alpha = \sqrt{4}$). But beyond four factors, that way of visualizing the

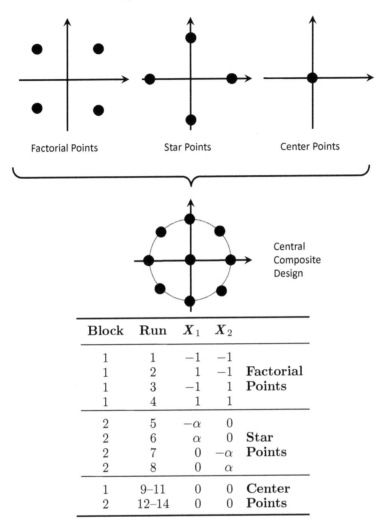

FIGURE 9.9: Central Composite Design for Two Factors

Block	Run	X_1	X_2	
1	1	-1	-1	
1	2	1	-1	**Factorial**
1	3	-1	1	**Points**
1	4	1	1	
2	5	$-\alpha$	0	
2	6	α	0	**Star**
2	7	0	$-\alpha$	**Points**
2	8	0	α	
1	9–11	0	0	**Center**
2	12–14	0	0	**Points**

design breaks down. Once α is settled, the actual values of the factors are determined from the coded values as discussed in Section 4.3. An example will be given in the next section (Section 9.5) to make this more clear.

The number of factorial points, star points, and center points necessary to run a central composite design is given in the second column of Table 9.2, along with the total of all three types of points. The total number of runs can be seen, via comparison to Table 9.1, to be much less than that for a 3^k design for three or more factors. The number of runs per estimated coefficient is seen to be in the desired range of 1.5 to 2.0, except for the design for seven factors (k=7) which is not of great practical importance, since running a Response Surface design with seven factors is extremely unusual. The five-factor design is particularly efficient, since it requires only three more runs than the four-factor design.

Thus, central composite designs are looking good so far. They have been seen to meet the first two of our design considerations: they allow fitting the full quadratic equation in a small number of runs. The third consideration (to allow sequential buildup of the design) is also met—the designs were constructed by starting with a first order design, the 2^k factorial design, and adding to that some star and center points to obtain the full design. The first-

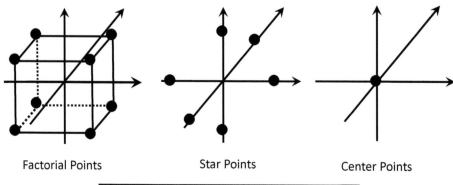

Block	Run	X_1	X_2	X_3	
1	1	-1	-1	-1	
2	2	1	-1	-1	
2	3	-1	1	-1	
1	4	1	1	-1	**Factorial**
2	5	-1	-1	1	**Points**
1	6	1	-1	1	
1	7	-1	1	1	
2	8	1	1	1	
3	9	$-\alpha$	0	0	
3	10	α	0	0	
3	11	0	$-\alpha$	0	**Star**
3	12	0	α	0	**Points**
3	13	0	0	$-\alpha$	
3	14	0	0	α	
1	15,16	0	0	0	**Center**
2	17,18	0	0	0	**Points**
3	19,20	0	0	0	

FIGURE 9.10: Central Composite Design for Three Factors

order design can be run first, and the second-order piece can be run next if needed. This is also related to the consideration of blocking, discussed below.

The fourth consideration (to allow detection of lack-of-fit) is also met and will be discussed in more depth in Section 10.4. The last consideration (to allow blocking) is also met. All the designs can be run in blocks. The factorial portion with center points constitutes one or more of the blocks, and the "star" points with some more center points is another block. This gives two big advantages to the design. First, the factorial portion can be run and analyzed first. If curvature is found to be negligible, then the star points need not be run at all (a big savings). Secondly, blocking is an important tool whose main purpose is to increase the precision of the results. This has been discussed in depth elsewhere in this book.

So, central composite designs are seen to be very good indeed. They have no major shortcomings, which explains why they are the most commonly used response surface designs in industry. The actual designs for two to seven factors, with the runs and blocking totally spelled out, are given in Appendix B.3.

TABLE 9.2
Number of Runs for a Central Composite Design

Number of Factors, k	Number of Runs in Central Composite Design (Factorial + Star + Center)	Number of Coefficients in Quadratic Equation
2	$2^2 + 4 + 6 = 14$ $(\alpha = 1.41;$ Two Blocks)	6
3	$2^3 + 6 + 6 = 20$ $(\alpha = 1.68;$ Three Blocks)	10
4	$2^4 + 8 + 6 = 30$ $(\alpha = 2.00;$ Three Blocks)	15
5	$2^4 + 10 + 7 = 33$ $(\alpha = 2.00;$ Two Blocks)	21
6	$2^5 + 12 + 10 = 54$ $(\alpha = 2.38;$ Three Blocks)	28
7	$2^6 + 14 + 12 = 90$ $(\alpha = 2.83;$ Nine Blocks)	36

9.5 Central Composite Design Example

This example is based loosely on work that was done at the Energy and Environmental Research Center at the University of North Dakota. The study was real, but the data and results were fabricated to protect confidentiality. Broadly, this example is concerned with the economic recycling of plastics. Specifically, polypropylene pyrolysis (i.e., heating) to crude fuel oil had been studied extensively in a fluidized bed process, and the thought was to see if a cheaper, kiln process could be used. The main difference between the two reactors is the fluidization velocity. In order to see if fluidization velocity was important, and thereby determine the feasibility of using a kiln, the liquid yield from pyrolysis needed to be studied at different fluidization velocities. Since reactor temperature was known to be very important, it was also included in the tests, primarily to see if the effect of fluidization velocity depended on temperature.

Since the effects of two factors needed to be determined, the central composite design for two factors was used (given in Table B.3-1). The design was run in blocks; the factorial portion was run first and the star points were to be added only if needed. The first block of the design is also shown in Table 9.3. The designs in the back of this book are always given in terms of the coded (and scaled) factors, denoted as X_1 and X_2 here. In order to run the experiments, the nominal ranges for the actual variables had to be decided. These were chosen by the engineers based on prior work to be 425 to 625 °C for reactor temperature and 0.25 to 0.75 ft/sec for fluidization velocity. Once the actual factor levels were determined, the experiments were run. It is worth repeating that the experiments are not to be run in the order listed in the table, but should be randomized to minimize the chance of bias in the conclusions. That is why Run 5 was the first actual experiment, Run 3 was the second, etc. The resulting yields are given in Table 9.3 and also shown graphically in Figure 9.11.

The results were analyzed using the methods described in detail in Chapters 3 and 4. A summary of the analysis is given in Table 9.4. Both reactor temperature (X_1) and fluidization velocity (X_2) were found to be statistically significant. This was determined by comparing their t-statistics to the critical two-sided t-statistic for 2 degrees of freedom and

TABLE 9.3
First Block of Central Composite Design for Two Factors: Polypropylene Pyrolysis

Run Number	Run Order	X_1	X_2	Temp. °C	Veloc. ft/sec	Yield
1	5	-1	-1	425	0.25	73.2
2	3	1	-1	625	0.25	51.1
3	2	-1	1	425	0.75	76.8
4	6	1	1	625	0.75	64.1
5	1	0	0	525	0.50	74.7
6	7	0	0	525	0.50	76.8
7	4	0	0	525	0.50	73.2

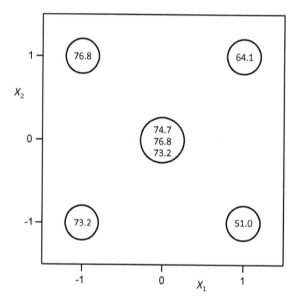

FIGURE 9.11: First Block of Central Composite Design for Two Factors: Polypropylene Pyrolysis

95% confidence of 4.303 (denoted by $t_{2,.05}$). As expected, curvature was found to be present, and the star point block was needed in order to adequately describe the response.

The star point block was therefore run (in random order, of course). In order to run the experiments, the settings for temperature and velocity had to be determined. This was done using the relationships between the uncoded and coded factors discussed in Section 4.3 (Results in Equation Form):

$X_i \equiv$ Coded and scaled value for Factor$_i$
$\quad =$ [Factor value − Center]/["High" value − Center].

Therefore,

$X_1 = (\text{Temp} - 525)/(625 - 525) = (\text{Temp} - 525)/100$

and

$X_2 = (\text{Veloc} - 0.50)/(0.75 - 0.50) = (\text{Veloc} - 0.50)/0.25.$

TABLE 9.4
Analysis of Factorial Experiments for Polypropylene Pyrolysis Example

Run Number	X_1	X_2	X_1X_2	Curvature X^2	Yield	
1	-1	-1	1	1	73.2	
2	1	-1	-1	1	51.1	Factorial
3	-1	1	-1	1	76.8	Avg = 66.3
4	1	1	1	1	64.1	
5	0	0	0	0	74.7	Center Point
6	0	0	0	0	76.8	Avg = 74.9
7	0	0	0	0	73.2	

Effects:	-17.40	8.30	4.70	-8.60	
t-Statistics:	-9.62	4.59	2.60	-6.23	$t_{2,0.05} = 4.303$

$s =$	1.81	(from center points)
$s_E =$	1.81	$(= s\sqrt{1/2 + 1/2})$
$s_C =$	1.38	$(= s\sqrt{1/4 + 1/3})$

Since we know that the X's are to be $+\alpha$ and $-\alpha$ (or 0), and we want to know the appropriate values for temperature and velocity, we must rearrange the equations. When we do so, we get

Temperature $= 525 + 100\ X_1$ and
Velocity $= 0.50 + 0.25\ X_2$.

So, for example, when we want X_1 to be $-\alpha$ (or -1.41), we must set the temperature to $525 + 100(-1.41) = 525 - 141 = 384\ °C$. The whole design (including the first block) is given in Table 9.5. The second block of experiments was run, and the crude oil yields were recorded. The yields are also given in Table 9.5 and shown graphically in Figure 9.12.

Once the data have been collected, a quadratic equation must be fit to the response. This task, called **regression analysis**, can be done by hand, but it is a tedious task much better suited to computers. The details of what these computer programs do is discussed later in this chapter, but we will go ahead (somewhat blindly for now) and obtain the quadratic equation that best fits our data.

The most rudimentary, and also probably the most available, programs that do regression analysis are spreadsheets. Of course, numerous statistical packages are available that do a more thorough job and do it virtually automatically. But the regression equation (or more precisely, the set of coefficients in the quadratic equation) is the same regardless of which program is used. The input for Microsoft Excel is shown in Table 9.6 and the output is shown in the top half of Figure 9.13. The output includes not just the coefficients, but the t-statistics (which were calculated by dividing the coefficients by their respective standard deviations) and their significance as well. It can be seen that all the coefficients in the equation were important (i.e., statistically significant) except for a block effect. Therefore, blocks were ignored in the final equation. In other words, a second regression analysis was performed with the blocking column omitted, and it is shown in the bottom half of Figure 9.13. As an aside, the reader may want to notice what happened to the other coefficients when blocking was deleted—absolutely nothing! This was not a fluke. The blocking was

TABLE 9.5

Complete Central Composite Design for Two Factors:
Polypropylene Pyrolysis

Run Number	Run Order	X_1	X_2	Temp. °C	Veloc. ft/sec	Yield
1	5	−1	−1	425	0.25	73.2
2	3	1	−1	625	0.25	51.1
3	2	−1	1	425	0.75	76.8
4	6	1	1	625	0.75	64.1
5	1	0	0	525	0.50	74.7
6	7	0	0	525	0.50	76.8
7	4	0	0	525	0.50	73.2
8	13	−1.41	0	384	0.50	80.6
9	9	1.41	0	666	0.50	61.0
10	12	0	−1.41	525	0.15	57.2
11	10	0	1.41	525	0.85	66.5
12	11	0	0	525	0.50	72.2
13	14	0	0	525	0.50	74.5
14	8	0	0	525	0.50	76.9

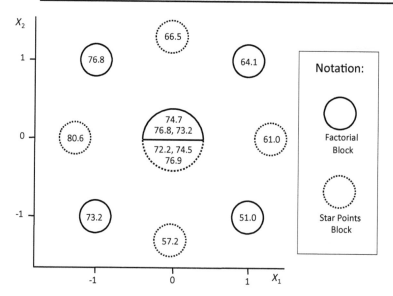

FIGURE 9.12: Complete Central Composite Design for Two Factors:
Polypropylene Pyrolysis

deliberately chosen so that even if there was a major shift in response from one block to the other, it would only impact the estimate of the mean (and the variability) not the other coefficients in the model. This is called ***orthogonal blocking***. For other designs (more factors), the blocks cannot always be made to be orthogonal, but the blocking is still chosen to make it as nearly orthogonal as possible.

A natural question at this point is what would have been done if other coefficients had turned out not to be statistically significant—would they have been dropped also? Some statisticians prefer to leave all the terms in the model, regarding the entire equation as a mathematical French curve. Others (the authors included) prefer to simplify the model by

TABLE 9.6

Input to Excel to Fit Quadratic Equation to Yield
Data: Polypropylene Pyrolysis

X_1	X_2	X_1^2	X_2^2	X_1X_2	Block	Yield
−1	−1	1	1	1	−1	73.2
1	−1	1	1	−1	−1	51.1
−1	1	1	1	−1	−1	76.8
1	1	1	1	1	−1	64.1
0	0	0	0	0	−1	74.7
0	0	0	0	0	−1	76.8
0	0	0	0	0	−1	73.2
−1.41	0	2	0	0	1	80.6
1.41	0	2	0	0	1	61.0
0	−1.41	0	2	0	1	57.2
0	1.41	0	2	0	1	66.5
0	0	0	0	0	1	72.2
0	0	0	0	0	1	74.5
0	0	0	0	0	1	76.9

deleting all terms that have insignificant t-ratios (then refitting the model). This question
will be covered fully in Chapter 10.

Before accepting this equation as gospel, the data should be checked for bad points
(called **outliers**) via residual plots, and the equation should be checked for lack of fit. Both
these diagnostics are very important, and are discussed fully in Chapter 10 as well. For
now, suffice it to say that the residual plots (not shown) looked fine, so all the data were
assumed to be representative. Likewise, the lack-of-fit ratio showed absolutely no problem
with our quadratic equation.

9.6 Graphical Interpretation of Response Surfaces

Once we are satisfied that the equation is a good representation of our system (over the
region studied), then the only thing that remains is to use it to come to some conclusions.
The final equation for the polypropylene pyrolysis example discussed in the previous section
is:

$$\hat{Y} = 74.72 - 7.83\,X_1 + 3.73\,X_2 - 1.96\,X_1^2 - 6.47\,X_2^2 + 2.35\,X_1X_2.$$

This equation "says it all"; all we have to do is understand what it is "saying." It is a
rare individual that can visualize the relationships between the response and the factors directly from the equation. Therefore, the equation is typically reported graphically (at least
for a small number of factors).

The best way to summarize the equation graphically is largely a matter of taste. A commonly used graph to show a response as a function of two factors is a contour plot. Figure
9.14 is a contour plot of yield versus the two factors in the polypropylene pyrolysis example.
It very clearly shows the nature of the relationships and the location of the highest yield in

SUMMARY OUTPUT   (Including Blocking Term in Model)

Regression Statistics	
Multiple R	0.987
R Square	0.974
Adjusted R Square	0.952
Standard Error	1.899
Observations	14

Term in Model	Coefficients	Standard Error	t-Stat	P-value
Intercept	74.72	0.78	96.39	0.00
X_1	-7.83	0.67	-11.64	0.00
X_2	3.73	0.67	5.54	0.00
$X_1{}^*X_1$	-1.96	0.70	-2.80	0.03
$X_2{}^*X_2$	-6.47	0.70	-9.21	0.00
$X_1{}^*X_2$	2.35	0.95	2.48	0.04
Block	-0.09	0.51	-0.17	0.87

SUMMARY OUTPUT   (Excluding Blocking Term in Model)

Regression Statistics	
Multiple R	0.987
R Square	0.974
Adjusted R Square	0.958
Standard Error	1.780
Observations	14

Term in Model	Coefficients	Standard Error	t-Stat	P-value
Intercept	74.72	0.73	102.83	0.00
X_1	-7.83	0.63	-12.42	0.00
X_2	3.73	0.63	5.91	0.00
$X_1{}^*X_1$	-1.96	0.66	-2.99	0.02
$X_2{}^*X_2$	-6.47	0.66	-9.83	0.00
$X_1{}^*X_2$	2.35	0.89	2.64	0.03

FIGURE 9.13: Regression Output for Polypropylene Pyrolysis Example

the region of experimentation. The limit of the experimental region is shown on the graph as a reminder that the equation is a mathematical French curve, and therefore it should not be extrapolated outside of the experimental region. That is not tempting in this example, but it often is in other situations. If a region beyond the limits of the current experimentation looks attractive, then further experiments must be conducted in that region. To extrapolate is at best risky, and at worst foolhardy.

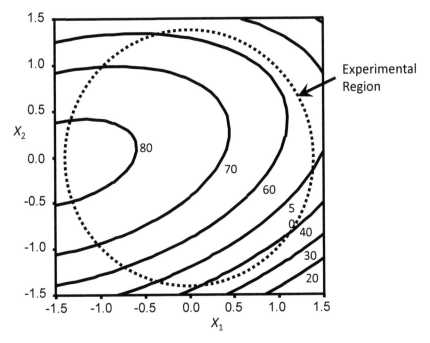

FIGURE 9.14: Contours of Yield vs. Coded Temperature (X_1) and Coded Fluidization Velocity (X_2) for Polypropylene Pyrolysis

In this example, however, the relationship between yield and fluidization velocity was the main focus. Temperature was brought into the study because it was thought that an interaction may exist. Therefore, a better summary graph for this situation is a simple yield versus velocity graph. Since temperature was important, the graph is actually a collection of curves, each one at a different temperature. These are shown in Figure 9.15. The curves were obtained by taking the equation for yield as a function of X_1 and X_2, putting in a specific value for X_1 (the coded temperature) and then simplifying. For example when temperature is 525, $X_1 = 0$, and the equation simplifies to:

$$\hat{Y} = 74.72 + 3.73\ X_2 - 6.47\ X_2^2.$$

Likewise, when temperature is 625, $X_1 = +1$, and the equation becomes:

$$\hat{Y} = 74.72 - 7.83(1) + 3.73X_2 - 1.96(1)^2 - 6.47\ X_2^2 + 2.35(1)X_2$$

or $\hat{Y} = 64.93 + 6.08\ X_2 - 6.47\ X_2^2.$

When there are more than two factors, it becomes increasingly difficult to visualize the response surface equation by using response curves like Figure 9.15; contour plots are usually better. Since there are more than two independent variables in the equation, several contour plots can be made at fixed levels of the independent variables not represented on the axes. For example, let us say that we studied the effect of catalyst concentration, reaction time, and temperature on the yield of a chemical reaction, and we obtained the following response surface equation:

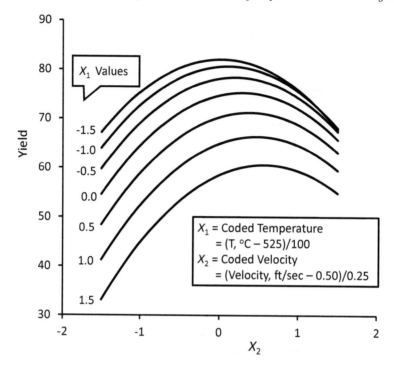

FIGURE 9.15: X-Y Plots of Yield vs. Coded Fluidization Velocity (X_2) at Several Temperatures

$$\tilde{Y} = 85.83 - 0.48X_1 - 0.75X_2 - 0.85X_3 - 1.29X_1^2 + 0.06X_2^2 + 0.66X_3^2$$
$$- 0.30\ X_1X_2 - 0.10\ X_1X_3 + 1.10\ X_2X_3$$

where:
X_1 = (catalyst, mole % − 0.03)/0.01,
X_2 = (time, min − 90)/30,
X_3 = (temperature, C − 65)/15, and
Y = yield of the reaction.

Figures 9.16–9.18 show contour plots with X_2 fixed at −1, 0, and 1, respectively (i.e., time = 60, 90, and 120 minutes). These can be thought of as three slices through the three-dimensional response contour space.

Contour plots can be used for the same purposes as simple graphs of the response versus each independent variable. When more than one response variable results from each experiment (e.g., Y_1 and Y_2), contour plots of the equations for each response are often overlaid in order to graphically determine a constrained maximum or minimum. For example, if the cost of the chemical reaction experiments could be expressed as a function of the three independent factors (catalyst, time, and temperature), a contour plot for cost could be drawn and overlaid on Figures 9.16–9.18. Then the maximum yield for a fixed cost level could be determined. If only the yield were important, then its maximum is readily apparent from the contour plots as being at short times, low temperatures, and middle catalyst amounts. For a larger number of factors, we can still take one- or two-dimensional slices through the surface and show them as curves or contour plots. But we end up with an exponentially

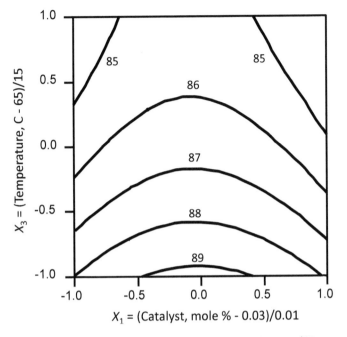

FIGURE 9.16: Yield Contours for Time = 60 minutes ($X_2 = -1$)

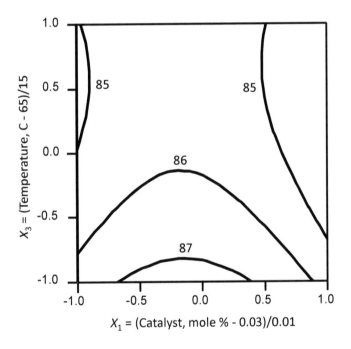

FIGURE 9.17: Yield Contours for Time = 90 minutes ($X_2 = 0$)

increasing number of plots. If we are looking for a maximum (or minimum) or we just want an overall feeling for what the response looks like, analytical or numerical methods may come to our rescue. These methods are discussed in Chapter 10.

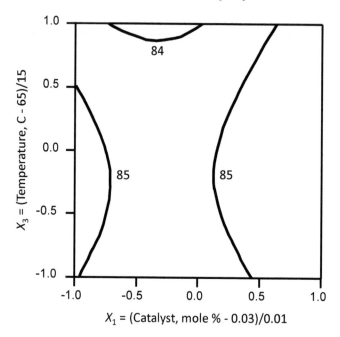

FIGURE 9.18: Yield Contours for Time = 120 minutes ($X_2 = 1$)

9.7 Other Response Surface Designs

9.7.1 Box–Behnken Designs

Another class of commonly used designs for full response surface estimation is called Box–Behnken designs, and this class has two advantages over central composite designs. First of all, they are more sparing in the use of runs, particularly for the very common three- and four-factor designs. The number of runs and the breakdown between center points and other points for the Box–Behnken designs are shown in the third column of Table 9.7. The exceptions to the rule can be seen from the table to be the five-factor design (the Box–Behnken design actually requires 13 more runs) and the six-factor design (which does not save any runs).

The second advantage is that the Box–Behnken designs are only three-level designs (i.e., each factor is controlled at only -1, 0, or $+1$), whereas the central composite designs are five-level designs (i.e., each factor will have been set at $-\alpha$, -1, 0, $+1$, or $+\alpha$ over the course of the set of experiments). On the surface this may seem to be totally inconsequential. But in an industrial setting, keeping the number of levels down to the bare minimum can often be a great help in the practical administration of an experimental program.

The Box–Behnken design for three factors is shown in Figure 9.19. The design consists of running all possible pairs of 2^2 designs, with the factor not considered being held at zero or its mid-level. Replicated center points are also added to complete the design. This pattern is the same for three to five factors; for six to nine factors the designs consist of sets of 2^3 designs plus center points. The specific designs for three to seven factors are given in Appendix B.4.

It should also be stated that Box–Behnken designs also meet the criteria of rotatability

TABLE 9.7
Number of Experiments Required for Response Surface Designs

Number of Factors	Central Composite Designs Section 9.4	Box–Behnken Designs Section 9.7.1	Small Composite Composite Section 9.7.2	Number of Coefficients in Quadratic Models
2	4 + 4 + 6 = 14 (Two Blocks)	No Design	6	6
3	8 + 6 + 6 = 20 (Three Blocks)	12 + 3 = 15 (One Block)	10	10
4	16 + 8 + 6 = 30 (Three Blocks)	24 + 3 = 27 (Three Blocks)	16	15
5	16 + 10 + 7 = 33 (Two Blocks)	40 + 6 = 46 (Three Blocks)	22	21
6	32 + 12 + 10 = 54 (Three Blocks)	48 + 6 = 54 (Three Blocks)	28	28
7	64 + 14 + 12 = 90 (Nine Blocks)	56 + 6 = 62 (Three Blocks)	40	36

Central Composite Designs: Factorial + Star + Center = Total
Box–Behnken Designs: Pattern + Center = Total

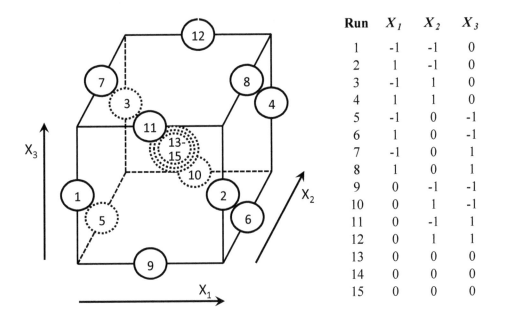

Run	X_1	X_2	X_3
1	-1	-1	0
2	1	-1	0
3	-1	1	0
4	1	1	0
5	-1	0	-1
6	1	0	-1
7	-1	0	1
8	1	0	1
9	0	-1	-1
10	0	1	-1
11	0	-1	1
12	0	1	1
13	0	0	0
14	0	0	0
15	0	0	0

FIGURE 9.19: Box–Behnken Design for Three Factors

(or nearly do), uniform precision of predictions, and can be broken down into two or more blocks when k, the number of factors, is between four and seven.

In fact, Box–Behnken designs meet all the considerations for a good design discussed in Section 9.3 except for one—they do not build upon the 2^k factorial design. This means that the full response surface design must always be run, whether or not it is needed. Remember that a big advantage of central composite designs is that the limited response surface portion (factorial + center points) can be run first and analyzed to determine if further work to separate the quadratics for each of the factors is justified. The star point block may never need to be run. That, of course, is a hefty savings! No such potential exists for the Box–Behnken designs. Conversely, if a limited response surface design was run first, there are no runs that could be added to it to make it into a Box–Behnken. You would have to start all over again. Naturally, you would never do that. Rather, you would just add star points (and center points) and be happy with a central composite design.

In summary, unless you are sure that a full response surface is necessary before beginning, or unless there is a distinct advantage to having only three levels of each factor, the central composite design is the design of choice.

9.7.2   Small Composite Designs

Sometimes the number of runs required to complete a central composite or Box–Behnken design may seem prohibitive. If the researcher is willing to assume his experimental region is restricted enough to ensure that the quadratic model will provide a good approximation to the underlying function, then *small composite designs* are available.

One situation where small composite designs would be desirable is when the experiments or runs are very expensive or time consuming. Another case is when there is little or no experimental error, so extra runs to increase precision are not needed. An example of this case is where the response is being calculated via a computer simulation. If this deterministic function is very time consuming to compute, it could be approximated by a quadratic model in order to reduce computations necessary to locate the approximate maximum or minimum.

The approach of fitting a quadratic model to a few data points works well whenever the underlying physical relationship is smooth and the experimental region is not too large (unlike that shown in Figure 9.8). As shown in Table 9.7, the number of experiments (or function evaluations in deterministic situations) required for a central composite or Box–Behnken design is approximately 1.5 to 2 times as large as the number of coefficients fit in the empirical quadratic model, while the number of experiments for the small composite designs listed in the fourth column of Table 9.7 is seen to be equal or nearly equal to the number coefficients in the model to be fit.

These designs were derived by taking a fraction of a factorial design and adding star points on the boundary of the experimental region for each factor as shown in Figure 9.20 for the three-factor case. Compare this to Figure 9.10 in Section 9.4, which shows the central composite design for three factors, and you will see that there are only two differences: a fractional factorial was used, and there are no center points. A listing of the small composite designs is given in Appendix B.5.

The huge disadvantage to using the small composite designs is the fact that the goodness of fit of the model, residual plots, etc., cannot be checked. Therefore, the researcher must be sure that his experimental region is small enough to ensure that the quadratic model will provide a good approximation. And, of course, the number of experiments used must be sufficient to provide the precision desired when experimental error is present.

9.7.3   Example—Coal Gasification Modeling

At the combustion research lab at BYU a computer program had been written to simulate the local fluid flow, chemical properties, and species concentrations in a coal gasification

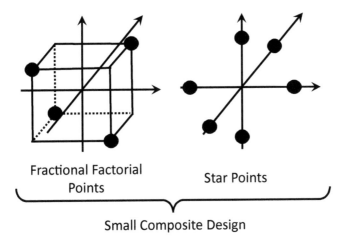

Fractional Factorial Points

Star Points

Small Composite Design

FIGURE 9.20: Small Composite Design for Three Factors

or combustion chamber. This simulation model required specification of 75 or more inputs or factors (X's) and took 10 to 15 CPU hours on a computer to make one prediction. Even though the simulations were quite repeatable at the same input conditions (i.e., there was essentially no experimental error) it would have been very difficult to locate the optimum conditions through direct simulation, because of the long computations. Therefore, a response surface approach was used to maximize the cold gas efficiency (CGE), which is defined as gross caloric value of the product gas divided by the gross caloric value of the coal fed, in a gasification simulation. Three factors, presumed to be the most critical, were selected for study. They were: (1) the system pressure, X_1, in the range 1 to 10 atmospheres, (2) the oxygen to coal ratio, X_2, in the range 0.70 to 1.15, and (3) the steam to coal ratio, X_3, in the range 0.0 to 0.3. Table 9.8 shows the conditions and results of 13 simulation experiments conducted at conditions defined by a Box–Behnken experimental design. Notice that the center point was not repeated, since the same result would have been obtained each time.

TABLE 9.8
Simulation Design and Results for
Coal Gasifier Modeling

Run	X_1	X_2	X_3	CGE
1	1	0.70	0.15	0.6375
2	10	0.70	0.15	0.8274
3	1	1.15	0.15	0.6167
4	10	1.15	0.15	0.6555
5	1	0.925	0.00	0.6734
6	10	0.925	0.00	0.7826
7	1	0.925	0.30	0.6396
8	10	0.925	0.30	0.7887
9	5.5	0.70	0.00	0.7275
10	5.5	1.15	0.00	0.6570
11	5.5	0.70	0.30	0.7843
12	5.5	1.15	0.30	0.6560
13	5.5	0.925	0.15	0.7882

After a quadratic model was fit to the data in Table 9.8, the maximum CGE was predicted to be 0.8428 at $X_1 = 10$ atm, $X_2 = 0.787$, and $X_3 = 0.2322$. This was confirmed by running an actual simulation at these conditions which resulted in CGE = 0.8486. In this particular case (three factors) there were 10 coefficients in the quadratic model and 13 simulation experiments were required. If the researchers had been totally confident in the quadratic model approximation, only 10 simulation experiments would really have been needed using the small composite design instead of the Box–Behnken, and 30–45 hours of CPU time could have been saved.

9.8   Summary

9.8.1   Procedure for Design of Experiments for RSM

The following is a summary in outline form of the procedure for running a response surface experimental design.

- List the important independent variables and their usual ranges (it should be remembered that if a central composite design will be used, the star points will extend beyond these ranges).

- List the dependent variables and their units.

- Select a class of RSM designs, keeping in mind that central composite designs allow building on a 2^k design which may be run first, that Box–Behnken designs require only three levels which may be simpler to administer, and that small composite designs should be used in deterministic situations (or other cases with very little variability in results) when no estimate of experimental error is needed.

- Obtain the design in coded X's from Appendix B.3 for central composite designs, B.4 for Box–Behnken designs, or B.5 for small composite designs.

- Decide whether or not the design should be run in blocks. This should normally be done in the case of central composite designs if the factorial part is to be analyzed first to determine if it is necessary to complete the second block of experiments. Blocking should be done for either central composite or Box–Behnken if it is likely that experimental uniformity is better within a block. The blocking is indicated in the appendices along with the design.

- Write out the experimental design in uncoded X's as a data collection worksheet. To do this, the coded X's are first copied from the appropriate table in Appendix B. For Example 1, Columns 3 and 4 of Table 9.3 were copied from Table B.3-1. Next, the high, low, and mid-level of the uncoded factor ranges are equated to +1, −1, and 0, respectively. For Factor 1 (temperature) this becomes:
High level: Temperature = 625 $\Longrightarrow X_1 = 1$
Low Level: Temperature = 425 $\Longrightarrow X_1 = -1$
Center Level: Temperature = 525 $\Longrightarrow X_1 = 0$
Lastly, the uncoded factor levels of the star points are determined, if needed, as follows:
The uncoded and coded factor levels are related by the general formula,

$X = $ (Factor $-$ Mid) / (Half of Difference between High and Low)

which becomes, for this example,

$X_1 = $ (Temperature $-$ 525)/[(625 $-$ 425)/2] $=$ (T $-$ 525)/100.

Solving for the uncoded factor level results in

T $=$ 525 + 100 X_1.

This equation is used to determine the uncoded factor levels of the star points, $\pm\alpha$. For Factor 1, the $+\alpha$ level is

T $=$ 525 + 100($+\alpha$) $=$ 525 + 100(1.414) $=$ 666

rounded to nearest integer.

- Review each of the experimental conditions for feasibility. If one or more runs does not appear operable, move them toward the center of the design by cutting down on all factors that are not zero (in coded units) in the same proportions. If the run is moved appreciably, it should be replicated to maintain the balance and precision of the design. If an experiment that was originally thought to be operable is found not to be, the same procedure applies.
  (Note: If a point, j, was moved by reducing all the coded factor levels for that point by an amount c ($0 < c < 1$), then the number of replicates that should be run at the new conditions, r_j , is given approximately by: $r_j = 1/c$. Of course, this number must be rounded to the nearest integer.)

- Assign random run order numbers to each run within each block. In cases where complete randomization is impractical (very expensive), it may be compromised a bit, but each level of the factors should be experienced at least twice. The exception to randomization is the center points. It may be preferable to spread them uniformly throughout the block, rather than to randomly run them, in order to get as wide a variation as possible and to facilitate checking for time trends.

- Run the experiments and collect the data. Often, as a practical matter, it is wise to use a data collection worksheet in which the experiments to be run are listed in the order to be run with the factor levels given in real (uncoded) units. This worksheet minimized the chance of mistakes, particularly if the runs are to be made by someone else.

9.8.2 Important Terms and Concepts

The following is a list of the important terms and concepts covered in this chapter.

response surface methodology (RSM)
empirical model
quadratic model
second order experimental design
3^k design
central composite design
star points

rotatability
uniform precision
orthogonal blocking
contour plot
X-Y plot
Box–Behnken design
small composite design

9.9   Exercises

1. You studied the effects of three factors of interest on the yield of a chemical reaction using a full (two-level) factorial design plus three center points. You found that curvature was important (as well as all three factors), and want to augment the design with "star" points and three more center points to make it a full central composite design. List the additional runs that are needed in the table below, in both coded and uncoded form. Note: $\alpha = 1.68$ for this design.

2. You have the same system described in Problem 1, but you have severe budget constraints. Therefore, you are thinking about running a Small Composite Design with the three factors.

   (a) Write out the whole central composite design in three factors using coded factors. Then put a check after each run that would still be required for the Small Composite Design. How many runs would be saved?

   (b) Under what conditions is the Small Composite Design justified?

   (c) If you use the Small Composite Design, how many degrees of freedom are available to estimate error once you fit the full quadratic equation?

Chapter 10

Response Surface Model Fitting

10.1 Introduction

As was explained in the last chapter, the goal of response surface methodology is to describe in detail the relationship between a response variable (Y) and the independent variables (X_1, X_2, ..., X_k). A general quadratic equation is used as the mathematical French curve to accomplish this goal once data have been collected using one of the RSM designs already discussed in Chapter 9. The primary objectives of this chapter are to find the best equation to describe the system under study, check its adequacy, and use it to see what the surface looks like.

Remember that the general quadratic model for k independent variables (discussed in Chapter 9, Equation 9.1) is:

$$\hat{Y} = b_0 + \sum_{i=1}^{k} b_i X_i + \sum_{i=1}^{k} b_{ii} X_i^2 + \sum_{i=1}^{k-1} \sum_{j=i+1}^{k} b_{ij} X_i X_j. \tag{10.1}$$

Finding the best equation means that we must determine the best values for the coefficients in the quadratic equation (b_o, b_1, ..., b_k, b_{11}, ..., b_{kk}, b_{12}, ..., $b_{k-1,k}$) for a particular set of experimental data. We should also simplify our model to the extent that is warranted by dropping out terms that are not justified (i.e., significant). We will use the method of least squares, which was described extensively in Section 8.4.2, to find the coefficients. Once you have your equation, you should not use it for anything until you check to see if it actually describes your data adequately. This was discussed in Sections 8.5 and 8.6, but we will add one more check (Section 10.4). Trimming the model down to the smallest adequate equation is discussed in Section 10.5; our philosophy will always be "the simpler, the better" unless greater complexity is actually shown to be needed. Some techniques for exploring the fitted equation (to determine optimum conditions, etc.) are discussed in Section 10.6. Finally, quantifying the precision of the model predictions is revisited in Section 10.7.

10.2 Estimation of Coefficients in a Quadratic Model

As stated in the introduction, our goal now is to find the best equation that describes a set of data. The measure of goodness we use (as discussed in detail in Chapter 8) is the sum of the squared vertical distances from the data points to the line, or the *SSE* (Sum of the Squared Errors) for short. In equation form:

$$SSE = \sum_i (Y_i - \hat{Y}_i)^2.$$

Remember that since we are making the sum of the squared distances as small as possible, the whole procedure is often called the *method of least squares*.

We found in Section 8.4 that the values of the coefficients that minimize SSE are given by Equation 8.36 which is repeated here for ease of reference as Equation 10.2.

$$\hat{\mathbf{B}} = [(\mathbf{X}^T\mathbf{X})^{-1}\,\mathbf{X}^T]\mathbf{Y}. \tag{10.2}$$

As previously stated in Chapter 8, Equation 10.2 is a simple but very powerful matrix equation which defines the least squares estimates for all the regression coefficients simultaneously. We saw how it was used for a factorial model in Chapter 8. To see how it is used for a RSM model, we will calculate the coefficients in the quadratic equation describing the polypropylene pyrolysis example in Chapter 9. The data are given in Table 9.5. Since we want to fit a full quadratic equation in two X variables to some data, the equation we will use to describe the system is:

$$\hat{Y} = b_0 + b_1 X_1 + b_2 X_2 + b_{11} X_1^2 + b_{22} X_2^2 + b_{12} X_1 X_2. \tag{10.3}$$

Using the coded values of the X's, we get the $\mathbf{X}$ and $\mathbf{Y}$ matrices from Table 9.6:

I	X_1	X_2	X_1^2	X_2^2	$X_1 X_2$	Yields
1	−1	−1	1	1	1	73.2
1	1	−1	1	1	−1	51.1
1	−1	1	1	1	−1	76.8
1	1	1	1	1	1	64.1
1	0	0	0	0	0	74.7
1	0	0	0	0	0	76.8
1	0	0	0	0	0	73.2
1	−1.41	0	2	0	0	80.6
1	1.41	0	2	0	0	61.0
1	0	−1.41	0	2	0	57.2
1	0	1.41	0	2	0	66.5
1	0	0	0	0	0	72.2
1	0	0	0	0	0	74.5
1	0	0	0	0	0	76.9

$\mathbf{X}=$ (left matrix), $\mathbf{Y}=$ (right matrix).

The least squares normal equation (Equation 8.35) for this example is:

$$
\begin{array}{ccc}
\mathbf{X}^T\mathbf{X} & \hat{\mathbf{B}} & \mathbf{X}^T\mathbf{Y}
\end{array}
$$

$$
\begin{bmatrix}
14 & 0 & 0 & 8 & 8 & 0 \\
0 & 8 & 0 & 0 & 0 & 0 \\
0 & 0 & 8 & 0 & 0 & 0 \\
8 & 0 & 0 & 12 & 4 & 0 \\
8 & 0 & 0 & 4 & 12 & 0 \\
0 & 0 & 0 & 0 & 0 & 4
\end{bmatrix}
\begin{bmatrix}
\hat{b}_0 \\
\hat{b}_1 \\
\hat{b}_2 \\
\hat{b}_{11} \\
\hat{b}_{22} \\
\hat{b}_{12}
\end{bmatrix}
=
\begin{bmatrix}
978.80 \\
-62.52 \\
29.75 \\
548.40 \\
512.60 \\
9.40
\end{bmatrix}
$$

and the solution (Equation 8.36 or Equation 10.2) is

$$
\hat{\mathbf{B}} \qquad\qquad (\mathbf{X}^T\mathbf{X})^{-1} \qquad\qquad\qquad \mathbf{X}^T\mathbf{Y}
$$

$$
\begin{bmatrix} \hat{b}_0 \\ \hat{b}_1 \\ \hat{b}_2 \\ \hat{b}_{11} \\ \hat{b}_{22} \\ \hat{b}_{12} \end{bmatrix} = \begin{bmatrix} 0.167 & 0 & 0 & -0.080 & -0.080 & 0 \\ 0 & 1/8 & 0 & 0 & 0 & 0 \\ 0 & 0 & 1/8 & 0 & 0 & 0 \\ -0.083 & 0 & 0 & 0.135 & 0.010 & 0 \\ -0.083 & 0 & 0 & 0.010 & 0.135 & 0 \\ 0 & 0 & 0 & 0 & 0 & 1/4 \end{bmatrix} \begin{bmatrix} 978.8 \\ -62.52 \\ 29.75 \\ 548.4 \\ 512.6 \\ 9.40 \end{bmatrix} = \begin{bmatrix} 74.72 \\ -7.83 \\ 3.73 \\ -1.96 \\ -6.47 \\ 2.35 \end{bmatrix}.
$$

These values can be compared to those given in the bottom half of Figure 9.13 which were calculated using a spreadsheet computer program. They should be the same, and they are.

Generally the first question of interest is: "How well does the equation fit?" As discussed in Section 8.5, the criterion that is most often used is R^2, the fraction of the total variability in the data that is explained by the model. This was calculated for the polypropylene pyrolysis example (see Table 10.1), and found to be $R^2 = 0.974$. This value of R^2 is extremely high (close to 1.0). This means that there is only a small bit of variation in the data (namely 2.6% of it) that is not explained by the model.

TABLE 10.1
Detailed Calculation of R^2 for Polypropylene Pyrolysis Example

Y_i	$\bar{Y}$	$Y_i-\bar{Y}$	$(Y_i-\bar{Y})^2$	$\hat{Y}_i$	$Y_i-\hat{Y}_i$	$(Y_i-\hat{Y}_i)^2$
73.2	69.91	3.29	10.80	72.74	0.46	0.212
51.1	69.91	-18.81	353.98	52.38	-1.28	1.650
76.8	69.91	6.89	47.41	75.49	1.31	1.715
64.1	69.91	-5.81	33.81	64.53	-0.43	0.189
74.7	69.91	4.79	22.90	74.72	-0.02	0.000
76.8	69.91	6.89	47.41	74.72	2.08	4.340
73.2	69.91	3.29	10.80	74.72	-1.52	2.300
80.6	69.91	10.69	114.18	81.85	-1.25	1.562
61.0	69.91	-8.91	79.46	59.78	1.22	1.500
57.2	69.91	-12.71	161.65	56.61	0.59	0.348
66.5	69.91	-3.41	11.66	67.12	-0.62	0.378
72.2	69.91	2.29	5.22	74.72	-2.52	6.333
74.5	69.91	4.59	21.03	74.72	-0.22	0.047
76.9	69.91	6.99	48.80	74.72	2.18	4.767
		SST =	969.12		*SSE =*	25.342

$$
\mathbf{R}^2 = (SST\text{-}SSE)/SST = (969.12 - 25.34)/969.12 = 0.974
$$

Figure 10.1 shows graphically how the data and the model compare for the polypropylene pyrolysis example by plotting the data, Y (on the y-axis), vs. the model predictions, $\hat{Y}$ (on the x-axis). The 45 degree line on the graph represents the perfect fit (where $Y = \hat{Y}$). As expected with a high R^2 value, the points are quite close to the 45 degree line.

Since the R^2 for our example says that the regression equation matches the data quite well, it is tempting to simply go ahead and use it for whatever purpose it was intended. But we should be cautious. The equation should not be used until we have verified that all our assumptions are correct.

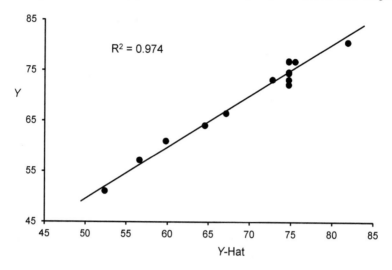

FIGURE 10.1: Data Values vs. Model Predictions for the Polypropylene Pyrolysis Example

10.3   Checking Model Assumptions (Residual Plots)

As discussed in detail in Chapter 8 (Section 8.6), we need to check our model's assumptions before going off and using it to improve our system. If any of our assumptions are not correct, our model could be quite wrong. The four assumptions are (again):

1. The errors are Normally distributed.

2. The errors have the same variance.

3. The errors are independent (or the experiment was randomized).

4. The errors have a mean of zero (i.e., the model is adequate).

As a review, let us check our assumptions for the polypropylene pyrolysis example in the order listed. The residuals are given again in Table 10.2 (along with the calculations for Normal plots which we use to check the first assumption).

To check that our errors are Normally distributed, a dot plot, a Normal plot and a Half-Normal plot are shown in Figures 10.2 (a), (b), and (c). All the plots look good, indicating that our assumption is fine.

Next, we checked that each of the data points has the same precision, and therefore each of the data points should be given equal weight. To check that assumption, the residuals were plotted against $\hat{Y}$ as shown in Figure 10.3. In this example it may appear as if the scatter is bigger at higher values of $\hat{Y}$, but there are more data at higher values, and so there would naturally be more scatter. The other plots of residuals shown in Section 8.6 should also be checked for evidence of non-constant variance, as well as for trends (which is the main goal of that section).

Our third assumption, that the errors are independent, is difficult to check (at best). However, since the run order (within blocks) was randomized, that removes the need to check that assumption.

TABLE 10.2

Residuals for the Polypropylene Pyrolysis Example and Detailed
Calculations Required for the Full-Normal and Half-Normal Plots

| Run | Y_i | $\hat{Y}_i$ | For Full-Normal Plot | | | | For Half-Normal Plot | | | |
			ϵ_i	$\epsilon_{(i)}$	$\frac{(i)-0.5}{n}$	z	$|\epsilon_i|$	$|\epsilon_{(i)}|$	$\frac{1}{2}+\frac{(i)-0.5}{2n}$	z
1	73.2	72.74	0.46	-2.52	0.036	-1.80	0.46	0.02	0.518	0.04
2	51.1	52.38	-1.28	-1.52	0.107	-1.24	1.28	0.22	0.554	0.13
3	76.8	75.49	1.31	-1.28	0.179	-0.92	1.31	0.43	0.589	0.23
4	64.1	64.53	-0.43	-1.25	0.250	-0.67	0.43	0.46	0.625	0.32
5	74.7	74.72	-0.02	-0.62	0.321	-0.46	0.02	0.59	0.661	0.41
6	76.8	74.72	2.08	-0.43	0.393	-0.27	2.08	0.62	0.696	0.51
7	73.2	74.72	-1.52	-0.22	0.464	-0.09	1.52	1.22	0.732	0.62
8	80.6	81.85	-1.25	-0.02	0.536	0.09	1.25	1.25	0.768	0.73
9	61.0	59.78	1.22	0.46	0.607	0.27	1.22	1.28	0.804	0.85
10	57.2	56.61	0.59	0.59	0.679	0.46	0.59	1.31	0.839	0.99
11	66.5	67.12	-0.62	1.22	0.750	0.67	0.62	1.52	0.875	1.15
12	72.2	74.72	-2.52	1.31	0.821	0.92	2.52	2.08	0.911	1.35
13	74.5	74.72	-0.22	2.08	0.893	1.24	0.22	2.18	0.946	1.61
14	76.9	74.72	2.18	2.18	0.964	1.80	2.18	2.52	0.982	2.10

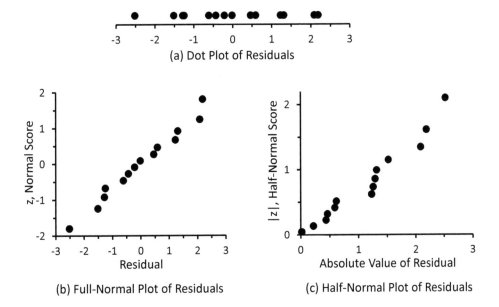

(a) Dot Plot of Residuals

(b) Full-Normal Plot of Residuals

(c) Half-Normal Plot of Residuals

FIGURE 10.2: Plots of Residuals to Check Normality of Errors for the Polypropylene
Pyrolysis Example

The last assumption is essentially that our model is adequate. So we plotted the residuals
versus the two independent variables and any other factors that may influence our response,
and we looked for trends. The extra factors were the run order and the blocks for this example. The suggested plots were made, and the plots are shown in Figure 10.4. None of the
plots show any trends, and so the model appears to be adequate in this regard. Although
the plots all look fine, there is a quantitative test we can and should use in addition to the

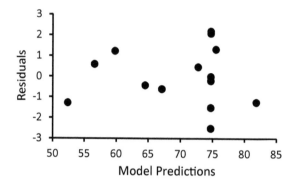

FIGURE 10.3: Plots of Residuals Versus Model Predictions for the Polypropylene Pyrolysis Example

plots to check model adequacy. This test is discussed in the next section.

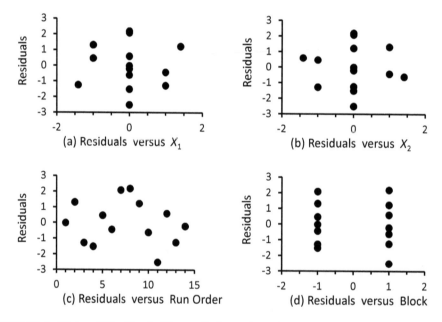

FIGURE 10.4: Plots of Residuals Versus Independent Variables, Run Order, and Blocks to Check Model Adequacy

10.4   Statistical Check of Model Adequacy –
Lack of Fit (LoF)

An objective statistical test is available (in addition to the residual plots) to check whether or not a quadratic model is giving an adequate representation of the data, but it requires some runs at the same conditions (replicates). The goal is exactly the same as the curvature test used for factorial designs. But the impact is even more severe. If the model is not adequate, the (quadratic) equation cannot be used reliably *anywhere* in the experimental region. Remember that for factorial designs, if the model was poor, it was still valid at the corners, but could not be used for interpolation inside the region.

The test is based on the sum of the squared residuals, *SSE*. The philosophy of the test is very simple and very basic. If the model is adequate, then *SSE* contains only experimental variability, and so it can be used to estimate the experimental error variance, σ^2. That is,

$$s_R^2 = SSE/\nu_R = SSE/[n - (p+1)] \approx \sigma^2. \tag{10.4}$$

(The subscript, R, on s_R^2 is used to denote that it is the estimate of σ^2 based on the regression model.) If the model is not adequate, at least some of the residuals will be too big. That will make *SSE* bigger than it ought to be, and s_R^2 will be an overestimate of σ^2. Thus s_R^2 can be compared to σ^2, and if it is significantly bigger (tested using a χ^2 distribution), then the model is deemed to be inadequate. If s_R^2 is not significantly bigger than σ^2, then the model is presumed to be fine.

This test looks great on paper, but there is one big catch—we usually do not know the value of σ^2. But if we have some replication, we can estimate σ^2 without using any model (other than the average at each set of conditions). We denote this estimate of σ^2 as s_{PE}^2. The subscript, PE, stands for pure error, because the only reason that replicates deviate from each other is the pure random variation in the data. Since all response surface designs have replication (at the center point at a minimum), the test described in this section should be used routinely with these designs.

The first step in using the test is to calculate s_{PE}^2. This is done by first identifying all replicates and calculating the average value at each replicated set of conditions. The sum of squared deviations from the average is then calculated at any point, j, at which there is replication (m_j is the number of replicates at point, j).

$$SS_{PE,j} = \sum_{i=1}^{m_j} (Y_{ij} - \bar{Y}_j)^2. \tag{10.5}$$

If only the center point was replicated, then the sum of squares of the center points around their average is the total SS_{PE}. If there is replication at several points, the various sums of squares are simply added together to get SS_{PE}.

$$SS_{PE} = \sum_j SS_{PE,j}. \tag{10.6}$$

In order to calculate s_{PE}^2, we also need to determine the degrees of freedom for pure error, DF_{PE} (also denoted by ν_{PE}). The degrees of freedom for any set of replicates is $\nu_{PE,j} = (m_j - 1)$. The total degrees of freedom for pure error is just the sum of the degrees of freedom for each set of replicates.

$$\nu_{PE} = \sum_j \nu_{PE,j}. \tag{10.7}$$

Once both SS_{PE} and ν_{PE} are calculated, we can determine s_{PE}^2 from the ratio,

$$s_{PE}^2 = SS_{PE}/\nu_{PE}. \tag{10.8}$$

And, since s_{PE}^2 is a perfectly valid estimate of σ^2, we could compare s_R^2 to it. But that is not what is usually done. The reason is that s_R^2 is based on a total sum of squared errors which includes a contribution from SS_{PE}. This contribution will mitigate any model inadequacies. We can do better.

In order to make the test more sensitive, what is typically done is to break the total sum of squares into two pieces—SS_{PE} and whatever is left over, which we denote by SS_{LoF}. The subscript, LoF, stands for Lack of Fit (of the model). We, likewise, break the total degrees of freedom into two pieces—ν_{PE} and whatever is left over, which we denote by ν_{LoF}. Since we already have the totals, and the pure error pieces, the remaining calculation is to get the lack-of-fit pieces by difference.

$$SS_{LoF} = SSE - SS_{PE} \tag{10.9}$$

$$\nu_{LoF} = \nu_R - \nu_{PE}. \tag{10.10}$$

It is important to realize that both parts of the original sum of squares, SSE, can be used to estimate the experimental error variance, σ^2. It has already been stated that SS_{PE}/ν_{PE} $(= s_{PE}^2)$ is a pure estimate of σ^2. But if the model is adequate, so is the quantity, SS_{LoF}/ν_{LoF} $(= s_{LoF}^2)$. As justification of this statement, remember that SSE is used to estimate σ^2 by dividing by ν_R. Therefore SSE is approximately equal to $\nu_R \sigma^2$. If we remove the amount, $\nu_{PE}\sigma^2$ $(= SS_{PE})$ from SSE, then what remains is $(\nu_R - \nu_{PE})\sigma^2$. Therefore, $SS_{LoF} \approx \nu_{LoF}\sigma^2$, or $\sigma^2 \approx SS_{LoF}/\nu_{LoF}$.

Thus, our test becomes a comparison of s_{LoF}^2 to s_{PE}^2 (which is our best estimate of σ^2). If our model is adequate, s_{LoF}^2 should be the same as s_{PE}^2 within experimental variability. If our model is not adequate, s_{LoF}^2 will be too big. Therefore our test asks the question: is s_{LoF}^2 significantly bigger than s_{PE}^2? Since two variances are being compared, the ratio of the two estimates of the (same) variance is distributed according to an F-Distribution,

$$F = \frac{s_{LoF}^2}{s_{PE}^2} = \frac{SS_{LoF}/\nu_{LoF}}{SS_{PE}/\nu_{PE}}, \tag{10.11}$$

and the yardstick is the tabled F-Distribution (with ν_{LoF} and ν_{PE} degrees of freedom).

The lack of fit calculations for our polypropylene pyrolysis example are shown in Table 10.3. The pure error estimate of variance, s_{PE}^2, was found to be 4.40, while the lack of fit estimate, s_{LoF}^2, was found to be 1.94. Since s_{LoF}^2 came out to be smaller than s_{PE}^2 (strictly due to chance), the question about s_{LoF}^2 being significantly bigger than s_{PE}^2 becomes mute, and the conclusion is that our model is fine. If s_{LoF}^2 had been bigger, we would have used the F-statistic (with 4 and 4 degrees of freedom) to tell us how much bigger than 1.0 the ratio, s_{LoF}^2/s_{PE}^2, can be just due to chance. In this case, s_{LoF}^2 would need to be 6.39 times bigger than s_{PE}^2 before we have enough evidence to say that the model is not adequate.

In our example, it should be noted that the center points were treated as two groups of replicates, one group in each block. This was done to be on the safe side in case there was actually a shift in the response from one block to the next (even though no shift was found to be significant in our regression analysis).

TABLE 10.3

Detailed Calculations Required for the Lack of Fit Test for the
Polypropylene Pyrolysis Example

Run	X_1	X_2	Block	Yield	j	Avg	d	d^2	ν
1	-1	-1	-1	73.2					
2	1	-1	-1	51.1					
3	-1	1	-1	76.8					
4	1	1	-1	64.1					
5	0	0	-1	74.7	1	74.90	-0.20	0.04	
6	0	0	-1	76.8	1	74.90	1.90	3.61	2
7	0	0	-1	73.2	1	74.90	-1.70	2.89	
8	-1.41	0	1	80.6					
9	1.41	0	1	61.0					
10	0	-1.41	1	57.2					
11	0	1.41	1	66.5					
12	0	0	1	72.2	2	74.53	-2.33	5.44	
13	0	0	1	74.5	2	74.53	-0.03	0.00	2
14	0	0	1	76.9	2	74.53	2.37	5.60	
							Totals =	17.59	4

$$SSE = 25.34 \qquad \nu_R = 8$$
$$SS_{PE} = 17.59 \qquad \nu_{PE} = 4 \qquad s^2_{PE} = 4.40$$
$$SS_{LoF} = 7.75 \qquad \nu_{LoF} = 4 \qquad s^2_{LoF} = 1.94$$

$$F(\text{Lack of Fit}) = s^2_{LoF}/s^2_{PE} = 1.94/4.40 = 0.44$$
$$F^*(4, 4, 0.95) = 6.39$$

10.5 Trimming Insignificant Terms from a Model

10.5.1 Justification

The overall goal of this section is to achieve model parsimony. The underlying principle is that the best model is the simplest model that "does the job." Although this philosophy is open to some debate, it is the firm belief of the authors.

To justify this position a bit, let us look at the simple situation of a response as a function of a single independent variable, and let us suppose that we have the data shown in Figure 10.5. We can fit numerous models of varying complexity to this data. If we limit ourselves to polynomials, we can still fit anything from a simple average to a complex ninth order polynomial. Since there seems to be a trend in the data, the simplest model actually used was a straight line. The least squares fits for that line and several higher order polynomials are shown in Figure 10.6. Which is best?

We can see from the figure that as we increase the complexity of the model it fits the data more and more closely. In fact, the eighth order polynomial fits the data almost exactly (and a ninth order polynomial *would fit exactly*), as the R^2 value indicates. Does that mean that we can conclude that the bigger a model is, the better it is? No! The problem is that the big models have gone past the point of describing the underlying relationship between X and Y, and they are describing the errors in the data. Since the errors are random, they will be different the next time some data are collected. So the big models are actually worse. Another way of stating the conclusion is that a model that is too big is better at predicting

Run	X	Y
1	0.1	5.90
2	0.3	5.84
3	0.5	7.61
4	0.7	10.10
5	0.9	8.57
6	1.1	8.40
7	1.3	9.97
8	1.5	8.62
9	1.7	7.38
10	1.9	7.95

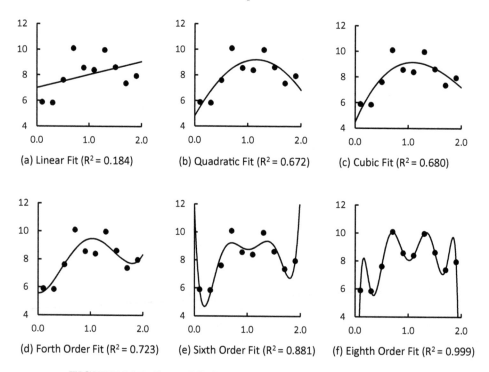

FIGURE 10.5: A Sample Set of X–Y Data

(a) Linear Fit ($R^2 = 0.184$)   (b) Quadratic Fit ($R^2 = 0.672$)   (c) Cubic Fit ($R^2 = 0.680$)

(d) Forth Order Fit ($R^2 = 0.723$)   (e) Sixth Order Fit ($R^2 = 0.881$)   (f) Eighth Order Fit ($R^2 = 0.999$)

FIGURE 10.6: Several Polynomial Fits to a Set of X–Y Data

the past (the data that we are analyzing), but what we want is a model that is good at predicting the future (new data that we may collect). The simplest adequate model is the best for that purpose; that is the quadratic fit in our example.

The next section describes how to trim away excess terms in a model in order to arrive at the simplest adequate equation.

10.5.2   Deleting Statistically Non-Significant Coefficients from Model

The general procedure for trimming a response surface model is to start by fitting the full quadratic equation (via least squares, of course). The residuals from the full model are checked via residual plots to ensure that the assumptions are valid before investing any more time in the model. If the residual plots look about right, then the model is trimmed down. Each coefficient in the model is examined for significance by comparing its t-statistic (the ratio of the coefficient to its standard error) to the critical t-value. These computations can be done manually using Equation 8.41, which is repeated below as Equation 10.12, remembering to replace σ^2 with s^2, its estimate.

$$Var(\hat{\mathbf{B}}) = (\mathbf{X}^T\mathbf{X})^{-1}\,\sigma^2. \tag{10.12}$$

This equation like Equation 10.2, is a very succinct yet very powerful formula. It gives the variability in our coefficients for any experimental design, $\mathbf{X}$. The result is a square symmetric matrix whose diagonal terms give the variances of the coefficients, and the off-diagonal terms give the covariances of pairs of the coefficients. It is the diagonal terms we will use when we test the statistical significance of our coefficients. Remember that to use the equation, we replace the unknown (but true value of the variance, σ^2, with its estimate, s^2. Then we must use the t-distribution (with the degrees of freedom in s^2) to test for statistical significance, rather than the Normal Distribution.

As a demonstration of the use of Equation 10.12, let us continue with the polypropylene pyrolysis example. The estimate of s^2 is SSE/ν_R. $SSE = 25.342$ as calculated in detail in Table 10.1, and the degrees of freedom for error, ν_R, is 14 (the number of data points) − 6 (the number of coefficients in our model) = 8. Therefore, our estimate of $s^2 = 25.342/8 = 3.168$, and $s = \sqrt{3.168} = 1.780$. This value of s can be compared to the value obtained by using the least squares regression function of a spreadsheet program to do the calculations. The output is shown at the bottom of Figure 9.13 and reproduced in Figure 10.7 for convenience. The estimate of s is labeled "Standard Error" and it is seen to be identical to the value we calculated.

SUMMARY OUTPUT   (Excluding Blocking Term in Model)

Regression Statistics	
Multiple R	0.987
R Square	0.974
Adjusted R Square	0.958
Standard Error	1.780
Observations	14

Term in Model	Coefficients	Standard Error	t-Stat	P-value
Intercept	74.72	0.73	102.83	0.00
X_1	-7.83	0.63	-12.42	0.00
X_2	3.73	0.63	5.91	0.00
$X_1{}^*X_1$	-1.96	0.66	-2.99	0.02
$X_2{}^*X_2$	-6.47	0.66	-9.83	0.00
$X_1{}^*X_2$	2.35	0.89	2.64	0.03

FIGURE 10.7: Spreadsheet Regression Output for Polypropylene Pyrolysis Example (Reproduced from the Bottom of Figure 9.13)

Since we already know $(\mathbf{X}^T\mathbf{X})^{-1}$ (which was calculated when we estimated the coefficients in our quadratic model—see Section 10.2), we can now use Equation 10.12 to estimate $Var(\hat{\mathbf{B}})$:

$$\text{Var}(\hat{\mathbf{B}}) = \begin{bmatrix} 0.167 & 0 & 0 & -0.080 & -0.080 & 0 \\ 0 & 0.125 & 0 & 0 & 0 & 0 \\ 0 & 0 & 0.125 & 0 & 0 & 0 \\ -0.083 & 0 & 0 & 0.135 & 0.010 & 0 \\ -0.083 & 0 & 0 & 0.010 & 0.135 & 0 \\ 0 & 0 & 0 & 0 & 0 & 0.250 \end{bmatrix} \times 3.168$$

or

$$\text{Var}(\hat{\mathbf{B}}) = \begin{bmatrix} 0.528 & 0 & 0 & -0.263 & -0.263 & 0 \\ 0 & 0.396 & 0 & 0 & 0 & 0 \\ 0 & 0 & 0.396 & 0 & 0 & 0 \\ -0.263 & 0 & 0 & 0.428 & 0.032 & 0 \\ -0.263 & 0 & 0 & 0.032 & 0.428 & 0 \\ 0 & 0 & 0 & 0 & 0 & 0.792 \end{bmatrix}.$$

As mentioned earlier, the diagonal terms in this matrix give us the variances of each of the coefficients in our model. We actually want the standard deviations, so we must take the square root of each diagonal element. If we do this, we get the standard deviations listed in Figure 10.7 in the column labeled "Standard Error" immediately after the coefficients. For example, the standard deviation of $\hat{b}_1$ is $\sqrt{0.396} = 0.629$. This computation was actually performed by the spreadsheet program, Excel, as part of the regression analysis.

Once we have all of the standard deviations of the coefficients, all that remains is to calculate the t-statistics, $t_i = \hat{b}_i/s_{\hat{b}_i}$, and compare them to the critical (tabled) value with the appropriate degrees of freedom. The t-statistics for our example (which were also computed by the spreadsheet program) are listed in Figure 10.7 in the column labeled "t-Stat." For example $t_1 = \hat{b}_1/s_{\hat{b}_1} = -7.83/0.629 = $ -12.42. The critical value of t for 8 degrees of freedom and a 95% confidence level (two-sided) is 2.306. Since all of the t-statistics are greater than this value, all of the coefficients are significant and should be kept in the model. It should be noted that the spreadsheet program even computed the significance level, denoted by "P-value."

If any coefficients are not significantly different from zero, then those coefficients should be dropped, *one-at-a-time* starting with the coefficient that has the smallest t-value. The regression is repeated until only statistically significant coefficients remain. If there are many coefficients that are not significant, this procedure will require many iterations, and it will be tedious. But doing the analysis in steps is necessary. The reason that only one term is dropped at a time (and the regression analysis repeated) is that the remaining coefficients may change and their significance will likely change (as well as the estimate of s). This may affect your conclusions about the significance of the remaining terms. It should be noted that most statistical software packages will do this type of stepwise regression automatically. Therefore, a statistics package is probably a worthwhile investment if this type of analysis is done reasonably often.

One further bit of advice is in order on dropping terms from the model that are not statistically significant. It may happen that a linear term in a model is not significant even though a quadratic term or an interaction term with the same variable is significant. In this event, which is not very common, it is recommended that you do not drop the linear term. Again, this recommendation is open to debate, but it is the opinion of the authors that it is a good one. The best analogy is to trimming a tree: you trim the branches not the trunk. The same reasoning applies to a mathematical model. You trim a particular variable in a model starting with its higher order terms (the quadratic and interactions), not with the main term (the linear term).

This procedure was carried out for our polypropylene pyrolysis example, and the output from the regression analysis was shown in Figure 9.13 in Section 9.5. After fitting the full model, including a term for the block, the only coefficient that was not significant was for the block term. This term was dropped from the model and the regression repeated. All the coefficients in the reduced model were significant, so the regression analysis was complete after only one term was deleted.

Sometimes, in the analysis of experimental data, there is a variable that looks questionable. In other words, most of the terms in the model for that variable are not significant, but there is a term (or two) that is just barely statistically significant, and you wonder if the variable is really needed. After all, if a variable appears in four terms in the model, it is roughly four times as likely that one of the terms will be statistically significant, so α is not really 0.05 for that variable. In a case like that, the whole variable should be tested for statistical significance. This is done by fitting the model with and without the variable under scrutiny. If m terms are dropped when the variable is deleted (i.e., all interactions involving the variable in addition to the linear and quadratic terms), the model will no longer fit as closely to the data, and *SSE* will get bigger. This happens whether the variable is significant or not. The statistical test is based on the fact that *SSE* increases by σ^2 (on the average) for every unnecessary term in the model that is dropped. When m terms are dropped, *SSE* should increase by $m\sigma^2$. If the terms are needed in the model, then *SSE* will increase by more than that. Thus, the increase in *SSE* is our gauge of how important the terms are to our model.

The statistical test is to estimate σ^2 by dividing the increase in *SSE* by m. This estimate is compared to our estimate of σ^2 from the regression of our bigger model. The ratio of the two estimates is distributed as an F with m and ν_R degrees of freedom. An example of this procedure is shown in Figure 10.8 for a response surface study with three independent variables. It can be seen in the example that X_3 was found to be unimportant, even though two of the terms involving X_3 were marginally significant. A final regression (not shown in the figure) was run deleting the not significant X_1^2 term (in addition to all X_3 terms).

10.6 Exploring the Response Surface

After all of the regression analysis and model checking has been done for a given set of data, you are left with a mathematical model that (hopefully) describes your system adequately. What remains to be done is to use that model to understand the system under study and perhaps optimize it. The best way to get an intuitive feeling for what the model is saying about how your system behaves is to make a plot or plots of the response. People are much more capable of grasping the content of a graph than an equation. The graphical interpretation of responses was already discussed in Section 9.6, and so it will not be delved into again.

On the other hand, when there are four or more factors or independent variables in a response surface equation, it becomes increasingly difficult to visualize the relationships with contour plots or other graphs, since there would be so many possible "slices" of the full surface to show. In this situation, a mathematical interpretation of the response surface may yield more satisfactory results.

REGRESSION ANALYSIS (Full Model)

R Squared			0.957	
Standard Error			2.934	Variance of Y = 8.611
Observations			25	$SSE = 15(8.611) = 129.16$
Degrees of Freedom			15	

Term in Model	Coefficients	Standard Error	t-Stat	t(15,0.95) = 2.131
Intercept	15.14			
X_1	-7.70	0.85	-9.09	
X_2	-11.70	0.85	-13.80	
X_3	-1.87	0.85	-2.20	
X_1*X_1	0.72	1.30	0.55	
X_2*X_2	5.17	1.30	3.96	
X_3*X_3	-0.28	1.30	-0.22	
X_1*X_2	5.21	1.04	5.02	
X_1*X_3	2.41	1.04	2.33	
X_2*X_3	0.81	1.04	0.78	

REGRESSION ANALYSIS (X3 Deleted)

R Squared			0.925	
Standard Error			3.428	Variance of Y = 11.748
Observations			25	$SSE = 19(11.748) = 223.22$
Degrees of Freedom			19	

Term in Model	Coefficients	Standard Error	t-Stat	t (19,0.95)= 2.093)
Intercept	15.08			
X_1	-7.70	0.99	-7.78	
X_2	-11.70	0.99	-11.80	
X_1*X_1	0.64	1.47	0.44	
X_2*X_2	5.09	1.47	3.46	
X_1*X_2	5.21	1.21	4.30	

Extra Sum of Squares (Delete X_3) =   94.06 $F(X_3) = (94.06/4) / 8.611 = 2.73$
Extra Degrees of Freedom, m =   4 $F (4,15,0.95) = 3.06$

FIGURE 10.8: Checking the Significance of X_3 in a Three-Factor Response Surface Study

10.6.1   Analytical Interpretation of Response Surfaces

It is known from elementary calculus that the maxima or minima of a simple quadratic equation in one variable, x, can be obtained by differentiating the equation with respect to x, setting the result equal to zero and solving for x. This procedure can be extended to higher dimensional quadratic equations also. If the quadratic equation is written in the form

$$Y = b_0 + \mathbf{x}\mathbf{B}_L + \mathbf{x}\mathbf{B}_Q\mathbf{x}^T \qquad (10.13)$$

where $\mathbf{x} = [\, X_1 \;\; X_2 \;\; \cdots \;\; X_k \,]$ is the vector of the coded level of each of the independent variables,

$$\mathbf{B}_L = \begin{bmatrix} b_1 \\ b_2 \\ \vdots \\ b_k \end{bmatrix} \text{ is the vector of coefficients of the linear terms, and}$$

$$\mathbf{B}_Q = \begin{bmatrix} b_{11} & b_{12}/2 & \cdots & b_{1k}/2 \\ & b_{22} & \cdots & b_{2k}/2 \\ & & \ddots & \\ sym. & & & b_{kk} \end{bmatrix}$$

is a symmetric matrix with the coefficients of the squared terms on the diagonal and half the coefficients of the cross product terms on the off-diagonals. Then the solution for X_1, X_2, ..., X_k, after differentiating Equation 10.13 with respect to each X (X_1, X_2, ..., X_k), and setting each result equal to zero, is obtained by solving the k simultaneous linear equations for the stationary point. The result, in matrix form, is:

$$\mathbf{x}_S = -(1/2)\mathbf{B}_Q^{-1}\mathbf{B}_L. \qquad (10.14)$$

As an example, consider the two variable quadratic equation

$$Y = 74.33 + 1.99\, X_1 + 0.45\, X_2 - 3.15\, X_1^2 - 2.12\, X_1 X_2 - 1.88\, X_2^2$$

where: Y = Yield of a chemical reaction,
$\quad\;\; X_1$ = (Aqueous to organic wt. ratio $-$ 0.5)/0.5,
and $\quad X_2$ = (Temperature $-$ 113 °C)/18.

In this case $\mathbf{B}_L = \begin{bmatrix} 1.99 \\ 0.45 \end{bmatrix}$, $\quad \mathbf{B}_Q = \begin{bmatrix} -3.15 & -1.06 \\ -1.06 & -1.88 \end{bmatrix}$, $\quad$ and

$$\mathbf{x}_S = -\frac{1}{2}\begin{bmatrix} -3.15 & -1.06 \\ -1.06 & -1.88 \end{bmatrix}^{-1}\begin{bmatrix} 1.99 \\ 0.45 \end{bmatrix} = \frac{1}{2}\begin{bmatrix} -0.3918 & 0.2209 \\ 0.2209 & -0.6565 \end{bmatrix}\begin{bmatrix} 1.99 \\ 0.45 \end{bmatrix} = \begin{bmatrix} 0.340 \\ -0.072 \end{bmatrix}.$$

These are the coordinates of the maximum as can be seen in Figure 10.9.

For a three-variable example, consider the chemical reaction example that was discussed in Section 9.6. For convenience, the equation derived experimentally is repeated here:

$$\hat{Y} = 85.83 - 0.48X_1 - 0.75X_2 - 0.85X_3 - 1.29X_1^2 + 0.06X_2^2 + 0.66X_3^2$$
$$- 0.30\, X_1 X_2 - 0.10\, X_1 X_3 + 1.10\, X_2 X_3.$$

In this case $\mathbf{B}_L = \begin{bmatrix} -0.48 \\ -0.75 \\ -0.85 \end{bmatrix}$ and $\mathbf{B}_Q = \begin{bmatrix} -1.29 & -0.15 & -0.05 \\ -0.15 & -0.06 & 0.55 \\ -0.05 & 0.55 & 0.66 \end{bmatrix}$

so that $\mathbf{x}_S$ is calculated to be:

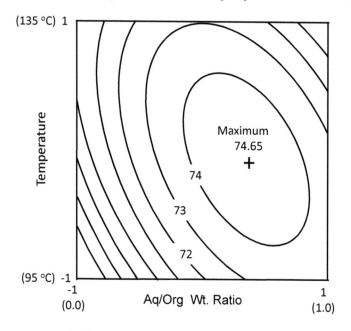

FIGURE 10.9: Contour Plot of Yield

$$\mathbf{x}_S = -\frac{1}{2} \begin{bmatrix} -1.29 & -0.15 & -0.05 \\ -0.15 & -0.06 & 0.55 \\ -0.05 & 0.55 & 0.66 \end{bmatrix}^{-1} \begin{bmatrix} -0.48 \\ -0.75 \\ -0.85 \end{bmatrix}$$

$$\mathbf{x}_S = \frac{1}{2} \begin{bmatrix} -0.7908 & 0.2140 & -0.2383 \\ 0.2140 & -2.5558 & 2.1461 \\ -0.2383 & 2.1416 & -0.2913 \end{bmatrix} \begin{bmatrix} -0.48 \\ -0.75 \\ -0.85 \end{bmatrix} = \begin{bmatrix} -0.2090 \\ 0.0045 \\ 0.6244 \end{bmatrix}.$$

Notice (by comparison with Figures 9.16–9.18) that this is not the coordinates of the maximum yield, which occurs at approximately $X_1 = 0$, $X_2 = -1$, and $X_3 = -1$. In general, $\mathbf{x}_S$ is called the stationary point of the response surface, and it must be determined if it is a maximum, minimum, or saddle point by taking the second derivatives of the quadratic equation. Examination of these second derivatives is equivalent to examination of the roots of the characteristic equation for $\mathbf{B}_Q$. The characteristic equation for $\mathbf{B}_Q$ is:

$$\left| \mathbf{B}_Q - \lambda \mathbf{I} \right| = 0 \tag{10.15}$$

where the $|\ |$ refers to the determinant of the matrix $(\mathbf{B}_Q - \lambda \mathbf{I})$. The roots of this equation are the eigenvalues, $\lambda_1, \lambda_2, \ldots, \lambda_p$. If all of the roots (eigenvalues) are positive, the stationary point, $\mathbf{x}_S$, represents a minimum of the response surface. If all of the roots are negative, it indicates that the stationary point, $\mathbf{x}_S$, is the maximum of the response surface. Finally, if some of the roots are positive while others are negative, it indicates that the stationary point, $\mathbf{x}_S$, is a saddle point in the response surface (which is neither a minimum nor a maximum).

For the two-factor example, the characteristic equation is:

$$\begin{vmatrix} -3.15 - \lambda & -1.06 \\ -1.06 & -1.88-\lambda \end{vmatrix} = (-3.15-\lambda)(-1.88-\lambda)-(-1.06)^2 = 0$$

The roots of this quadratic equation are $\lambda_1 = -3.75$ and $\lambda_2 = -1.28$, indicating that $\mathbf{x}_S$ is a maximum. The eigenvalues of the matrix, $\mathbf{B}_Q$, for the second example (three-factor system) discussed in Section 9.6 and repeated above are $\lambda_1 = -1.31$, $\lambda_2 = -0.26$, and $\lambda_3 = +0.99$, which indicates that the stationary point, $\mathbf{x}_S$, is a saddle point. Computer programs for fitting quadratic equations of the form of Equation 10.13 automatically compute $\mathbf{x}_S$ and the eigenvalues of $\mathbf{B}_Q$.

If the purpose of constructing a response surface is to locate the maximum or minimum response, the stationary point may be very helpful. If $\mathbf{x}_S$ is within the experimental region (i.e., each component is within -1 to $+1$ for a Box–Behnken design or within $-\alpha$ to $+\alpha$ for a central composite design), and it is a maximum or minimum (whatever is sought), the purpose has been accomplished. This was the case when computing $\mathbf{x}_S$ for the two-factor example in this section. If the stationary point is out of the region, or not the maximum or minimum being sought, like it was for the three-factor example in this section, additional work may be necessary. If the maximum is on a boundary of the experimental region as in Figure 9.16, the stationary point of the reduced equation (after fixing time, $X_2 = 1$ and temperature, $X_3 = -1$) may be useful; otherwise numerical techniques described in the next subsection can be used.

10.6.2 Numerical Methods for Interpreting Response Surfaces

With the availability of modern computers, function minimization is often accomplished by numerical methods rather than analytically (as described in the last section). If the response is described as a linear function of the independent variables or factors, minimization or maximization within the experimental region is a linear programming problem which can be solved using the well-known simplex algorithm. When the response is described by a quadratic equation, then maximization, minimization, or constrained maximization or minimization can be accomplished using nonlinear programming. Most of these algorithms will work well on quadratic functions like those developed from response surface designs. Some commercially available computer programs for response surface analysis also have nonlinear programming algorithms.

10.7 Precision of Predictions

Determining the precision of a prediction calculated using our model was discussed in Chapter 8, Section 8.4.5. In summary, our model prediction at some set of conditions, $\mathbf{X}_o$, is calculated by $\hat{\mathbf{Y}} = \mathbf{X}_0\hat{\mathbf{B}}$. Although an equation has the appearance of being very precise, it clearly has some potential error because $\hat{\mathbf{B}}$ is not known exactly. The variability in $\hat{\mathbf{Y}}$ is given by Equation 8.47, which is repeated below as Equation 10.16:

$$Var(\hat{\mathbf{Y}}) = \mathbf{X}_0(\mathbf{X}^T\mathbf{X})^{-1}\mathbf{X}_0^T\sigma^2. \tag{10.16}$$

Remember that $\mathbf{X}$ is the n sets of conditions at which experiments were run, and thus it does not change, and it is known precisely (presumably). On the other hand, $\mathbf{X}_o$ is the collection of points at which we would like to calculate the variance of $\hat{\mathbf{Y}}$. That can be as little as only a single point or a very large number of points, and the matrix changes accordingly. It is totally up to us; we can choose any set of conditions of interest to us. The only thing to keep in mind is that the danger is great when extrapolating outside of the region of experimentation. The polynomial model is merely a flexible curve that should only be expected to be useful in the vicinity of our data.

Continuing with our polypropylene pyrolysis example, if we are interested in the curve of $\hat{Y}$ versus fluidization velocity, when the temperature is 525°C, we can calculate predictions (the curve) using the regression equation and the variability in those predictions using Equation 10.16. To make sure that the use of Equation 10.16 is clear, let us calculate the variability of $\hat{Y}$ for our polypropylene pyrolysis example at a fluidization velocity of 0.15 and a temperature of 525°C. Remembering our coding:

$$X_1 = [\text{temperature} - 525] / 100 = (525 - 525) / 100 = 0,$$
$$X_2 = [\text{velocity} - 0.50] / 0.25 = (0.15 - 0.50) / 0.25 = -1.4,$$
$$X_3 = X_1^2 = (0)^2 = 0,$$
$$X_4 = X_2^2 = (-1.4)^2 = 1.96, \text{ and}$$
$$X_5 = X_1 X_2 = (0)(-1.4) = 0.$$

The $\mathbf{X}_o$ matrix (being careful to include a "1" for the constant term) then becomes:

$$\mathbf{X}_o = \begin{bmatrix} 1 & 0 & -1.40 & 0 & 1.96 & 0 \end{bmatrix}.$$

When we use this $\mathbf{X}_o$ matrix in Equation 10.16, and substitute s^2 for σ^2, we get:

$$\text{Var}(\hat{\mathbf{B}}) = \begin{bmatrix} 1 & 0 & -1.4 & 0 & 1.96 & 0 \end{bmatrix} \begin{bmatrix} 0.53 & 0 & 0 & -.26 & -.26 & 0 \\ 0 & 0.4 & 0 & 0 & 0 & 0 \\ 0 & 0 & 0.4 & 0 & 0 & 0 \\ -.26 & 0 & 0 & 0.43 & 0.03 & 0 \\ -.26 & 0 & 0 & 0.03 & 0.43 & 0 \\ 0 & 0 & 0 & 0 & 0 & 0.79 \end{bmatrix} \begin{bmatrix} 1 \\ 0 \\ -1.4 \\ 0 \\ 1.96 \\ 0 \end{bmatrix}$$

or $\text{Var}(\hat{\mathbf{B}}) = [1.932].$

The error limits on the equation values are typically reported as plus or minus twice the standard deviation of $\hat{Y}$, which are approximate 95% confidence limits on the curve. For these values of X_1 and X_2, the error limits are $\pm 2\sqrt{1.932} = \pm 2.78$. Error limits for other values of fluidization velocity were calculated in a similar fashion, changing $\mathbf{X}_o$ appropriately. The results of those calculations are shown numerically in Table 10.4 and graphically in Figure 10.10.

TABLE 10.4
Calculated Error Limits for Polypropylene Pyrolysis Example

Fluidization Velocity	X_2	$\hat{Y}$	$s_{\hat{Y}}^2$	$s_{\hat{Y}}$	$\hat{Y} - 2s_{\hat{Y}}$	$\hat{Y} + 2s_{\hat{Y}}$
0.15	−1.4	56.83	1.93	1.39	54.05	59.61
0.25	−1.0	64.53	0.83	0.91	62.70	66.35
0.35	−0.6	70.15	0.54	0.74	68.68	71.62
0.45	−0.2	73.71	0.53	0.73	72.26	75.17
0.55	0.2	75.20	0.53	0.73	73.75	76.66
0.65	0.6	74.62	0.54	0.74	73.15	76.09
0.75	1.0	71.98	0.83	0.91	70.15	73.80
0.85	1.4	67.26	1.93	1.39	64.48	70.04

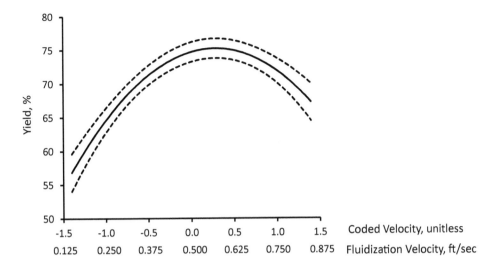

FIGURE 10.10: Predicted Yield and Error Limits for Polypropylene Pyrolysis Example (at a Temperature of 525 °C)

The other thing to notice from this example is that the predictions get less precise as you move away from the center of the experimental region. This is strictly true for linear models (including models with interactions). But for response surfaces (quadratic models), the relationship between the precision of predictions and the distance from the center of the experimental region is more complex. Remember that the number of center points in an RSM design is chosen to make the precision in the center of the experimental region roughly equal to the precision around the periphery of the region. Making predictions outside the experimental region runs into the issue of the precision of the predictions getting worse, but more importantly, it runs into the problem of the quadratic model no longer being a good approximation of the real response surface.

10.8 Summary

10.8.1 General Procedure for Analysis of Data from RSM Design

- FIT FULL QUADRATIC MODEL

  Use a computer program that is based on minimizing the residual sum of squares.

- CHECK DATA FOR OUTLIERS

  1. Calculate residuals.
  2. Construct a Half-Normal Plot of the residuals.
  3. If the residuals look OK (fall on a straight line), proceed with the next step (bullet).
  4. If any residuals are too big, throw out the worst data point and go back to the

first step (fit the full quadratic equation again to the reduced set of data). If at all possible, the data point that is being eliminated should be rerun in order to determine whether the result was spurious, or whether the model simply does not fit the data at those conditions.

- TRIM MODEL DOWN TO THE SMALLEST ADEQUATE FUNCTION

  1. Check each independent variable to see if it can be deleted. This is done by fitting the model with and without the variable under scrutiny. If m terms are dropped when the variable is deleted, you must see if *SSE* increased by significantly more than m×s². Note: s² is estimated from the fit of the bigger model (with ν degrees of freedom), and significance is judged by comparing the ratio of $(SS_{Increase}/\text{m})/s^2$ to an *F*-Distribution with m and ν degrees of freedom.

  2. Delete any *high order terms* (for the important variables) that are not significant. This should be done one term at a time, starting with the term that has the smallest *t*-value. Note: It may happen that a linear term is not significant even though an interaction or a quadratic term with the same variable is significant. In this case (not very common), it is recommended that you **do not drop** the linear term.

- CHECK MODEL ADEQUACY

  1. Check lack of fit to ensure that a quadratic model is adequate. This is done by breaking the residual sum of squares, *SSE*, and its associated degrees of freedom, ν_R, into two pieces: (1) the sum of squares from pure error (from replicates), SS_{PE}, and (2) the sum of squares due to lack of fit, SS_{LoF}, by difference ($SS_{LoF} = SSE - SS_{PE}$). Note: $\nu_{LoF} = \nu_R - \nu_{PE}$. The lack of fit variance, s^2_{LoF}, can then be calculated ($s^2_{LoF} = SS_{LoF}/\nu_{LoF}$) and checked to see if it is significantly bigger than $s^2 = s^2_{PE} = SS_{PE}/\nu_{PE}$. The significance is determined by comparison to an *F*-distribution with ν_{LoF} and ν_{PE} degrees of freedom.

  2. Check residuals for trends. This is done by plotting the residuals against any variable that makes sense (e.g., run order, X_1, etc.). If there are any trends, it is an indication that something should be added to the model.

- DISPLAY FINAL MODEL IN GRAPHICAL FORM

This is usually best done via contour plots if there are only two (or perhaps three) important variables.

10.8.2   Important Equations

Estimation of Coefficients, $\hat{\mathbf{B}}$ (Equation 8.36 and Equation 10.2)

$$\hat{\mathbf{B}} = (\mathbf{X}^T\mathbf{X})^{-1}\,\mathbf{X}^T\mathbf{Y}$$

Variability of Coefficients, $\hat{\mathbf{B}}$ (Equation 8.41 and Equation 10.12)

$$Var(\hat{\mathbf{B}}) = (\mathbf{X}^T\mathbf{X})^{-1}\,\sigma^2$$

Variability in Model Predictions, $\hat{\mathbf{Y}}$ (Equation 8.47 and Equation 10.16)

$$Var(\hat{\mathbf{Y}}) = \mathbf{X}_0(\mathbf{X}^T\mathbf{X})^{-1}\mathbf{X}_0^T\sigma^2$$

Approximate Variability in Model Predictions, $\hat{\mathbf{Y}}$ (Equation 8.48)

$$Var(\hat{\mathbf{Y}}) \approx c\sigma^2/n$$

10.8.3 Important Terms and Concepts

The following is a list of the important terms and concepts covered in this chapter:

quadratic equation
model adequacy
assumptions verified
lack-of-fit test
model trimming
$SS_{Increase}$ test (for a factor)
stationary point on surface
eigenvalues of $\mathbf{B}_Q$
error limits on predictions

10.9 Exercises

1. The Y data below were fit using a linear model in X. The regression output is also given below. Is the model adequate? Use a lack-of-fit Test to determine your answer.

X	Y	ANOVA			
-2	31		df	*SS*	*MS*
-2	33	Regression	1	4500	4500
-1	53	Residual	8	118	14.75
-1	53	Total	9	4618	
0	71				
0	69	**REGRESSION OUTPUT**			
1	83				
1	83		**Coeff**	**Std Error**	*t*-**Stat**
2	93	Intercept	66	1.214	54.344
2	91	X	15	0.859	17.467

2. The following data were taken to study the effects of three factors on the yield of a chemical reaction. Note: the goal is to minimize the % impurity.

Run	Run Order	Block	X_1	X_2	X_3	% Impurity
1	6	−1	−1	−1	−1	8.55
2	2	−1	1	−1	−1	31.33
3	4	−1	−1	1	−1	8.24
4	9	−1	1	1	−1	30.89
5	7	−1	−1	−1	1	27.79
6	5	−1	1	−1	1	29.13
7	8	−1	−1	1	1	27.37
8	1	−1	1	1	1	30.77
9	11	−1	0	0	0	19.27
10	3	−1	0	0	0	17.37
11	10	−1	0	0	0	17.76
12	14	1	−1.68	0	0	10.93
13	17	1	1.68	0	0	31.25
14	20	1	0	−1.68	0	20.74
15	18	1	0	1.68	0	19.61
16	12	1	0	0	−1.68	25.52
17	13	1	0	0	1.68	37.06
18	15	1	0	0	0	20.1
19	19	1	0	0	0	18.53
20	16	1	0	0	0	20.83

(a) Determine the best model for describing the system based on the data.

(b) Verify that any assumptions in your analysis are reasonable (via plots of residuals), and that there is no lack of fit for your model.

(c) Use that model to determine optimum operating conditions via graphical display of the equation.

(d) What is your expected yield at the optimum conditions, and what are its error limits?

3. Yan et al. [2011] performed a central composite experiment to determine the optimal conditions for enzymatic saccharification of food waste. The coded factors were $X_1=((\text{Glucoamylase load(U/g)} - 120)/20)$, $X_2=((\text{Incubation time}-2.0)/0.5)$, $X_3=((\text{Temperature} - 55)/5))$, and $X_4=((\text{pH}-5)/0.5)$, and the response was $y=$ Reducing sugar concentration (g/L). The objective was to identify conditions that would maximize the response. The design and resulting response data in standard order are shown in the table below.

(a) Fit the full quadratic model to the data.

(b) Identify the conditions that maximize the response (Reducing sugar concentration).

Run	X_1	X_2	X_3	X_4	Y
1	−1	−1	−1	−1	79.451
2	1	−1	−1	−1	119.125
3	−1	1	−1	−1	115.220
4	1	1	−1	−1	154.421
5	−1	−1	1	−1	103.253
6	1	−1	1	−1	125.526
7	−1	1	1	−1	112.332
8	1	1	1	−1	146.967
9	−1	−1	−1	1	105.214
10	1	−1	−1	1	117.396
11	−1	1	−1	1	127.318
12	1	1	−1	1	138.615
13	−1	−1	1	1	114.216
14	1	−1	1	1	128.998
15	−1	1	1	1	127.814
16	1	1	1	1	136.412
17	−2	0	0	0	96.132
18	2	0	0	0	154.218
19	0	−2	0	0	97.425
20	0	2	0	0	152.396
21	0	0	−2	0	99.229
22	0	0	2	0	114.392
23	0	0	0	−2	114.511
24	0	0	0	2	134.697
25	0	0	0	0	152.397
26	0	0	0	0	152.759
27	0	0	0	0	152.641
28	0	0	0	0	153.985
29	0	0	0	0	153.723
30	0	0	0	0	154.367

Chapter 11

Sequential Experimentation

11.1 Introduction

The purpose of experimentation is to increase knowledge about a phenomenon. The choice of an appropriate experimental design to be used in experimentation will be influenced by how much knowledge is already available and what additional knowledge is sought. Figure 11.1 (a compression of Figure 1.4) represents the state of knowledge as a continuum ranging from a very low state of knowledge on the left to a very high state of knowledge on the right. Under the knowledge line some appropriate stages of experimentation and their purposes are indicated.

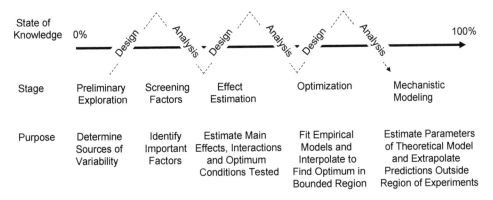

FIGURE 11.1: Experimentation and the State of Knowledge

When at a very low state of knowledge (on the far left of Figure 11.1), where there is little idea about what potential factors may influence the responses of interest, preliminary experiments to determine sources of variability (as described in Chapter 6) are useful. The example in Section 6.8 describes a situation where discovering the sources of variability led to ideas about what factors may have the most influential effect on the response of interest. Further experimentation with these factors led the experimenters to develop an equation that helped them determine conditions to produce catalyst supports appropriate for different applications.

When planning experiments where there may be many potential factors, Occam's razor indicates that changes in the response can likely be explained by changes in a small subset of the potential factors. In the context of experimental design, this principle is often called *effect sparsity* and it can help us in planning experiments at the screening stage (indicated as the second stage in Figure 11.1). At this stage it is generally wasteful to plan large, full factorial designs or resolution V fractions with many factors. It is better to start small. Begin with an economical screening design and add to it later if needed. With this in mind, we can plan a conditional sequence of experimental designs that will allow us to estimate the important main effects and two-factor interactions with the minimal number of experiments. The designs useful at this stage were described in Chapter 7.

Usually no more than 25% of the budget for a research program should be expended on the initial screening experiments. In initial screening experiments, things may not work out as planned due to poor choice of factor levels or other procedural problems. Even after successful completion of a set of screening experiments, follow up experiments may be needed to identify significant interaction effects and allow for fitting a prediction equation. For this reason, plenty of time should be left to repeat botched experiments and conduct follow-up trials with important variables. The example in Chapter 1, Section 6, illustrates a sequence of experiments in chemical process development. It started with a 12 run screening design, and was followed by two additional sets of experiments that led to optimization of the process.

Although many of the examples in this book have illustrated specific experimental designs (and the associated analysis and interpretation of data) appropriate at one stage of the knowledge continuum shown in Figure 11.1, in practice experimentation often consists of a sequence of experimental designs and analysis as knowledge about the phenomenon increases.

11.2   Augmenting Screening Designs to Resolve Confounding

Since the confounding patterns for fractional factorials can be used for identifying the interactions confounded with any large, unassigned effects, these designs are ideal for defining a conditional sequence of experiments. Figure 11.2 is a flow diagram that illustrates a simple sequence that starts with the resolution IV, half fraction (2^{4-1}).

The initial experiment in this sequence of designs (shown in Table B.1-6) only involves four factors. If, after running this design, the data analysis shows that no unassigned effects (that represent strings of confounded two-factor interactions) are significant, experimentation stops and only the main effects are considered when interpreting the effects and defining a prediction model. If, on the other hand, one or more unassigned or interaction strings appear to be significant, a different approach is required.

Experience has shown that significant two-factor interactions are more likely when at least one of the main effects involved in the interaction is also significant. For example, when an experiment is conducted using the 2^{4-1} plan in Table B.1-6 and the largest effects found are main effects X_1, X_3, and the unassigned effect E_1 (that represents the confounded interactions $X_1 \times X_3$ and $X_2 \times X_4$), then E_1 is more likely to represent the $X_1 \times X_3$ interaction than the $X_2 \times X_4$ interaction. The fact that interactions usually occur between factors that have significant main effects is called *effect heredity*. When an interaction occurs between two factors that do not have significant main effects, it means the effect of one factor is equal and opposite depending on the level of the other factor (as shown in the bottom row and second and third columns of Figure 3.7). This situation is rare.

In Figure 11.2, when a confounded string of interactions appears to be significant, it should be checked to see if there is a clear interpretation that can be described by effect heredity. If so, the significant main effects and interactions should be interpreted, and a prediction model formed containing those significant effects and interactions. If predictions from this model match the results of confirmatory experiments, the sequence is complete.

If the prediction model is not accurate, or no simple interpretation of effects and interactions

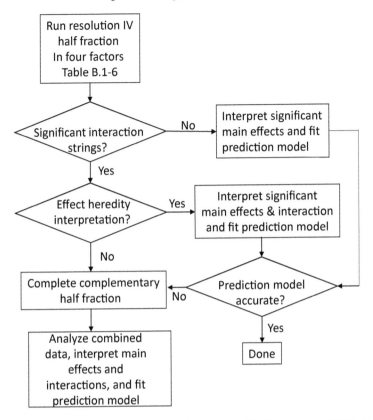

FIGURE 11.2: Sequence of Experiments Starting with Resolution IV Half Fraction

following effect heredity can be proposed, a second step consisting of running the remaining experiments needed for a full factorial should be completed. The sequence shown in Figure 11.2 can be illustrated by returning to the example used in Section 7.6. This was a blocked design, but the runs made on day 2 are exactly the resolution IV half-fraction design for four factors listed in Table B.1-6.

If the runs on day 2 had been completed on day 1, Step 1 from Figure 11.2 would have been followed. The next step would be to check to see if any of the unassigned effects were large. Table 11.1 shows the analysis of the day 2 data. The standard error of the effects, s_E, was calculated by pooling the three unassigned effects (none of which appear large) as was done in Section 3.10. All four main effects appear to be significant, while none of the interaction strings are significant. Therefore, following the flow diagram in Figure 11.2 no further experimentation would be necessary and the main effects could be interpreted alone.

Notice that the main effects calculated from the half fraction in Table 11.1 are very similar to the effects calculated from the full factorial in Table 7.22. The conclusions and interpretation after eight experiments would be the same as that given in Section 7.6, and we see that by assuming effect sparsity and using a sequential approach the experimental effort could have been cut in half.

Admittedly, the plan for a sequence of experiments that follows the flow diagram in Figure 11.2 is a simple case because it starts with only four factors. However, Figure 11.3 is a similar diagram that is much more general. It can be followed to plan a sequence of experiments starting with any number of factors.

Figure 11.3 has two entry points. The first entry point is Step 1, where the initial design is a resolution III fractional factorial like those shown in Tables B.1-5, B.1-7, and B.1-10 or a

TABLE 11.1

Analysis of Half Fraction Consisting of Day
2 Data from Table 7.22

Term	Effect	t-value	P-value
Constant	80.00		
X_1	64.50	11.37	0.001
X_2	-24.00	-4.23	0.024
X_3	-62.50	-11.02	0.002
X_4	-21.00	-3.70	0.034
$E_1=12+34$	3.50		
$E_2=13+24$	-8.00		
$E_1=14+23$	-4.50		

$$s_E=5.67$$

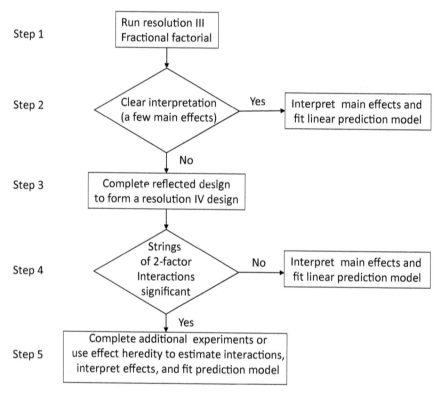

FIGURE 11.3: Sequence of Experiments Starting with Resolution III Screening Design

resolution III Plackett–Burman design like those shown in Tables B.1-1 to B.1-4. The second
entry point is Step 3, where the initial experiment is a resolution IV fractional factorial like
those shown in Tables B.1-8 or B.1-11.

The decision on what entry point to use depends upon the number of factors under study
and the number of runs, n_F, which is calculated on the basis of information and precision
described in Section 4.2. We will first discuss the use of the flow diagram as if beginning at
Step 3. Later we will show an example of following the flow diagram from Step 1.

The chemical process experiment presented in Section 7.4.6 was a 1/8th fractional factorial of resolution IV (from Table B.1-8). Planning a sequence of experiments by starting with a resolution IV experiment like this would follow the flow diagram in Figure 11.3 by entering at Step 3. The next step would be to analyze the data and determine if any unassigned effects (that represent strings of two-factor interactions) are significant. In the example chemical process experiment, there were no apparent interactions, and therefore the next step according to Figure 11.3 would be to stop experimentation and interpret the main effects and develop a prediction equation. This is exactly what was done in Section 7.4.6. Because there were no significant interactions, the sequential plan allowed essentially the same information to be obtained with a 16-run 1/8th fraction as would have been obtained with a 128-run 2^7 full factorial.

If some of the interaction strings appear to be significant in a resolution IV design, the next step on the flow diagram would be number 5. At Step 5, additional experiments are completed to allow separate estimation of each of the interactions confounded with any significant unassigned factor, or the effect heredity principle is used to determine which interaction would most plausibly represent each significant string of confounded interactions. An example of how to choose additional experiments to allow separate estimation of confounded interactions was shown in Table 7.16 and discussed in Section 7.4.7. On the other hand, if the experimenter feels that he can use the effect heredity principle to guess which interaction (in each significant string of confounded interactions) is the important one, a set of confirmation experiments could be run to verify the guess. Guessing may seem to be contradictory to the scientific principles we are trying to establish with experimentation, but experience has shown that effect heredity usually holds in experimental data.

Another principle that can help an experimenter make an educated guess as to which interaction best represents a significant unassigned effect is called the *hierarchical ordering principle*. By hierarchical ordering we mean that if several interactions are confounded with a significant unassigned effect, one of the lowest order interactions will normally be the important one. A two-factor interaction means the effect of one factor depends upon the level of another factor, while a three-factor interaction means the effect of one factor depends on the combination of levels of two other factors. A four-factor interaction means the effect of one factor depends on the combination of levels of three other factors. The experience summarized by the hierarchical ordering principle would lead one to believe that lowest order interactions confounded with a significant unassigned effect are more likely to be important than higher order interactions.

To illustrate the use of the hierarchical ordering principle and effect heredity paradigm to choose a set of confirmatory experiments, let's return to the chemical process example shown in Section 7.4.6 and based on the design in Table B.1.8. In that example, a resolution IV 2^{7-3} experiment was completed finding that the significant main effects (in order of magnitude) were X_4 = Stir, X_1 = pH, X_6 = RPM, and X_7 = Pump. Suppose that the unassigned effect E_1 also appeared to be significant. Ignoring the unused factor 8, the defining relation for Table B.1-8 is I = 1235 = 2346 = 1347 = 1456 = 2457 = 1267 = 2457. Multiplying on both sides of the defining relation by 12, we find all the interactions confounded with unassigned Effect E_1 to be 12 + 35 + 67 + 1346 + 2347 + 2456 + 1457. (Note: Only the two-factor interactions were listed in the CONFOUNDINGS below Table B.1-8.)

First the hierarchical ordering principle would lead one to believe that E_1 represents one of the two-factor interactions confounded with it (i.e., $X_1 \times X_2$, $X_3 \times X_5$, or $X_6 \times X_7$) rather than one of the four-factor interactions. Next, the effect heredity paradigm would rule out the $X_3 \times X_5$ interaction since neither main effect X_3 = Addition Time, nor X_5 = H$_2$O level, was significant. Therefore, E_1 could represent either the $X_1 \times X_2$ interaction or the $X_6 \times X_7$.

If the $X_1 \times X_2$ was guessed to be the important interaction, the prediction equation would be:

$$Y(\%impurities) = 1.09 - 0.38X_1 - 0.54X_4 - 0.24X_6 - 0.23X_7 + 0.165X_1X_2, \quad (11.1)$$

where 0.165 is one half of the effect for the unassigned E_1. With this prediction equation, the levels of the process variables that are predicted to yield the lowest predicted impurities are ($X_1 = +$, $X_2 = -$, $X_4 = +$, $X_6 = +$, and $X_7 = +$).

If, on the other hand, the $X_6 \times X_7$ was guessed to be the important interaction, the prediction equation would be:

$$Y(\%impurities) = 1.09 - 0.38X_1 - 0.54X_4 - 0.24X_6 - 0.23X_7 + 0.165X_6X_7. \quad (11.2)$$

With this prediction equation, another combination of levels of the process variables ($X_1 = +$, $X_2 = +$, $X_4 = +$, $X_6 = +$, and $X_7 = +$) is predicted to result in an equally low level of impurities as the levels predicted from Equation 11.1.

Finally, to determine whether the $X_1 \times X_2$ interaction or the $X_6 \times X_7$ best represents the unassigned effect E_1, a set of confirmatory experiments would be conducted at both the combinations of factor levels shown above. If the confirmatory experiments at the factor levels predicted to be best by Equation 11.1 resulted in the lowest impurities, it can be assumed that the $X_1 \times X_2$ interaction best represents E_1. On the other hand, if the two sets of confirmatory experiments resulted in equally low impurity levels, it can be assumed that the $X_6 \times X_7$ interaction best represents E_1. In either case no further experiments would be needed, and an explanation or interpretation of the assumed interaction should be made.

Because the hierarchical ordering principle and the effect heredity principle have been shown to be true in the majority of cases, their use often reduces the number of follow-up experiments needed. Recall the simple example of a half fraction presented in Section 7.4.3. In that example the hierarchical ordering principle was followed in assuming the main effects were responsible rather than the two-factor interactions. The confirmation experiments showed the assumption was correct, eliminating the need for further experiments.

Now that we have discussed the sequence of experiments that can be planned by starting at Step 3 in Figure 11.3, let's continue by discussing how a sequence of experiments can be planned starting at Step 1. At Step 1, we start with a resolution III fractional factorial (i.e., Table B.1-5, B.1-7, or B.1-10 or Tables B.1-1 to B.1-4.). Resolution III designs are the most economical possible. After running a resolution III design, if there appears to be only one or two main effects significant, we move to the right at Step 2. At that point, it is usually safe to assume that no interactions are important. So experimentation stops, the main effects are interpreted, and a simple linear prediction equation is written. When more than two main effects appear significant in a resolution III design, or if the main effects model does not accurately predict confirmation experiments, the assumption that interactions are negligible may not be valid. To see why, let's consider a specific case.

In resolution III designs, each main effect that we can estimate is confounded with a string of two-factor interactions. For example, looking at the confoundings from Table B.1-5, we see:

$$X_1 = 1 + 24 + 35 + 67$$
$$X_2 = 2 + 14 + 36 + 57$$
$$X_3 = 3 + 15 + 26 + 47$$
$$X_4 = 4 + 12 + 37 + 56$$
$$X_5 = 5 + 13 + 27 + 46$$
$$X_6 = 6 + 17 + 23 + 45$$
$$X_7 = 7 + 16 + 25 + 34$$

If Factors 1 and 2 had the largest effects but Factor 4 also had what seemed to be a significant effect, then by the effect heredity paradigm, it would be hard to say whether the three main effects were important or whether Factors 1 and 2 were important and the fourth factor looked significant because it was confounded with the interaction of Factors 1 and 2. The only way we can answer a question like this is to complete additional experiments.

If a resolution III fractional factorial is augmented by its mirror image or reflected design, [Box et al., 1978] have shown that the main effects can be de-confounded from the strings of two-factor interactions they are confounded within the original design. A mirror image or reflected design is a design with all the signs reversed from the original design. In other words, replace each − with a + in the original design and each + with a −. When we combine an eight-run resolution III fractional factorial with its mirror image, we will have a 16-run design. Likewise when we combine a 16-run or 32-run resolution III design with their mirror images we get a 32- or 64-run design, respectively. A combined design with double the number of runs in the original resolution III design will be a resolution IV design that allows estimation of: 1) the main effects (clear of two-factor interactions), 2) the strings of two-factor interactions that were previously confounded with main effects, and 3) the block effect that represents the average difference between the two sets of experiments (original resolution III design and mirror image design).

After the reflected design is completed and combined with the data from the original resolution III design, we are again at Step 3 in Figure 11.3. To illustrate the whole sequence in Figure 11.3 starting at Step 1, we will present (in the next section) an excellent example of sequential experiments published by [Box and Hunter, 1961].

11.3   Application of Sequential Experimentation in a Process Start-up

In the start up of a new manufacturing unit, considerable difficulty was experienced at the filtration stage. Similar units operated satisfactorily at other sites, but this particular unit, although apparently similar in most major respects to the other units, gave a crude product which required much longer filtration times. Meetings were held to discuss possible explanations and to consider ways of curing the trouble. The following variables were proposed as possibly being responsible:

1. **Water Supply**

   The new plant used piped water from the local municipal reservoir. An alternate but somewhat limited supply of water was available from a local well. It was proposed that the effect of changing to the well water should be tried since it was argued that the well water corresponded more closely to the water used at other sites.

2. **Raw Material**

   The raw material used was manufactured on the site and it was suggested that this might be in some way deficient. It was proposed that raw material which had been satisfactorily used in manufacturing the product at another site should be shipped in and tested locally.

3. **Temperature of Filtration**

This was not thought to be a critical factor over the range involved, and no special attempt to control this temperature had been made. However, the physical arrangement of the new process was such that filtration was accomplished at a somewhat lower temperature than had been experienced at other plants. By temporarily covering pipes and equipment, provision could be made to raise the temperature to the level experienced elsewhere.

4. **Recycle**
The only major difference between production facilities at the other plants and the present one lay in the introduction of a recycle stage which slightly increased conversion of the reagents prior to precipitation and filtration. Arguments were advanced which accounted for the longer filtration time in terms of this recycle stage. Arrangements could be made to temporarily eliminate the recycle stage.

5. **Rate of Addition of Caustic Soda**
Immediately prior to filtration, a quantity of caustic soda liquor was added resulting in precipitation of the product. The addition rate was somewhat faster with the new plant, but it was possible to reduce the rate of addition.

6. **Type of Filter Cloth**
The filter cloths employed in this plant were very similar to those used at the other sites. However, they did come from a more recently supplied batch, and it was suggested that their performance should be compared with cloths from previously supplied batches which were still available.

7. **Hold-up Time**
Prior to filtration, the product was held in a stirred tank. The average period of hold-up in the new plant was somewhat less than that used in the other plants, but it could easily be increased.

In the following list of factors, the minus version corresponds to the usual operation for the new plant and the plus version to the proposed level. Thus, we have:

TABLE 11.2
Factors for Eight-Run Resolution III Factorial Design

Factor	− Level	+ Level
X_1 = Water	Town	Well
X_2 = Raw material	On site	Other
X_3 = Temperature of filtration	Low	High
X_4 = Recycle	Included	Omitted
X_5 = Rate of NaOH addition	Fast	Slow
X_6 = Filter Cloth	New	Old
X_7 = Hold-up time	Low	High

A series of tests were run wherein each of these variables was changed to the proposed level while all other variables were held constant. However, the results were disappointing; the improvement sought was not found in any of the tests. There was increased pressure from management to solve the problem and the process engineers, being out of ideas, called in a consultant on the project. The consultant suggested they vary all of the seven variables or factors simultaneously in a 2^{7-3} 16-run fractional factorial design. The engineers were not willing to commit to 16 experiments since they were seeking a quick solution; however, with no other ideas, they compromised and agreed to run a 2^{7-4} 8-run fractional factorial design.

The 1/16 replicate of the 2^7 that was used is shown in Figure 11.4. It is the same design as that given in Table B.1-5 in Appendix B, obtained by using the generators: $4 = 12$, $5 = 13$, $6 = 23$, and $7 = 123$. The defining relations from the generators are: $I = 124 = 135 = 236 = 1237$. The pair-wise products are 2345, 1346, 347, 1256, 257, and 167. The three-way products are 456, 1457, 2467, and 3567. The four-way product is 1234567. So, the complete defining relation is:

$$I = 124 = 135 = 236 = 1237 = 2345 = 1346 = 347 = 1256 = 257 = 167 = 456 = 1457 = 2467 = 3567 = 1234567.$$

The eight experimental runs to be made were run in random order, and gave the filtration times listed in Figure 11.4.

	A	B	C	D	E	F	G	H	I	J
1	Run	Mean	X1	X2	X3	X4	X5	X6	X7	Filt. Time
2	1	1	-1	-1	-1	1	1	1	-1	68.4
3	2	1	1	-1	-1	-1	-1	1	1	77.7
4	3	1	-1	1	-1	-1	1	-1	1	66.4
5	4	1	1	1	-1	1	-1	-1	-1	81.0
6	5	1	-1	-1	1	1	-1	-1	1	78.6
7	6	1	1	-1	1	-1	1	-1	-1	41.2
8	7	1	-1	1	1	-1	-1	1	-1	68.7
9	8	1	1	1	1	1	1	1	1	38.7
10	SUMPRODUCT	520.7	-43.5	-11.1	-66.3	12.7	-91.3	-13.7	2.1	
11	Effect		65.09	-10.88	-2.775	-16.58	3.175	-22.83	-3.425	0.525	

FIGURE 11.4: 2^{7-4}(8-Run) Design and Results

The filtration times for Runs 6 and 8 were comparable to those experienced at other similar units and the process engineers were ready to conclude the project and recommend running at the conditions listed in Run 8. However, the consultant argued that maybe not all factors were important and the plus level of each factor was generally more expensive than the minus level (usual operation). In the long run, it would pay them to find out which factors were really important and which were not since they could then set the unimportant variables to their $(-)$ cheaper level.

The analysis of the data is shown at the bottom of Figure 11.4. The calculated effects along with the confounding patterns shown in Table B.1-5 are put together in Table 11.3 for convenience.

TABLE 11.3
Calculated Effects and Confounding Patterns for Filtration Experiments

Water	$X_1 = 1 + 24 + 35 + 67 =$	-10.9
Raw material	$X_2 = 2 + 14 + 36 + 57 =$	-2.8
Temperature of filtration	$X_3 = 3 + 15 + 26 + 47 =$	-16.6
Recycle	$X_4 = 4 + 12 + 37 + 56 =$	3.2
Rate of NaOH addition	$X_5 = 5 + 13 + 27 + 46 =$	-22.8
Filter Cloth	$X_6 = 6 + 17 + 23 + 45 =$	-3.4
Hold-up time	$X_7 = 7 + 16 + 25 + 34 =$	0.5

From Table 11.3 (and more clearly from the Lenth plot of effects in Figure 11.4) it can be seen that the estimates −10.9, −16.6, and −22.8 are beyond the ±ME limits and would be considered significant. The simplest interpretation of the results would be that the main effects of the factors 1, 3, and 5 were important. However, due to the confounding and the effect heredity paradigm, other interpretations are possible. For example, possibly the main effects of factors 3 and 5 along with their interaction (which is confounded with 1) were significant and responsible for the observed results. If that were true, and Factor 1 was not important, then it could be run at its cheaper level. An equivalent interpretation is that the main effects of factors 1 and 5 are significant along with their interaction. And yet another very reasonable interpretation is that factors 1 and 3 as well as their interaction are significant (and not Factor 5).

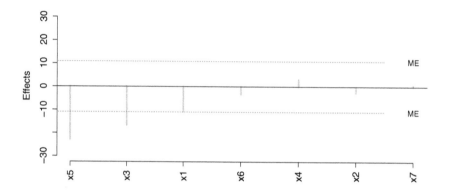

FIGURE 11.5: Lenth Plot of Effects in Figure 11.4

There was no way to determine which of these interpretations was correct without additional experiments. Therefore, the answer to the question at Step 2, in the flow diagram in Figure 11.3, is no, and the next step was Step 3, where a reflected design would be run to de-confound the strings of two-factor interactions.

In this experiment, unlike the copper plating example presented in Section 7.5.3, there was a higher cost to run at the high level of each factor, as stated earlier. Therefore, it would not have been economical in the long run to run the filtration unit with factors 1, 3, and 5 set at their high levels, if that were unnecessary. So, it was worth further investigation to find out which of the three factors (1, 3, or 5) were significant and must be run at their high levels and which (if any) were insignificant and could be run at their low levels.

Figure 11.6 shows the factor settings and results of the reflected design that was run to de-confound the main effects from the strings of two-factor interactions. Once again, if the two groups of data are combined, the main effects can be estimated free of two-factor interactions. And the groups of two-factor interactions listed in Table 11.3 can be estimated as well. In addition a block effect, which is the difference between the two groups of data, can be estimated.

Figure 11.7 is a worksheet for calculating the effects. The first seven columns are copied from Figures 11.4 and 11.6. The next seven columns, representing interaction strings, were obtained by multiplying (element-wise) the first two pairs of columns represented in each string of interactions. For example, the column labeled 12 was obtained by multiplying the X_1 column by the X_2 column, and it represents the sum of the 12, 37, and 56 interactions. The last contrast column labeled Block is −1 for each of the eight runs in the original

	A	B	C	D	E	F	G	H	I
1	Run	X1	X2	X3	X4	X5	X6	X7	Filt. Time
2	1	1	1	1	-1	-1	-1	1	66.7
3	2	-1	1	1	1	1	-1	-1	65.0
4	3	1	-1	1	1	-1	1	-1	86.4
5	4	-1	-1	1	-1	1	1	1	61.9
6	5	1	1	-1	-1	1	1	-1	47.8
7	6	-1	1	-1	1	-1	1	1	59.0
8	7	1	-1	-1	1	1	-1	1	42.6
9	8	-1	-1	-1	-1	-1	-1	-1	67.6

FIGURE 11.6: Reflected Design for Filtration Unit

resolution III design, and +1 for each run in the reflected design. The effects are calculated by the usual spreadsheet method using the SUMPRODUCT function. Remember that it gives the sum of products of the column of contrasts and the column of response values (the filtration times). For this reason, we had to put +1 and −1 as the factor settings in the spreadsheet rather than just + and −. The effects were calculated by dividing the sum of products by eight.

Similar results could be obtained using a regression program. The dependent variable for regression is filtration time, and the 15 columns of contrasts in Figure 11.7 are the independent variables. The calculated regression coefficients would be exactly one-half the effects shown in Figure 11.7.

Figure 11.8 is a normal plot of the effects from Figure 11.7, where it can be seen that Factors 1, 5, and the interaction $X_1 \times X_5$, that represents the sum of the interactions 15 + 26 + 47, appear to be the only significant effects. In the flow diagram in Figure 11.3, the answer to the question at Step 4 is yes. Step 5 is next, where either additional experiments will be run to de-confound the interactions or the effect heredity paradigm will be employed to define acceptable operating conditions for confirmation experiments.

It is possible the interaction effect was due to either 26 (the interaction between X_2 and X_6) or 47 (the interaction between (X_4 and X_7)), but in this case the plant engineers thought it was easier to believe that main effects 1 and 5 and their two-factor interaction were important (as the effect heredity paradigm would lead them to believe). Assuming the interaction was due to 15, no additional experiments were needed to de-confound the three interactions, and the two-way table of average values shown in Figure 11.9 helped the engineers discover an acceptable operating condition.

According to the representation of the results in Figure 11.9, if the high level of Factor 1 (well water) and the high level of Factor 5 (slow addition of caustic) were used, the filtration time should be equivalent to what it was at other similar plants. All other factors could be left at their lower (less expensive) level. This representation also explains why the desirable operating condition was not discovered with the earlier one-at-a-time trials that were performed. The two-way diagram shows the interaction to be very important and both Factor 1 and Factor 5 need to be at their high levels in order to achieve the desired result.

	A	B	C	D	E	F	G	H	I	J	K	L	M	N	O	P	Q
Run		X1	X2	X3	X4	X5	X6	X7	12+37 +56	13+27 +46	14+36 +57	15+26 +47	16+25 +34	17+23 +45	35+24 +67	Block	Filtration Time
1		-1	-1	-1	1	1	1	-1	1	1	-1	-1	-1	1	-1	-1	68.4
2		1	-1	-1	-1	-1	1	1	-1	-1	-1	-1	1	1	1	-1	77.7
3		-1	1	-1	-1	1	-1	1	-1	1	1	-1	1	-1	-1	-1	66.4
4		1	1	-1	1	-1	-1	-1	1	-1	1	-1	-1	-1	1	-1	81.0
5		-1	-1	1	1	-1	-1	1	1	-1	-1	1	1	-1	-1	-1	78.6
6		1	-1	1	-1	1	-1	-1	-1	1	-1	1	-1	-1	1	-1	41.2
7		-1	1	1	-1	-1	1	-1	-1	-1	1	1	-1	1	-1	-1	68.7
8		1	1	1	1	1	1	1	1	1	1	1	1	1	1	-1	38.7
9		1	1	1	-1	-1	-1	1	1	1	-1	-1	-1	1	-1	1	66.7
10		-1	1	1	1	1	-1	-1	-1	-1	-1	-1	1	1	1	1	65.0
11		1	-1	1	1	-1	1	-1	-1	1	1	-1	1	-1	-1	1	86.4
12		-1	-1	1	-1	1	1	1	1	-1	1	-1	-1	-1	1	1	61.9
13		1	1	-1	-1	1	1	-1	1	-1	-1	1	1	-1	-1	1	47.8
14		-1	1	-1	1	-1	1	1	-1	1	-1	1	-1	-1	1	1	59.0
15		1	-1	-1	1	1	-1	1	-1	-1	1	1	-1	1	-1	1	42.6
16		-1	-1	-1	-1	-1	-1	-1	1	1	1	1	1	1	1	1	67.6
SUMPRODUCT		-53.5	-31.1	-3.3	21.7	-153.7	-0.5	-34.5	3.7	-28.9	8.9	-129.3	38.7	-26.9	-33.5	-23.7	
Effect		-6.69	-3.89	-0.41	2.71	-19.21	-0.06	-4.31	0.46	-3.61	1.11	-16.16	4.84	-3.36	-4.19	-2.96	

FIGURE 11.7: Worksheet for Calculating Effects from Combined Resolution III and Reflected Designs

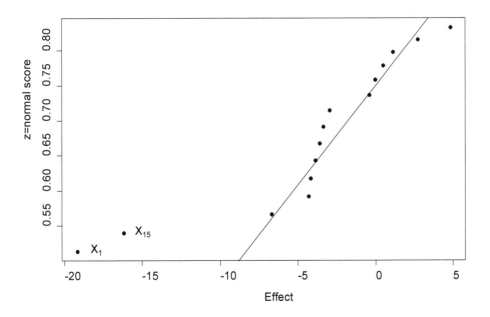

FIGURE 11.8: Normal Plot of Effects from Combined Filtration Experiments

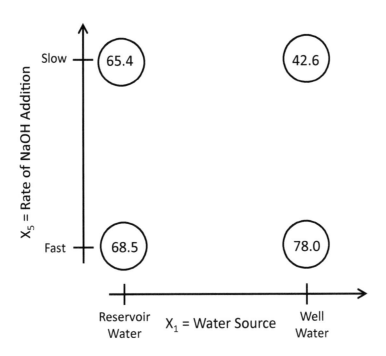

FIGURE 11.9: Two-Way Graph of X_1 and X_5 Means

Interaction effects cannot be seen in one-at-a-time trials. This benefit of factorial designs (allowing for the discovery of interactions) cannot be over emphasized.

Of the possible explanations for which one of the interactions in the significant string is important, the 15 interaction seemed by far the most likely. Nevertheless, the fact that none of the factors 2, 4, 6, and 7 have significant main effects does not, of course, preclude the possibility that their interactions exist. The crucial test was whether the trouble would be cured by using well water and the slow addition rate of caustic soda while leaving the other variables at their usual levels.

A number of confirmation trials were run in the plant where the only modifications made were the use of well water with a slow rate of addition of caustic. These runs did give satisfactorily short filtration times in the neighborhood of 40 minutes. It confirmed the assumption that the interaction was between factors 1 and 5 and validated the efficacy of the proposed change in operating conditions. The modification was adopted for all future production.

This example illustrates the sequential use of fractional factorial designs starting at Step 1 in Figure 11.3, and continuing at Step 3 by running a reflected second design to de-confound main effects from two-factor interactions. It also illustrates the use of the effect heredity paradigm for the interpretation of a confounded string of two-way interactions by picking the most likely single important interaction based on which main effects are significant. Confirmation runs validated the choice.

Another important concept illustrated by this example is the fact that finding out which variables or factors are not important may be just as valuable as finding out which variables are important. Unimportant variables can be set to their least expensive levels, since they don't significantly affect the results. This may identify great areas of cost savings when experimenting with prototype products and production process designs or the selection of materials and components.

11.4   Sequential Analysis without Follow-up Experiments

In some situations the cost of experimentation is high, and it would be preferable to avoid additional experiments, like the reflected design discussed in the last section. After screening experiments are conducted and the important main effects are discovered, it may be possible to re-analyze the same data to detect significant interactions without any additional experiments. This cannot be done when the design for the screening experiments is a fractional factorial. However, it is possible with Plackett–Burman designs and other designs to be discussed in this section.

11.4.1   Sequential Analysis of Data from a Plackett–Burman Design

In the example presented in Section 11.3, three main effects (X_1 = water source, X_3 = temperature of filtration, and X_5 = rate of addition of NaOH) appeared to be the important factors affecting filtration time after the analysis of an eight-run screening experiment. However, after augmenting the original eight experiments with eight more experiments (defined by the reflected design), analysis of the combined data showed that main effect X_3 was not important; instead, X_5 and the interaction between X_1 and X_5 appeared to be the important effects. The initial confusion was caused by the fact that the X_3 effect was completely confounded with the $X_1 \times X_5$ interaction in the initial eight-run fractional factorial design.

By completely confounded, we mean that if a column $X_1 \times X_5$ were added to the computation worksheet shown in Figure 11.4, it would be identical to the X_3 column in the worksheet, and the computed effect of X_3 and the computed $X_1 \times X_5$ interaction effect would be identical. For this reason, the additional eight experiments were required to separate the two effects.

The Plackett–Burman screening designs (described in Section 7.5) and certain other screening designs do not completely confound main effects with two-factor interactions. While these designs are normally used for detecting a subset of active main effects in a screening design, they can also be used to estimate the important main effects plus a few interactions. This can be done using regression analysis, and often simple models involving a few main effects and few interactions involving those main effects can be discovered in the analysis of data from the screening design without the need to run additional experiments. On the other hand, regression analysis cannot be used to fit models involving interactions if a resolution III fractional factorial is used for the screening experiment.

To illustrate the detection of main effects and a few interactions after a single screening experiment, consider the following example from Yan et al. [2011]. They investigated the effect of factors in the enzymatic saccharification of food waste to produce an economical process for transforming food waste hydrolysates to fuel ethanol. In China the major recycling methods for food waste are to use it for fertilizer or animal feed. However, excessive waste water is produced in desalting food wastes for fertilizer production, and animal feeds produced from food waste often cause hygiene problems. Therefore, an environmentally friendly method for converting food waste into high value fuel ethanol was investigated.

Enzymatic hydrolysis experiments were performed using combinations of two commercial enzyme solutions. The first was fungal α-amylase and the second was glucoamylase. The experiments were performed in 500 mL Erlenmeyer flasks, each containing 300 mL of minced food waste in a shaking incubator. The dry mass of the food waste was composed of mainly starch sugars, protein, fat, and cellulose. One unit of fungal α-amylase is defined as the amount of enzyme that hydrolyzes 1 mg water soluble corn starch per minute under assay conditions, and one unit of glucoamylase is defined as the amount of enzyme required to produce 1 mg of glucose in 1 hour under the assay conditions. The factors that varied from run to run were agitation speed, amounts of the two enzymes, incubation temperature, incubation time, pH, and the amount of additives CaCl and surfactant (Polysorbate). Table 11.4 shows the Factor labels and levels used in the experiments.

TABLE 11.4
Factors and Levels for Enzymatic Hydrolysis Experiments

Factor	Definition	Coded Factor	Low Level (−)	High Level (+)
A	Agitation Speed (rpm)	X_1	150	300
B	α-Amylase load (u/g)	X_2	10	20
C	Polysorbate 80 (w · w^{-1} %)	X_3	0	.25
D	Glucoamylase load (u/g)	X_4	80	160
E	CaCl$_2$ (w · w^{-1} %)	X_5	0	.5
F	Temperature (°C)	X_6	45	65
G	pH	X_7	4.0	6.0
H	Incubation time (h)	X_8	1.0	3.0

A Plackett–Burman design was used for the experiments. The response was the average of the reducing sugar concentration from three repeat experiments at each condition. The goal was to maximize the response. Table 11.5 shows the experimental runs and responses in a standard order. In this table X_1–X_8 represent the coded factor levels as described in Section 4.3. Note that in this experiment the authors used less runs than $n_F => \geq k + 5$ as recommended in Section 7.5.2.

TABLE 11.5

Plackett–Burman Design with Reducing Sugar
Concentration as Response

Run	X_1	X_2	X_3	X_4	X_5	X_6	X_7	X_8	Y (g/L)
1	+	−	+	−	−	−	+	+	78.891
2	+	+	−	+	−	−	−	+	129.902
3	−	+	+	−	+	−	−	−	83.153
4	+	−	+	+	−	+	−	−	120.217
5	+	+	−	+	+	−	+	−	105.393
6	+	+	+	−	+	+	−	+	116.350
7	−	+	+	+	−	+	+	−	119.712
8	−	−	+	+	+	−	+	+	101.587
9	−	−	−	+	+	+	−	+	135.468
10	+	−	−	−	+	+	+	−	78.134
11	−	+	−	−	−	+	+	+	109.516
12	−	−	−	−	−	−	−	−	91.997

The analysis of the data from the Plackett–Burman design in Table 11.5 was completed using a worksheet similar to those used in the analysis of filtration example discussed last section. The results are shown in Figure 11.10. This analysis shows the four main effects (X_4 = Glucoamylase load, X_6 = Temperature, X_7 = pH, and X_8 = Incubation time) all had significant effects upon the reducing sugar concentration. Examining the signs of the main effects would lead one to believe that the high level of X_4, or Glucoamylase load = 160; the high level of X_6, or Temperature = 65 °C; the low level of X_7, or pH = 4; and the high level of X_8, or Incubation time = 3, would result in the maximum response.

However, in a Plackett–Burman Design, the main effects are confounded with interactions as in a fractional factorial design. Therefore, it is possible that some of the main effects appear large because they are confounded with significant interactions, as was the case in the filtration example presented in the last section. If the interactions can be estimated, it is possible (as in the last example) that the higher (and more expensive) levels will not be required for all three of the factors: Glucoamylase load, Temperature, and Incubation time.

In a Plackett–Burman design there is no defining relation or simple formula to show what two-factor interactions are confounded with each main effect. Therefore, the job of determining which if any interactions are important is more difficult. In a Plackett–Burman design, each main effect is partially confounded with many two-factor interactions rather than being completely confounded with a few. When there is complete confounding, interaction effect columns in a worksheet like Figure 11.4 (with added columns for interactions) will be identical to the columns for main effects with which they are confounded. In that situation the main effects cannot be estimated separately from the interactions. By combining the original design with a reflected design as shown in Figure 11.7, the main effect columns and the interaction columns became orthogonal. Therefore, both the main effects and the strings of two-factor interactions they are confounded with could both be estimated independently.

The partial confounding of main effects and two-factor interaction present in Plackett–Burman designs lie somewhere in between complete confounding and orthogonality. In this situation the main effect columns are correlated (as described in Section 2.3) with many different two-factor interaction columns. The main effects and two-factor interactions cannot be estimated using worksheet calculations like those shown in Figure 11.7, but they can be estimated using least squares multiple regression, as long as there are fewer terms in the model than there are runs in the design.

	A	B	C	D	E	F	G	H	I	J	K	L	M
1	Mean	X1	X2	X3	X4	X5	X6	X7	X8	u1	u2	u3	Y
2	1	1	-1	1	-1	-1	-1	1	1	1	-1	1	78.891
3	1	1	1	-1	1	-1	-1	-1	1	1	1	-1	129.902
4	1	-1	1	1	-1	1	-1	-1	-1	-1	1	1	83.153
5	1	1	-1	1	1	-1	1	-1	-1	-1	1	1	120.217
6	1	1	1	-1	1	1	-1	1	-1	-1	-1	1	105.393
7	1	1	1	1	-1	1	1	-1	1	1	-1	-1	116.350
8	1	-1	1	1	1	-1	1	1	-1	-1	-1	-1	119.712
9	1	-1	-1	1	1	1	-1	1	1	-1	1	-1	101.587
10	1	-1	-1	-1	1	1	1	1	1	1	1	1	135.468
11	1	1	-1	-1	-1	1	1	-1	1	-1	1	-1	78.134
12	1	-1	1	-1	-1	1	1	1	1	-1	1	1	109.516
13	1	-1	-1	-1	-1	-1	1	-1	-1	-1	-1	1	91.997
14	SUMPRODUCT	-12.546	57.732	-30.500	154.238	-30.150	88.474	-83.854	73.108	-19.800	-25.302	-5.044	
15	Effects	-2.091	9.622	-5.083	25.706	-5.025	14.746	-13.976	12.185	-3.300	-4.217	-0.841	
16	t-statistic	-0.668	3.075	-1.624	8.214	-1.606	4.712	-4.466	3.894				
17	p-value	0.552	0.054	0.203	0.004	0.207	0.018	0.021	0.030				
18													
19	S =	3.12942											

FIGURE 11.10: Computation Worksheet for Plackett–Burman Design

For the hydrolysis experiment, the main effects X_4, X_6, X_7, X_8 were found to be significant, with X_4 being the most dominant. From effect heredity paradigm, we can guess that the most likely two-factor interactions to be significant would be those that involve these significant main effects, namely: $X_4 \times X_6$, $X_4 \times X_7$, $X_4 \times X_8$, $X_6 \times X_7$, $X_6 \times X_8$, and $X_7 \times X_8$. However, it would not be wise to fit the 10-term regression model:

$$Y = b_0 + b_4 X_4 + b_6 X_6 + b_7 X_7 + b_8 X_8 + b_{46} X_4 X_6 + b_{47} X_4 X_7 + b_{48} X_4 X_8 +$$
$$b_{67} X_6 X_7 + b_{68} X_6 X_8 + b_{78} X_7 X_8 \qquad (11.3)$$

because with only 12 experimental observations, there would not be enough data to accurately estimate the experimental error variance and determine significance of the coefficients.

The best way to identify significant interactions is to use some type of regression subsetting procedure. One way to do this is to use an all-subsets regression procedure. Another approach is to use a forward stepwise regression, and a third approach is to use a backwards elimination regression. This third approach can only be used when the number of runs in the design is at least two greater than the number of variables in the full regression model.

In an all-subsets regression procedure, the program fits a regression model with all possible subsets of variables and keeps the summary results from a specified number of models for each subset size. For example, if an all-subsets procedure was run using the variables X_4, X_6, X_7, X_8, $X_4 \times X_7$, $X_4 \times X_8$, $X_6 \times X_7$, $X_6 \times X_8$, and $X_7 \times X_8$, the program would fit all models with one variable, all possible models with 2 variables (like X_4 and X_6 or X_6 and $X_7 \times X_8$ etc.), and all models with 3, 4, 5, 6, 7, 8, 9 and 10 variables.

In a forward stepwise regression procedure, the program fits all models with one variable and picks the variable that results in the highest R^2. Next, the program finds a second variable that has the highest correlation with the residuals (i.e., actual response minus predicted value) from the best one-variable model, and then fits a two-variable model that includes this second variable with the variable from the best one-variable model. This procedure is continued, increasing the number of variables in the model one step at a time until some stopping rule is satisfied. A stepwise regression algorithm can be modified to incorporate the effect heredity principle in the following way. If a two-factor interaction is the next variable to enter the model, the procedure should also include the main effects involved in this interaction (if they are not already in the model).

Finally, a backwards elimination regression starts with the full model (again this is only possible if the the number of runs in the design is at least two greater than the number variables in the full regression model). After fitting the full model, the output is examined and the variable with smallest t-statistic in absolute value (or the largest P-value) is eliminated from the model and the regression is run again. This procedure of eliminating a term from the model is repeated until all remaining terms in the model are significant. This procedure can also be modified to incorporate effect heredity by simply not eliminating both main effects that define a significant two-factor interaction in the model.

The backward elimination procedure can be carried out with any statistical program that has a regression routine (including a spreadsheet program) by simply following the outline in the last paragraph. However, it would be tedious to carry out an all-subsets regression or the forward stepwise regression subset procedure with a spreadsheet program. These procedures can most easily be carried out using a statistical program. For example, the commercially available program Minitab© [MINITAB, 2015] contains an all-subsets routine called Best Subsets, and the `leaps` package [Lumley, 2009] for the public domain program R [R Development Core Team, 2003] contains a `regsubsets` function for doing all subsets regression. The commercially available program Minitab© has a **Forward stepwise** procedure where the effect heredity principle can be selected in its **Stepwise** template, and the `ihstep`, and `fstep` functions in the R package `daewr` [Lawson, 2015] can do a similar thing. While the commercially available statistical programs have a GUI user interface, the public domain program R has a command interface which is like a programming language. However the R package Rcmdr[Fox, 2005] provides a GUI interface for routines such as forward stepwise regression, backwards elimination, and all-subsets regression. Example input and output for all these programs, using the examples from this book, are available on the website.

Using the significant main effects X_4, X_6, X_7, and X_8 found in the worksheet computations (Figure 11.10) and all the interactions involving these significant factors (i.e., $X_4 \times X_6$, $X_4 \times X_7$, $X_4 \times X_8$, $X_6 \times X_7$, $X_6 \times X_8$, and $X_7 \times X_8$), a backwards elimination procedure was run. The first step involved fitting the full model in Equation 11.3. In the second step, the term with the highest P-value ($X_4 \times X_6$) was eliminated and the regression model was fit again. In the third step, the term with the highest P-value in the second step ($X_4 \times X_7$) was eliminated and the regression model was fit again. Finally, in the fourth step the term with the highest P-value in the third step ($X_7 \times X_8$) was eliminated and the regression model was fit again. After the fourth step, all the terms in the model were significant and there was nothing left to remove. The output of the lm procedure for fitting least squares regression in the program R is shown in Table 11.6.

TABLE 11.6
Regression Output from R lm function

```
Coefficients:
              Estimate Std. Error t value Pr(>|t|)
(Intercept)   105.860      1.105  95.840 7.11e-08 ***
X4             16.209      1.285  12.616 0.000227 ***
X6              5.703      1.198   4.760 0.008903 **
X7             -7.294      1.285  -5.677 0.004751 **
X8              8.084      1.198   6.748 0.002514 **
X4:X8          -5.009      1.392  -3.598 0.022794 *
X6:X7           5.976      1.392   4.293 0.012712 *
X6:X8           4.091      1.271   3.220 0.032291 *
---
Signif. codes:  0 *** 0.001 ** 0.01 * 0.05 . 0.1   1

Residual standard error: 3.826 on 4 degrees of freedom
Multiple R-squared:  0.9861,Adjusted R-squared:  0.9616
F-statistic:  40.4 on 7 and 4 DF,  p-value: 0.001497
```

Using the forward stepwise procedure in the program Minitab© with the *Hierarchy* option resulted in exactly the same model. This indicates that there are some significant interactions in addition to the main effects found in Figure 11.10. However, it is not at all clear that the three interactions identified in the output above are the most important ones. The all-subsets regression procedure in Minitab (a portion of the output shown in Table 11.7) indicates six models with 7–9 terms that all fit the data well and each one contains a different subset of interactions.

Figure 11.11 shows interaction plots for all the interactions among the variables X_4, X_6, X_7, and X_8. There appears mild interactions, as described in Section 3.4, between X_4 and X_6 (i.e., Temperature × Incubation time), X_4 and X_8 (i.e., Temperature × pH), and X_6 and X_7 (i.e., Incubation time × Glucoamylase load). However, the interaction between X_4 and X_6 is not in the model found by backwards elimination or forward stepwise regression.

TABLE 11.7
Minitab All-Subsets Regression Results

						X	X	X	X	X	X
						4	4	4	6	6	7
						*	*	*	*	*	
		R-Sq	R-Sq	Mallows		X X X X	X	X	X	X	X	X
Vars	R-Sq	(adj)	(pred)	Cp	S	4 6 7 8	6	7	8	7	8	8

Vars	R-Sq	R-Sq (adj)	R-Sq (pred)	Mallows Cp	S	X4	X6	X7	X8	4·6	4·7	4·8	6·7	6·8	7·8
7	98.6	96.2	90.7	5.5	3.8263	X	X	X	X				X	X	X
7	96.1	89.1	70.6	8.2	6.4355	X	X	X	X	X	X	X			
8	99.0	96.3	95.6	7.1	3.7554	X	X	X	X			X	X	X	X
8	98.8	95.6	91.2	7.3	4.1132	X	X	X	X	X			X	X	X
9	99.0	94.7	91.2	9.0	4.5043	X	X	X	X	X		X	X	X	X
9	99.0	94.7	91.1	9.0	4.5052	X	X	X	X		X	X	X	X	X
10	99.1	89.8	*	11.0	6.2537	X	X	X	X	X	X	X	X	X	X

It is in one of the eight variable models found in the all-subsets regression, but that model also contains the interaction between X_6 and X_8 (Incubation time × pH), which doesn't appear very strong in Figure 11.11.

The confusion results from the fact that the interactions are partially confounded with the main effects and other two-factor interactions. There appears to be at least three important interactions, and this is difficult to sort out with only 12 runs (less than the number recommended in Section 7.5.2). With at least three interactions, this is a rather complex relationship between the response and the factors. However, in just 12 experiments with a Plackett–Burman design with eight factors and a sequential analysis of the data, four significant main effects and three mild interactions have been identified. It is not completely clear which three interactions are important without follow-up experiments.

Usually in screening designs, the relationship between the response and the factors is not so complex. For example, in the filtration example presented in Section 11.3, only one main effect and one interaction were found to be significant after 16 experiments. This again is an illustration of the effect sparsity paradigm. In this more usual situation, sequential analysis of the data from a Plackett–Burman design may allow identification of a few important main effects and interactions without the follow-up experiments that would be required when using a resolution III fractional factorial screening design (as shown in the procedure in Figure 11.3).

As an illustration of this, Hunter et al. [1982] completed a 12-run Plackett–Burman design investigating seven potential factors in the process of weld repairing cracked castings. The response of interest was the fatigue life of the repaired parts when put on test. An initial analysis of the data (using a normal plot of the effects as described in Section 7.5.3) revealed only one significant main effect. A simple linear regression analysis of the response on that factor only accounted for 45% of the variation in the response (i.e., $R^2 = 0.4474$).

Several other researchers have re-analyzed the data from the article by [Hunter et al., 1982] using a sequential analysis approach. For example, a forward stepwise regression (that you can reproduce as one of the exercises in this chapter) identified a model with two main effects and a strong interaction between them. This model explained the data much better ($R^2 = 0.9104$) and identified factor settings that would lead to a 44% improvement in the fatigue life.

When performing a forward stepwise analysis of the data from the article by [Hunter et al., 1982], the single term that had the highest R^2 for predicting the response was a

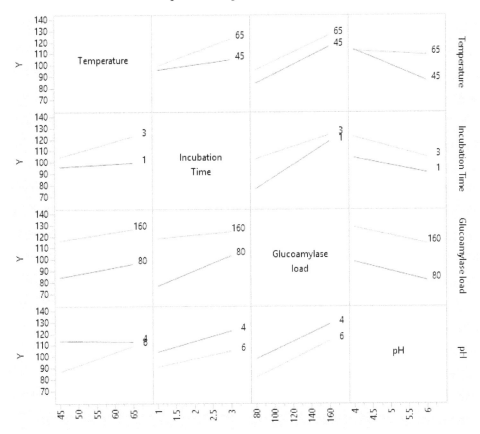

FIGURE 11.11: Interaction Plots for Enzymatic Hydrolysis Experiments

two-factor interaction. Therefore, to be consistent with the effect heredity paradigm, this interaction and both of the main effects involved in this interaction should be included in the model at the first step.

When there are only one or two main effects and two-factor interactions, the use of Plackett–Burman designs and sequential analysis may allow identification of the important effects without the need for follow-up experiments. However, the Plackett–Burman designs listed in Appendix B.1 are only available in run sizes of 12, 20, 24, or 28. If the number of runs required to obtain the necessary precision is 16, there is no Plackett–Burman design. However, Li and Nachtsheim [2000] and Jones and Montgomery [2010] have developed 16-run designs that have properties similar to the Plackett–Burman designs in that main effects and two-factor interactions are partially confounded. These designs are included on the website for the book, and data from these designs can be analyzed using subset regression techniques.

Returning to the enzymatic hydrolysis experiments, the interpretation of the mild interactions shown in Figure 11.11 would not change the fact that the high and more expensive levels of Glucoamylase load (160 u/g), Temperature (65 °C), and Incubation time (3 h) are predicted to result in the maximum response. The authors of the article felt that the optimal conditions were still unclear after the screening experiment. They elected to perform 30 follow-up experiments by varying the four factors X_4, X_6, X_7, X_8 (found to be significant in screening) according to a central composite design. This would allow for curvature in the prediction equation and the possibility of identifying higher response values within the bounds on the factors established in the screening experiment. The data for these 30 additional experiments are shown in Exercise 3 of Chapter 10.

A general quadratic model was fit to this data and maximum predicted value of the response (reducing sugar concentration = 164.3) was found at the stationary point (described in Section 10.6.1) where Glucoamylase load is 142.2 u/g, the Incubation time was 2.48 h, the Temperature was 54.45 °C, and pH was 4.82. This combination of factor levels is represented by the "+" in the contour plot shown in Figure 11.12. Identification of this combination of factor levels increased the predicted response by over 20% and did not require the highest and most expensive levels of Glucoamylase load, Temperature, and Incubation time.

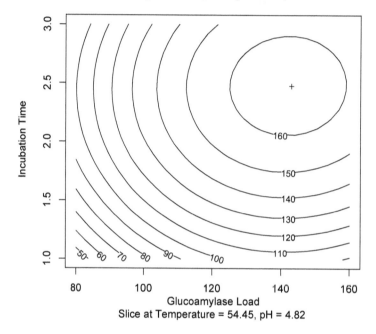

FIGURE 11.12: Region of Maximum Reducing Sugar Concentration

11.4.2 Sequential Analysis of Screening Designs to Find Quadratic Optima

In the enzymatic hydrolysis experiments, follow-up experiments were required to identify the optimal combination of factor settings that was in between the high and low factor levels chosen in the initial screening design. The 12 runs in the Plackett–Burman screening design plus the 30 runs in the follow-up central composite design took a total of 42 runs.

When experimenters anticipate that the optimal factor settings may be somewhere between the high and low levels chosen in a screening design, there is a class of screening designs that have three levels for each factor that can be varied over a continuous range. These designs are called *definitive screening designs* and were developed by Jones and Nachtsheim [2011, 2013].

In definitive screening designs, main effects are completely orthogonal to other main effects, two-factor interactions, and quadratic effects. Two-factor interactions and quadratic effects are only partially confounded with each other. Therefore, data from these designs can be analyzed using a subset regression procedure to identify a general quadratic model in a subset of the factors. Definitive screening designs are available with 4 to 12 three-level factors. They are shown in Appendix B in Tables B.8-1 to B.8-9, and electronic spreadsheets containing these designs in standard order can be found on the website for the book.

With only three-level factors, definitive screening designs require only $2k + 1$ runs, where k is the number of factors. For example, only 17 runs would be required if three levels were used for each of the eight factors in the enzymatic hydrolysis experiments. This would be a substantial savings compared to the 42 runs actually used. Definitive screening designs will allow fitting a full quadratic model to any subset of three factors.

To illustrate the use of a definitive screening design, consider the example from Libbrecht et al. [2015]. They studied the synthesis of soft template mesoporous carbons that are used as adsorbents. In their study, bisphenol A (BPA) was the absorbant. BPA has been classified as an endocrine disruptive chemical that puts human health at risk, and has been detected in industrial waste water, groundwater, surface water, and even drinking water. The study focused on optimizing the method of synthesizing mesoporous carbons and investigating the material properties of these carbons on their ability to absorb BPA. The synthesis method was a simple acid-catalyzed evaporation-induced self-assembly (EISA).

The synthesis parameters varied in the study are shown in Table 11.8. The important responses were the mesoporous surface area and the carbon content of the mesoporous carbons. These responses were identified as the main properties that influenced the ability of the carbons to absorb BPA. Since absorption capacity of the carbons increased as both the surface area and carbon content increased, the goal of the experimentation was to identify a combination of the synthesis parameters that would result in both high surface area and carbon content.

TABLE 11.8

Factors in Mesoporous Synthesis Experiments

				Levels	
Factor	Definition	Coded	Low (−)	Mid (0)	High (+)
A	Carbonization Temperature	X_1	600	800	1000
B	Ratio (precursor/surfactant)	X_2	0.7	1.0	1.3
C	EISA—time	X_3	6	15	24
D	Curing—time	X_4	12	18	24
E	EISA—Surface Area	X_5	65	143	306

The five factors were varied over three levels, each according to a definitive screening design. The design took $2 \times 5 + 1 = 11$ runs. The authors repeated one of the experiments to make sure the original response was not an outlier, and they repeated the center point three times so that there would be a measure of replicate error. This resulted in a total of 14 runs, compared to 26 runs that would be required for a central composite design, or 43 runs that would be required for a Box-Bhenken design. The design (in coded factor levels) and responses are shown in Table 11.9

A full quadratic model (including all five factors) could not be fit to the data because the number of terms in the model (21) was greater than the number of runs. Instead, the data were analyzed in a sequential manner using regression methods. In the first step, a regression model that included only the main effects X_1–X_5 (which are unconfounded with each other) was fit for both responses. This was done in order to identify insignificant factors that could be eliminated. In the second step, a full quadratic model was fit in the terms found to be significant in the first step. This was possible using regression because the main effects in this design are not correlated with the quadratic effects nor two-factor interactions, and quadratic effects and two-factor interactions are only partially confounded. In the final step, a backwards-elimination procedure was used to eliminate insignificant terms from the model.

TABLE 11.9

Definitive Screening Design for Mesoporous Synthesis Experiments

Run	\multicolumn{5}{c}{Coded Factor Levels}					\multicolumn{2}{c}{Responses}	
	X_1	X_2	X_3	X_4	X_5	Surface area	wt %C
1	0	+	+	−	−	241	94.26
2	0	−	−	+	+	295	93.76
3	+	0	−	−	+	260	95.75
4	−	0	+	+	−	338	90.22
5	+	−	0	+	−	320	97.50
6	−	+	0	−	+	265	90.16
7	+	−	+	0	+	275	94.79
8	−	+	−	0	−	248	89.44
9(1)	+	+	+	+	0	66	94.31
9(2)	+	+	+	+	0	119	-
10	−	−	−	−	0	383	90.23
11(1)	0	0	0	0	0	313	95.00
11(2)	0	0	0	0	0	305	-
11(3)	0	0	0	0	0	304	-

For the response, surface area, the first step of the sequential analysis revealed that main effects X_1, X_2, and X_4 were significant. Since the design supports fitting the full quadratic in any three factors, the second step was to fit the full quadratic model:

$$SA = \beta_0 + \beta_1 X_1 + \beta_2 X_2 + \beta_4 X_4 + \beta_{11} X_1^2 + \beta_{22} X_2^2 + \beta_{44} X_4^2 + \beta_{12} X_1 X_2 + \beta_{14} X_1 X_4 + \beta_{24} X_2 X_4$$

to the data. After the backwards elimination, the final equation found was:

$$SA = 304.343 - 49.43 X_1 - 79.04 X_2 - 20.51 X_4 - 45.19 X_2{}^2 - 22.12 X_2 X_4, \qquad (11.4)$$

Because of the interaction, surface area is maximized when $X_4 = -$, or Curing time = 12 h.

For the wt %C response, the first step of the sequential analysis showed that only X_1 was significant. After fitting the quadratic model in X_1, both terms were found to be significant; therefore, the final equation was identified as:

$$wt\%C = 94.34 + 2.7875 X_1 - 1.54 X_1{}^2. \qquad (11.5)$$

Figure 11.13 shows the contours of Equation 11.4 overlaid with the contours of Equation 11.5 (where X_4 is held constant at −1 or Curing time = 12 h). In this figure it can be seen that surface area (solid contours) is maximized with Ratio ≈ 0.84 (or X_2 ≈ 0.533). With respect to X_1, or Carbonization Temperature, there is a trade-off between surface area and wt %C. As the Carbonization Temperature increases, wt %C also increases, but surface area decreases. The figure, or a spreadsheet solver program, can be used to identify a level of Carbonization Temperature that will result in high surface area and adequately high wt %C.

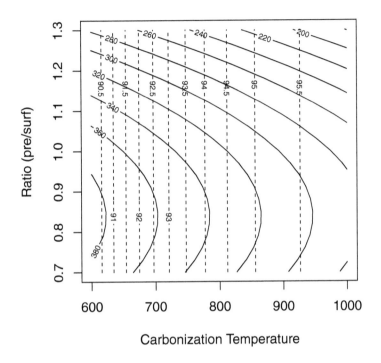

FIGURE 11.13: Contours of Predicted Mesoporous Surface Area (Solid) and Carbon Content (Dashed)

11.5 Summary

Throughout this book, we have described how various types of experimental designs are each tailored to a specific situation in terms of prior knowledge. Their use generally advances a user to the next state of knowledge as described in Figures 3.1, 7.1, 9.1 and 11.1. In this chapter we describe how a sequence of experiments using different designs can be used to move from a low state of knowledge to a higher state of knowledge. An example in Chapter 1 illustrated how a screening design followed by a factorial design and then a response surface design increased knowledge from the screening to optimization phase. An example in Chapter 6 illustrated how a preliminary exploration design followed by a factorial design advanced research from a low state of knowledge to crude optimization. In this chapter we show simple augmenting fractional factorial designs with a few more experiments can do the same thing. Finally, we show how repeated analysis of data from Plackett–Burman or definitive screening designs can again move a researcher from the screening to optimization phase without any further experiments.

11.5.1   Important Terms and Concepts

The following is a list of the important terms and concepts covered in this chapter.
reflected design
partial confounding of main effects and two-factor interactions
forward stepwise regression
backwards elimination
all-subsets regression
effect heredity principle
definitive screening Design

11.6   Exercises

1. Wastewater reclamation is one way to provide the world's growing population with potable water. Membrane technology, such as reverse osmosis (RO) systems, can be used in wastewater reclamation. However, periodic cleaning must be incorporated into the operation of a membrane plant as a preventative measure against severe and irreversible membrane fouling. Membranes can be cleaned by chemical and physical methods. Physical cleaning methods depend on mechanical forces to dislodge and remove foulants from the membrane surface. Physical methods used include forward flushing, reverse flushing, backwashing, vibrations, air sparge, and CO_2 back permeation. Chen et al. [2003] performed a series of experiments to find the optimal combination of factor levels in physical and chemical membrane cleaning processes. The table below shows the factors and levels they used in experiments to study the physical cleaning of RO membranes.

TABLE 11.10
Factors in RO Membrane Cleaning Experiment

	Levels	
Definition	Low $(-)$	High $(+)$
Production interval between cleaning	2 h	6 h
Duration of forward flush	1 min	5 min
Duration of backwash	1 min	5 min
Pressure during forward flush	6.21×10^5 Pa	1.24×10^6 Pa
Type of water used	RO permeate	Tap water
Sequence of forward flush and backwash	backwash first	forward flush first

The responses they measured were CWF recovery % and wash water usage. They are defined as:

$$\text{CFW recovery \%} = \frac{\text{CFW}_i}{\text{CFW}_c}$$

where CFW_i is clean water flux of membrane before water production and CFW_c is clean water flux after cleaning.

$$\text{wash water usage} = \frac{\text{volume of wash water use}}{\text{total volume of water produced}}.$$

The first set of experiments was an eight-run fractional factorial. The plan was the same as Table B.1-5 in Appendix B with the factors assigned to the first six columns of the table. The responses listed in the standard order of Table B.1-5 are:

CFW recovery %: 85.43, 79.17, 91.32, 79.27, 91.81, 90.95, 96.78, 87.54

wash water usage: 0.06, 0.03, 0.16, 0.04, 0.23, 0.08, 0.31, 0.11

(a) Analyze the data from the two responses and determine which main effects appear to be significant.

(b) Based on the confounding of main effects with two-factor interactions shown in Table B.1-5, are there alternate explanations for the significant effects in the model (i.e., could one of the main effects detected actually represent a two-factor interaction that involves at least one of the other significant main effects in the model?)

After the first set of experiments was complete, Chen et al. [2003] ran eight more experiments according to the reflected design of Table B.1-8. The responses for the reflected design in standard order are:

CFW recovery %: 91.61, 92.32, 84.29, 93.82, 80.49, 85.55, 74.48, 92.02

wash water usage: 0.11, 0.27, 0.10, 0.22, 0.07, 0.11, 0.03, 0.07

(c) Combine the data from the initial eight experiments and the reflected design and analyze the two responses in a way similar to that shown in Figure 11.7.

(d) What main effects and two-factor interactions appear to be significant when analyzing the combined data?

2. Hunter et al. [1982] completed a 12-run Plackett–Burman design investigating seven potential factors in the process of weld repairing cracked castings. The table below shows the factor settings and the response (log fatigue life).

Run	X_1	X_2	X_3	X_4	X_5	X_6	X_7	u_1	u_2	u_3	u_4	Y
1	+	+	−	+	+	+	−	−	−	+	−	6.058
2	+	−	+	+	+	−	−	−	+	−	+	4.733
3	−	+	+	+	−	−	−	+	−	+	+	4.625
4	+	+	+	−	−	−	+	−	+	+	−	5.899
5	+	+	−	−	−	+	−	+	+	−	+	7.000
6	+	−	−	−	+	−	+	+	−	+	+	5.752
7	−	−	−	+	−	+	+	−	+	+	+	5.682
8	−	−	+	−	+	+	−	+	+	+	−	6.607
9	−	+	−	+	+	−	+	+	+	−	−	5.818
10	+	−	+	+	−	+	+	+	−	−	−	5.917
11	−	+	+	−	+	+	+	−	−	−	+	5.863
12	−	−	−	−	−	−	−	−	−	−	−	4.809

(a) Use a worksheet like Figure 11.10 to calculate the effects of the seven factors X_1–X_7 and the unassigned factors u_1–u_4.

(b) Make a normal probability plot of the effects and determine if any of the main effects appear to be significant.

(c) Based on a linear model in the factor(s) you find to be significant in (b), predict the optimal factor level(s) and the average log fatigue life predicted at these level(s).

(d) Compute one step of a forward stepwise regression using the main effects and all two-factor interactions as candidate terms for the model. This can be easily accomplished in the following way using any regression program or a spreadsheet program. First, determine which main effect or two-factor interaction column in an expanded worksheet has the highest absolute correlation coefficient with the response. If the term that has the highest correlation with the response is a main effect, it should agree with what you found in (b) and then you are finished. If a two-factor interaction has the highest correlation with the response, fit a regression model that includes the interaction and both main effects that are involved in the interaction (to preserve effect heredity).

(e) If any of the three terms in the regression model you found in (d) are insignificant, eliminate it and refit the model using regression.

(f) Using the final model you find in (e), predict the optimal factor level(s) and the average log fatigue life predicted at these level(s), and compare this to what you found in (c).

3. Erler et al. [2013] used a definitive screening design as part of a process-characterization campaign of a candidate vaccine product. The six-input factors were X_1 = protein concentration, X_2 = lysine concentration, X_3 = reaction duration, X_4 = pH, X_5 = reaction temperature, and X_6 = formaldehyde to protein ratio. The response of interest was the extent of polymerization. The $2\times6 + 1 = 13$ experiments in the definitive screening design were augmented by an additional four center points resulting in the 17-run design (presented in random order) below.

Run	X_1	X_2	X_3	X_4	X_5	X_6	Y=Ext. of Poly.
1	−	−	−	−	+	0	2.9
2	+	−	+	−	0	−	14.9
3	0	0	0	0	0	0	14.4
4	+	+	0	−	+	+	27.5
5	0	0	0	0	0	0	14.6
6	0	+	−	−	−	−	3.6
7	−	+	+	0	+	−	2.5
8	−	+	−	+	0	+	8.5
9	0	0	0	0	0	0	14.1
10	+	0	−	+	+	−	22.0
11	−	0	+	−	−	+	4.2
12	0	0	0	0	0	0	14.2
13	−	−	0	+	−	−	2.7
14	0	0	0	0	0	0	15.3
15	0	−	+	+	+	+	28.2
16	+	+	+	+	−	0	30.7
17	+	−	−	0	−	+	31.9

(a) The authors of the article came up with a 10-term regression model that fit their data well. The terms included the six main effects, the $X_3 \times X_4$ interaction, the $X_1 \times X_6$ interaction, $X_5 \times X_6$ interaction, and the quadratic term in X_5. Use regression to fit this model and determine if all the coefficients are significant.

(b) Use six steps in a forward stepwise regression to identify the model you fit in (a). Define all six main effects, the 15 two-factor interactions, and six quadratic terms as candidates for your forward stepwise regression. In order to insure effect heredity, if an interaction enters the model at any step, also include the main effects that are involved in that interaction as well (note this can be accomplished by clicking the Hierarchy... button on the Forward selection template in the Regression:Stepwise menu of the commercially available program Minitab© MINITAB [2015] or the ihstep, and fstep functions in the *R* package daewr [Lawson, 2015]).

Chapter 12

Mixture Experiments

12.1 Introduction

Many products, such as textile fiber blends, explosives, paints, polymers, ceramics, etc., are made by mixing two or more ingredients together. The qualities of the product are not dependent upon the total amount of the ingredients in the mixture, but only on the relative proportions in which they are present.

If the proportion of the ith component is X_i and there are k components in the mixture, then the proportions must satisfy the constraints

$$0 \leq X_i \leq 1 \text{ for each component, and } \sum_{i=1}^{k} X_i = 1.$$

For example, in a two-component mixture problem, $0 \leq X_1 \leq 1$, $0 \leq X_2 \leq 1$, and $X_1 + X_2 = 1$. These constraints prevent us from using the experimental designs we have already described for studying the relationship of product quality to mixture components. The 2^2 factorial design would include all the corners of a square, while only two of these points would be permissible in studying mixture problems as shown below in Figure 12.1.

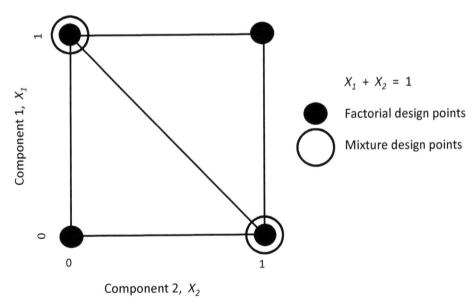

FIGURE 12.1: Experimental Region for Factorial and Mixture Experiments

The experimental region for a non-mixture experiment would consist of the entire area inside a square or rectangle while for the mixture design it consists of only the points on the line $X_1 = 1 - X_2$. Therefore, the experimental region for the mixture problem (a line) is one dimension less than the experimental region for the normal problem (which is a two-dimensional plane).

For the three-component mixture problem, the experimental region consists of a two-dimensional surface, namely the triangular plane shown in Figure 12.2, rather than the normal three-dimensional cube. In the four-component mixture problem, the region consists of the three-dimensional volume of a regular tetrahedron, and so on.

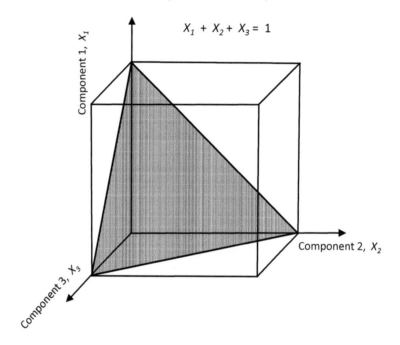

FIGURE 12.2: Experimental Region for Three-Component Mixture Experiments

The recommended experimental designs for mixture experiments differ depending on the purpose, just as they did for the screening and response surface designs discussed earlier in the book. Different designs for mixture experiments will be presented after a discussion of the linear, quadratic, and cubic models used for these problems.

12.2   Models for Mixture Problems

12.2.1   Overview of Modeling

As in all response surface analysis, the objective of the statistical analysis of mixture problems is to determine a model to predict the response as a function of the mixture components. This can be done with a linear model like:

$$\hat{Y} = b_0 + \sum_{i=1}^{k} b_i X_i. \tag{12.1}$$

where $\hat{Y}$ is the response and the k X_i's are the mixture components. Or, it can be done with a quadratic model like:

$$\hat{Y} = b_0 + \sum_{i=1}^{k} b_i X_i + \sum_{i=1}^{k} b_{ii} X_i^2 + \sum_{i=1}^{k-1} \sum_{j=i+1}^{k} b_{ij} X_i X_j. \tag{12.2}$$

The linear model would generally be used for cases where the mixture components are additive and the product quality is defined as a simple linear combination of the component proportions. The quadratic model would be used if there were some interaction (synergism or antagonism) between components, and the product quality or response from a mixture of components was better or worse than would be predicted by the straight linear combination of the mixing proportions. For the three-component case, the linear model is written as:

$$\hat{Y} = b_0 + b_1 X_1 + b_2 X_2 + b_3 X_3,$$

and the quadratic model is written as:

$$\hat{Y} = b_0 + b_1 X_1 + b_2 X_2 + b_3 X_3 + b_{11} X_1^2 + b_{22} X_2^2 + b_{33} X_3^2 \\ + b_{12} X_1 X_2 + b_{13} X_1 X_3 + b_{23} X_2 X_3.$$

At first glance these look like the normal linear model or quadratic response surface model. However, due to the mixture constraint $X_1 + X_2 + X_3 = 1$, it can be seen that once the value of two of the X's are known, the third one is fixed. Therefore, the prediction models are redundant, and $\hat{Y}$ is not really a function of three X's but only of two of them. Therefore, the models must be modified (i.e. reduced) to eliminate that redundancy.

12.2.2 Slack Variable Models

One approach that can be used is to simply eliminate one of the factors from the model. This method is most useful when one of the factors is the major component proportion-wise, and is the most inert. As an example, let us say that we have a three-component system, and of the three components, X_3 fits that description. Therefore, we could drop X_3 from the models, and the models could be written as:

$$\hat{Y} = b_0 + b_1 X_1 + b_2 X_2 \text{ (linear), or}$$

$$\hat{Y} = b_0 + b_1 X_1 + b_2 X_2 + b_{11} X_1^2 + b_{22} X_2^2 + b_{12} X_1 X_2 \text{ (quadratic)}$$

for the three-component case. This is called the slack variable form of the model. Figure 12.3 illustrates the meaning of the coefficients in the linear model. In the linear case, the coefficients can be easily interpreted. The intercept, b_0, is the response when only the slack variable is present. The linear coefficients give the amount that each of the other variables contribute to the response when they are added to the mixture.

The slack variable model in its general form for the quadratic case is:

$$\hat{Y} = b_0 + \sum_{i=1}^{k-1} b_i X_i + \sum_{i=1}^{k-1} b_{ii} X_i^2 + \sum_{i=1}^{k-2} \sum_{j=i+1}^{k-1} b_{ij} X_i X_j. \tag{12.3}$$

This model looks like the normal quadratic model used in response surface analysis, with the kth component omitted. However, unlike the coefficients in the linear model, those

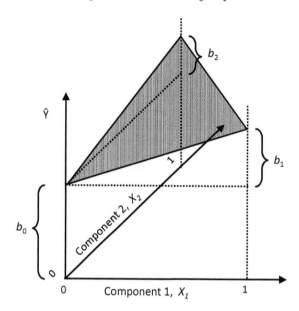

FIGURE 12.3: Graphical Interpretation of Coefficients in a Linear Slack Variable Model

in the quadratic model cannot be interpreted in the same way that they are for non-mixture models. The interpretation is actually very confusing. Because of that, insignificant coefficients, in general, should not be removed from a fitted equation. This is a major deficiency in this form of the model (for any but the linear case), and it led Sheffé to propose a different form which is much more widely used.

12.2.3   Scheffé Models

The Scheffé [1958] form of the mixture model, like the slack variable model, is not redundant. In most situations it is a more reasonable model to use since all the mixture components are represented, and the coefficients are more easily interpreted.

12.2.3.1   Scheffé Linear Model

The general form of the linear Scheffé model is:

$$\hat{Y} = \sum_{i=1}^{k} b_i X_i. \tag{12.4}$$

The interpretation of the linear coefficients is very straightforward. Each b_i is simply predicted response (the height of the vertex) at $X_i = 1$.

12.2.3.2   Scheffé Quadratic Model

The general form of the Scheffé quadratic model is:

$$\hat{Y} = \sum_{i=1}^{k} b_i X_i + \sum_{i<j=2}^{k} b_{ij} X_i X_j. \tag{12.5}$$

The coefficients, b_{ij}, in the quadratic Scheffé model also have a simple interpretation in that they represent the amount of curvature along the face of the line connecting the point $X_i = 1$ and $X_j = 1$. This is shown graphically for the three-component case in Figure 12.4.

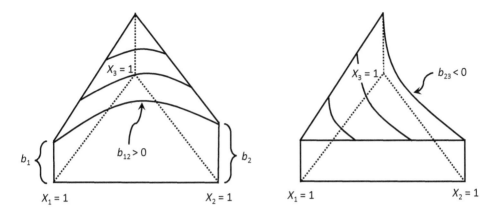

FIGURE 12.4: Graphical Interpretation of Cross-Product Coefficients in a Scheffé Quadratic Model

12.2.3.3 Scheffé Special Cubic Model

Frequently in mixture experiments a special cubic model is needed to represent the interaction between mixing components. The form of the special cubic model is:

$$\hat{Y} = \sum_{i=1}^{k} b_i X_i + \sum_{i<j=2}^{k} b_{ij} X_i X_j + \sum_{h<i<j=3}^{k} b_{hij} X_h X_i X_j. \tag{12.6}$$

An example of a special cubic model for k = 3 components is:

$$\hat{Y} = b_1 X_1 + b_2 X_2 + b_3 X_3 + b_{12} X_1 X_2 + b_{13} X_1 X_2 + b_{23} X_2 X_3 + b_{123} X_1 X_2 X_3,$$

and the cubic term b_{123} represents the difference in the response at the mixture $X_1 = 1/3$, $X_2 = 1/3$, $X_3 = 1/3$ and the value predicted at that point by the quadratic model.

12.2.3.4 Scheffé Full Cubic Model

If even the special cubic model is found to be inadequate, more terms can be added to complete the cubic model. The form of the full Scheffé cubic model is:

$$\hat{Y} = \sum_{i=1}^{k} b_i X_i + \sum_{i<j=2}^{k} b_{ij} X_i X_j + \sum_{h<i<j=3}^{k} b_{hij} X_h X_i X_j + \sum_{i<j=2}^{k} c_{ij} X_i X_j (X_i - X_j). \tag{12.7}$$

12.3 Experimental Designs for Mixture Problems

12.3.1 Unconstrained Problems

Unconstrained mixture problems are encountered when studying mixtures which may be 100 percent one component. For example, if studying the properties of a lubricant made up of mixtures of various types of mineral and synthetic oils, there would be no restriction on the proportion of one type oil in the mixture (i.e., it could be 100 percent mineral oil, 100 percent synthetic, or anything in between). For this type of problem, the best experimental designs available are the simplex designs.   The simplex designs for linear models consist of one observation taken at each vertex. For the three-component mixture, this would consist of the points shown in Table 12.1. The responses to be recorded at the three vertices are symbolically labeled Y_1, Y_2, and Y_3.

TABLE 12.1
Simplex Design in Three Components

Run Number	X_1	X_2	X_3	Symbolic Response
1	1	0	0	Y_1
2	0	1	0	Y_2
3	0	0	1	Y_3

Since there are three parameters to be estimated in the linear model, this design has no extra runs to estimate experimental error or test the goodness of fit. This can be compensated for by replicating the design or by adding a replicated center point. Adding a replicated center point allows for making a lack-of-fit test to determine if the linear model is adequate. The lack-of-fit test is determined by comparing the variation between the observed and predicted response at the center point to the variation among the replicate responses at the center point. An example of a three-component simplex design with a replicated center point is shown below in Table 12.2 and graphically in Figure 12.5.

TABLE 12.2
Simplex Design with Replicated Center Point

Run Number	X_1	X_2	X_3	Symbolic Response
1	1	0	0	Y_1
2	0	1	0	Y_2
3	0	0	1	Y_3
4	1/3	1/3	1/3	$Y_{123(1)}$
5	1/3	1/3	1/3	$Y_{123(2)}$
6	1/3	1/3	1/3	$Y_{123(3)}$

The simplex designs for quadratic models include points at the midpoint of the lines connecting each vertex in order to allow estimation of the non-linear effects. For example, the three-component simplex design for a quadratic model is shown in Table 12.3. The replicated center points in the quadratic simplex design allow for both testing the lack of fit from a quadratic model and fitting the special cubic model to the data.

This design is shown graphically in Figure 12.6 along with formulas for computing the coefficients in the Scheffé model. These formulas give the least squares estimates of the coefficients in Equation 12.6.

The generalization of these designs to more components is straight-forward. In Appendix B, Tables B.6-1 to B.6-5 show quadratic mixture designs for three to seven components along with formulas for estimating the coefficients in the special cubic model. A simplex linear design can be created from these tables by including only the vertex points and overall

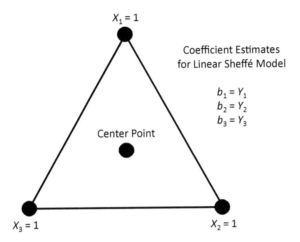

FIGURE 12.5: Linear Simplex Design and Estimates of the Scheffé Linear Model Coefficients

TABLE 12.3
Simplex Quadratic Design

Run Number	X_1	X_2	X_3	Symbolic Response
1	1	0	0	Y_1
2	0	1	0	Y_2
3	0	0	1	Y_3
4	1/2	1/2	0	Y_{12}
5	1/2	0	1/2	Y_{13}
6	0	1/2	1/2	Y_{23}
7	1/3	1/3	1/3	$Y_{123(1)}$
8	1/3	1/3	1/3	$Y_{123(2)}$
9	1/3	1/3	1/3	$Y_{123(3)}$

center points.

12.3.2 An Example

In agricultural field tests two or more herbicides are often mixed together in so-called "tank mixes" in an attempt to find a mixture which is more effective than individual components in controlling pest weeds. In a specific test various combinations of herbicide A (formulated to control broad leaf weeds), herbicide B (formulated to control seedling grasses), and a general purpose herbicide C were mixed together in the proportions shown in Table 12.4.

Each mixture was applied to a randomly selected plot of corn and the weed control index in percent was recorded as the response. The design can be seen to be a quadratic simplex design with three replicates of the center point. The coefficients in the quadratic Scheffé model were estimated and found to be:

$$\hat{b}_1 = 73, \ \hat{b}_2 = 68, \ \hat{b}_3 = 80, \ \hat{b}_{12} = 26, \ \hat{b}_{13} = 38, \ \hat{b}_{23} = 4$$

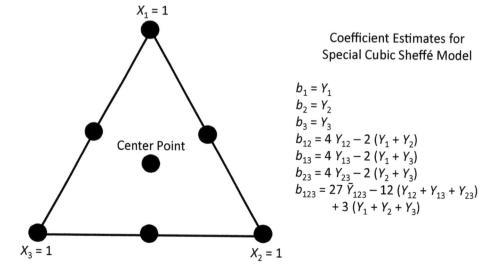

FIGURE 12.6: Quadratic Simplex Design and Estimates of the Scheffé Special Cubic Model Coefficients

TABLE 12.4

Herbicide Mixture Experiments

Mixture	A	B	C	% Weed Control
1	1	0	0	$Y_1 = 73$
2	0	1	0	$Y_2 = 68$
3	0	0	1	$Y_3 = 80$
4	1/2	1/2	0	$Y_{12} = 77$
5	1/2	0	1/2	$Y_{13} = 86$
6	0	1/2	1/2	$Y_{23} = 75$
7	1/3	1/3	1/3	$Y_{123(1)} = 92$
8	1/3	1/3	1/3	$Y_{123(2)} = 93$
9	1/3	1/3	1/3	$Y_{123(3)} = 88$

TABLE 12.5

Predictions Versus Results for Herbicide Mixture Experiments

Mixture	A	B	C	Actual % Weed Control	Predicted % Weed Control	Residual
1	1	0	0	73	73	0
2	0	1	0	68	68	0
3	0	0	1	80	80	0
4	1/2	1/2	0	77	77	0
5	1/2	0	1/2	86	86	0
6	0	1/2	1/2	75	75	0
7	1/3	1/3	1/3	92	81.22	10.78
8	1/3	1/3	1/3	93	81.22	11.78
9	1/3	1/3	1/3	88	81.22	6.78

using the formulas shown in Figure 12.6. The predicted values and differences between actual and predicted values are shown in Table 12.5. The sum of squares which represents the variation between the actual and predicted values at the center point is:

$$S^2_{LOF} = n_c(\bar{Y}_{123}\text{-}\hat{Y}_{123})^2 = 3\,(91 - 81.22)^2 = 286.815$$

and has 1 degree of freedom (the number of independent conditions, 7, minus the number of coefficients in the model, 6). The sum of squares which represents the variation in replicated center points is:

$$S^2_{PE} = \sum_{i=1}^{n_c}(Y_i - \bar{Y}_c)^2/(n_c - 1) = [(92{-}91)^2{+}(93{-}91)^2{+}(88{-}91)^2]/2 = 7$$

which has 2 degrees of freedom. These two sums of squares can be compared using the *F*-ratio to see if the deviation of the actual responses from the predicted response can be explained reasonably by experimental variation.

$$F_{1,2} = 286.8/7 = 40.97.$$

The tabulated *F*-ratio at the 5% point for 1 and 2 degrees of freedom is 18.51. In this case the computed *F*-ratio is larger than the critical value indicating there is a significant lack of fit to the quadratic Scheffé model. Therefore, the special cubic model is needed to provide a better fit to the data. The coefficient of the special cubic term is calculated as shown in Figure 12.6 as:

$$b_{123} = 27(91) - 12(77{+}86{+}75) + 3(73{+}68{+}80) = 264$$

and the special cubic equation is:

$$\hat{Y} = 73X_1{+}68X_2{+}80X_3{+}26X_1X_2{+}38X_1X_3{+}4X_2X_3{+}264X_1X_2X_3.$$

Figure 12.7 shows a contour plot of the fitted special cubic equation. In this case it can be seen that a mixture of the three components will provide more effective weed control than any one of the three components (vertex points) or any mixture of two (which would lie along the line connecting any two vertices). The exact coordinates of the best mixture is $X_1 = 0.374$, $X_2 = 0.234$, and $X_3 = 0.392$, where the weed control is predicted to be 91.8%. This point is indicated on the graph.

12.3.3   Constrained Mixture Problems

In some mixture problems it is impossible to have a product which is 100 percent pure in one component. For example, a rocket propellant, which is a mixture of binder, oxidizer, and fuel, cannot be 100 percent binder or any other component for that matter. The proportion of the three components can be varied, but must remain within certain constraints in order for the propellant to work at all. In this case, the experimental region is not the entire triangular plane of the unconstrained three-component mixture experiment. Instead, it is some subset of the plane. Kurotori [1966] described an example where the binder in a rocket propellant must be at least 20 percent, the oxidizer must be at least 40 percent, and the fuel must be at least 20 percent. In this case, the experimental region is the shaded area shown in Figure 12.8. It can be clearly seen that the simplex designs will be inappropriate in this

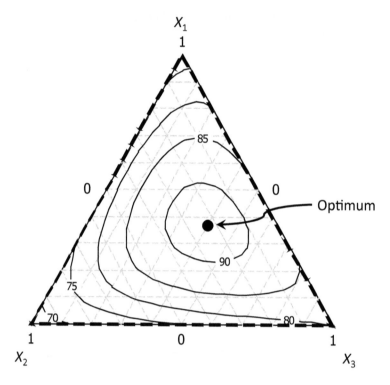

FIGURE 12.7: Contour Plot of Percent Weed Control

situation, since all design points except the center point are not even in the experimental region.

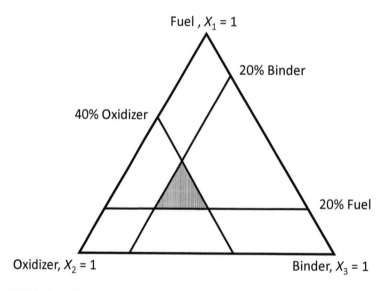

FIGURE 12.8: Experimental Region for Constrained Mixture Experiment

12.3.3.1   Pseudocomponents

In cases where each mixture component has only a lower constraint, appropriate experimental designs for constrained mixture problems can be defined by using pseudocomponents. pseudocomponents are not pure components but a mixture of several components. If the lower constraint for the mixture component X_i is L_i (i.e. $X_i \leq L_i$) then the ith pseudocomponent X_i' is defined by:

$$X_i' = (X_i - L_i)/(1 - \sum_{j=1}^{k} L_j), \qquad (12.8)$$

and ith pure component can be written in terms of the pseudocomponents

$$X_i = L_i + (1 - \sum_{j=1}^{k} L_j) X_i'. \qquad (12.9)$$

For example, in the rocket propellant example described in the last section the upper vertex of the experimental region shown in Figure 12.8 is

Fuel $X_1 = 0.40$
Oxidizer $X_2 = 0.40$,
Binder $X_3 = 0.20$

and the lower constraints for each component are

Fuel $L_1 = 0.20$
Oxidizer $L_2 = 0.40$.
Binder $L_3 = 0.20$

Therefore, the pseudocomponents for the upper vertex can be calculated as:

$$X_1' = (X_1 - L_1)/(1 - \sum_{j=1}^{k} L_j) = \frac{0.40 - 0.20}{1 - 0.80} = 1.00$$

$$X_2' = (X_2 - L_2)/(1 - \sum_{j=1}^{k} L_j) = \frac{0.40 - 0.40}{1 - 0.80} = 0.00$$

$$X_3' = (X_3 - L_3)/(1 - \sum_{j=1}^{k} L_j) = \frac{0.20 - 0.20}{1 - 0.80} = 0.00.$$

All of the pseudocomponents will be between 0 and 1 for any mixture in the constrained experimental region, and therefore a simplex design in the pseudocomponents can be defined. Once this is done, the actual experimental mixtures can be determined by using Equation 12.9 to convert the pseudocomponents back to the pure components. For example, a quadratic simplex design in three pseudocomponents is shown in Table 12.6. The pure component levels are obtained from the pseudocomponents, as already mentioned, using Equation 12.9. For example, for Run 1

$X_1' = 1.0$, so $X_1 = 0.20 + (1 - 0.80)(1.0) = 0.40$

TABLE 12.6

Experimental Design and Data

| | Pseudo-Component | | | Pure Component | | | |
Run	X'_1	X'_2	X'_3	X_1 Fuel	X_2 Oxidizer	X_3 Binder	Response Elasticity
1	1	0	0	0.400	0.400	0.200	2350
2	0	1	0	0.200	0.600	0.200	2450
3	0	0	1	0.200	0.400	0.400	2650
4	1/2	1/2	0	0.300	0.500	0.200	2400
5	1/2	0	1/2	0.300	0.400	0.300	2750
6	0	1/2	1/2	0.200	0.500	0.300	2950
7	1/3	1/3	1/3	0.267	0.467	0.267	3000

$X'_2 = 0.0$, so $X_2 = 0.40 + (1 - 0.80)(0.0) = 0.00$
$X'_3 = 0.0$, so $X_3 = 0.40 + (1 - 0.80)(0.0) = 0.00$.

And an experimental design in the pure components is created as shown on the right half of Table 12.6. The response for this experiment was the elasticity of the mixture. This design is shown graphically in Figure 12.9 (which is an enlargement of the shaded region in Figure 12.8). To analyze the data in Table 12.6 the special cubic model is fit to the pseudocomponents. This will be left for an exercise.

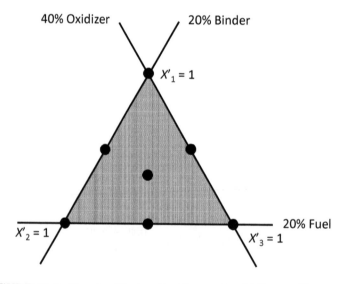

FIGURE 12.9: Simplex Design for Constrained Mixture Experiment

12.3.3.2 Extreme Vertices Design

In many mixture problems there are both upper and lower constraints on each component, i.e. $L_i \leq X_i \leq U_i$ for each component X_i. For example, coke briquettes, which are used as a reducing agent in metal production in an electric smelter furnace, are made by mixing certain proportions of coal calcinate, tar solids, and solids free of tar. These components are pressed into fixed size briquettes, and then baked at a constant temperature. It is known that the fixed carbon of the final briquettes is a function of the mixture components, and it is desirable to determine what mixture proportions will produce the highest fixed carbon.

This is a constrained mixture problem because the three components must be within the following limits in order for satisfactory coking to occur:

80 percent	$\leq$	calcinate	$\leq$	90 percent
8 percent	$\leq$	solids-free tar	$\leq$	15 percent
0 percent	$\leq$	tar solids	$\leq$	5 percent

In this problem the experimental region is irregular and the pseudocomponents cannot be used to define an experimental design. A useful experimental strategy for this situation is to use the extreme vertices designs of Mclean and Andersoni [1966]. These designs consist of all the extreme vertices of the irregular experimental region. To continue our example, the constrained experimental region and design points for the coke briquette problem are shown in Figure 12.10, and a listing of these vertices is given in Table 12.7. The centroid point is formed by averaging the components from all the vertices. Often midpoints of the sides are added to the list of experimental mixtures in order to fit the coefficients in the quadratic or special cubic model. These midpoints are found by averaging the components of the two corresponding vertices.

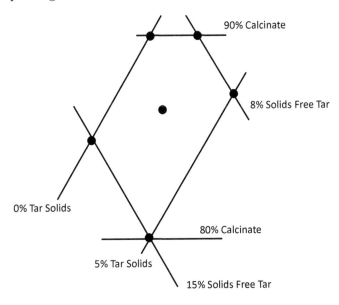

FIGURE 12.10: Constrained Experimental Region for Coke Briquette Experiment

TABLE 12.7
Extreme Vertices for Coke Briquette Experiment

Run	Calcinate	Tar	Tar Solids	
1	0.900	0.100	0.000	
2	0.900	0.080	0.020	
3	0.870	0.080	0.050	
4	0.800	0.150	0.050	
5	0.850	0.150	0.000	
6	0.864	0.112	0.024	(Centroid)

When there are more than three mixture components, it is difficult to visualize the experimental region or determine the extreme vertices graphically. Snee and Marquart [1974] have developed an algorithm for locating the extreme vertices in these more complicated situations. The algorithm is a five-step procedure which is illustrated with the coke briquette example.

1. Rank the components in order of increasing range, $U_i - L_i$.

Component	U_i	L_i	Range	Rank
X_1	0.90	0.80	0.10	3
X_2	0.15	0.08	0.07	2
X_3	0.05	0.00	0.05	1

2. Set up a two-level factorial design in the k−1 components with the smallest ranges using the upper and lower constraint values as factor levels.

Run	X_3	X_2
1	0.00	0.08
2	0.05	0.08
3	0.00	0.15
4	0.05	0.15

3. Calculate the value of the omitted component X_k (with largest range) as $1.0 - \sum_{i=1}^{k-1} X_i$ for each of the 2^{k-1} runs listed in Step 2.

Run	X_3	X_2	X_1
1	0.00	0.08	0.92
2	0.05	0.08	0.87
3	0.00	0.15	0.85
4	0.05	0.15	0.80

4. If the calculated value of X_k in Step 3 falls within its upper and lower constraint (as Runs 2, 3, and 4 do) these runs are extreme vertices. The other runs (in this case, only Run 1) must be adjusted.

5. For each of the runs where X_k is out of its lower to upper constraint range, adjust X_k to its nearest limit, and form k−1 new runs where each of the other k−1 components are in turn adjusted by the same magnitude but in the opposite direction as X_k, so that $\sum_{i=1}^{k-1} X_i = 1.0$.

The design is now complete, but the new values for the first k−1 components need to be checked to see if they fall within their constraints. If they all are within their constraints, the adjusted k−1 runs are added to the list of extreme vertices, in place of the original run where X_k was out of its constraint range. The new values replacing Run 1 are given below. They are both found to be acceptable.

This algorithm may duplicate some vertices, but it will find all of the extreme vertices. Piepel [1988] has published FORTRAN code and Lawson et al. [2015] have made an R-package for a similar algorithm that can be used for locating extreme vertices in up to 12

Run	X_3	X_2	X_1	
1a	0.02	0.08	0.90	(Adjusting X_1 and X_3)
1b	0.00	0.10	0.90	(Adjusting X_1 and X_2)
2	0.05	0.08	0.87	
3	0.00	0.15	0.85	
4	0.05	0.15	0.80	

components.

When fitting a quadratic or special cubic model in a constrained region, face centroids and the overall centroid are usually added to the extreme vertices design. The face centroids can be found by locating all extreme vertices that have k−3 components equal, and averaging the levels of the other three components. For example, Anderson and Mclean [1976] present a constrained mixture experiment where the formula for a flare is obtained by mixing four chemicals, X_1 = magnesium, X_2 = sodium nitrite, X_3 = strontium nitrate, and X_4 = binder, together with the following constraints:

$$
\begin{array}{ccccc}
0.40 & \leq & X_1 & \leq & 0.60 \\
0.10 & \leq & X_2 & \leq & 0.50 \\
0.10 & \leq & X_3 & \leq & 0.50 \\
0.03 & \leq & X_4 & \leq & 0.08
\end{array}
$$

In this case, k=4 and the constrained region is shown in Figure 12.11. Table 12.8 shows the eight extreme vertices (Runs 1–8) of the constrained region, which are also shown in Figure 12.11. It can be seen that Runs 1–4 all have the same values for k−3 = 4−3 = 1 component (namely X_2). Therefore the average of the components for these four runs is Run 9, the face centroid of the side defined by the vertices 1, 2, 3, and 4 (see Figure 12.11). Runs 5 through 8 all have a common value for X_3 (lower constraint) and therefore Run 10 (the average of the component values for these runs) is the face centroid for the side defined by the vertices 5, 6, 7, and 8. The other face centroids (not shown in Figure 12.11) are defined as shown in Table 12.8, and the overall centroid (Run 15, also not shown in Figure 12.11) is the average of the components for all the vertices.

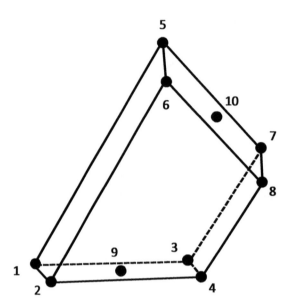

FIGURE 12.11: Constrained Experimental Region for Flare Experiment

TABLE 12.8
Extreme Vertices and Centroids for Flare Experiment

Run	X_1	X_2	X_3	X_4	Y	Type of Data Point
1	0.40	0.1000	0.4700	0.030	75	Extreme Vertex
2	0.40	0.1000	0.4200	0.080	180	Extreme Vertex
3	0.60	0.1000	0.2700	0.030	195	Extreme Vertex
4	0.60	0.1000	0.2200	0.080	300	Extreme Vertex
5	0.40	0.4700	0.1000	0.030	145	Extreme Vertex
6	0.40	0.4200	0.1000	0.080	230	Extreme Vertex
7	0.60	0.2700	0.1000	0.030	220	Extreme Vertex
8	0.60	0.2200	0.1000	0.080	350	Extreme Vertex
9	0.50	0.1000	0.3450	0.055	220	Face Centroid (1,2,3,4)
10	0.50	0.3450	0.1000	0.055	260	Face Centroid (5,6,7,8)
11	0.40	0.2725	0.2725	0.055	190	Face Centroid (1,2,5,6)
12	0.60	0.1725	0.1725	0.055	310	Face Centroid (3,4,7,8)
13	0.50	0.2350	0.2350	0.030	260	Face Centroid (1,3,5,7)
14	0.50	0.2100	0.2100	0.080	410	Face Centroid (2,4,6,8)
15	0.50	0.2225	0.2225	0.055	425	Overall Centroid

12.4 Data Analysis and Model Fitting for Constrained Mixture Problems

When an extreme vertices design is used there are no longer simple formulas to calculate estimates of the b's in Equation 12.3 or 12.5 from the observed experimental data. In this case, a linear regression routine must be used to estimate the b-parameters of the mixture model. To check how well the model fits the data, the same R^2, t- and F-statistics, the goodness-of-fit F-statistic and residual plots discussed in the second half of Chapter 10 are appropriate. Their application to mixture models will be discussed in Section 12.4.1. The methods of contour plotting and numerical analysis of a fitted model discussed in Section 10.6 are also applicable to the mixture model, and will be discussed in Section 12.4.2.

12.4.1 Least Squares Model Fitting Examples

The method of least squares can be used to fit both the slack variable form of the model (Equation 12.3) or the Scheffé form of the model (Equations 12.5 and 12.6). For example, assume the mixture experiments were run for the coke briquetting problem and the following crushing strengths were observed:

Run Number	X_1	X_2	X_3	Crushing Strength
1	0.90	0.10	0.00	1122
2	0.90	0.08	0.02	1109
3	0.87	0.08	0.05	1103
4	0.80	0.15	0.05	1118
5	0.85	0.15	0.00	1130
6	0.85	0.12	0.03	1114

To fit the slack variable linear model to this data, we will use the normal equations (given in matrix form as Equation 8.35), or more precisely the solution of them, Equation 8.36. The $\mathbf{X}$ and $\mathbf{Y}$ matrices used in these equations would be:

$$\mathbf{X} = \begin{bmatrix} 1 & 0.10 & 0.00 \\ 1 & 0.08 & 0.02 \\ 1 & 0.08 & 0.05 \\ 1 & 0.15 & 0.05 \\ 1 & 0.15 & 0.00 \\ 1 & 0.12 & 0.03 \end{bmatrix} \qquad \mathbf{Y} = \begin{bmatrix} 1122 \\ 1109 \\ 1103 \\ 1118 \\ 1130 \\ 1114 \end{bmatrix}$$

where the major component (calcinate $= X_1$) is designated as the slack variable. The normal equations are $(\mathbf{X}^T\mathbf{X})\hat{\mathbf{B}} = \mathbf{X}^T\mathbf{Y}$:

$$\begin{bmatrix} 6 & 0.68 & 0.15 \\ 0.68 & 0.0822 & 0.0167 \\ 0.15 & 0.0167 & 0.0063 \end{bmatrix} \begin{bmatrix} \hat{b}_0 \\ \hat{b}_2 \\ \hat{b}_3 \end{bmatrix} = \begin{bmatrix} 6696 \\ 760.04 \\ 166.65 \end{bmatrix}$$

and the solution is $\hat{\mathbf{B}} = (\mathbf{X}^T\mathbf{X})^{-1}\,\mathbf{X}^T\mathbf{Y}$:

$$\begin{bmatrix} \hat{b}_0 \\ \hat{b}_2 \\ \hat{b}_3 \end{bmatrix} = \begin{bmatrix} 3.0637 & -22.807 & -12.4872 \\ -22.807 & 196.1538 & 23.0769 \\ -12.4872 & 23.0769 & 394.872 \end{bmatrix} \begin{bmatrix} 6696 \\ 760.04 \\ 166.65 \end{bmatrix} = \begin{bmatrix} 1098.91 \\ 210 \\ -269 \end{bmatrix}.$$

Therefore the fitted slack variable linear model is:

$$y = 1098.91 + 210X_2 - 269X_3.$$

Using this equation the predicted values can be found for each observation as $(\hat{\mathbf{Y}} = \mathbf{X}\hat{\mathbf{B}})$:

$$\begin{bmatrix} 1119.93 \\ 1110.34 \\ 1102.26 \\ 1116.97 \\ 1130.44 \\ 1116.05 \end{bmatrix} = \begin{bmatrix} 1 & 0.10 & 0.00 \\ 1 & 0.08 & 0.02 \\ 1 & 0.08 & 0.05 \\ 1 & 0.15 & 0.05 \\ 1 & 0.15 & 0.00 \\ 1 & 0.12 & 0.03 \end{bmatrix} \begin{bmatrix} 1099 \\ 210 \\ -269 \end{bmatrix}.$$

To continue the analysis of this example, we need to test the significance of the equation. We start by examining the variation between the actual data and the predictions from the model compared to the variation of the data about simply its average. The calculation of both of these quantities is given in Table 12.9.

TABLE 12.9
Calculation of SSE and SST for Flare Experiment

Y	$\hat{Y}$	$Y-\hat{Y}$	$(Y-\hat{Y})^2$	$\bar{Y}$	$Y-\bar{Y}$	$(Y-\bar{Y})^2$
1122	1119.93	2.07	4.285	1116	6	36
1109	1110.34	−1.34	1.790	1116	−7	49
1103	1102.26	0.74	0.554	1116	−13	169
1118	1116.97	1.03	1.061	1116	2	4
1130	1130.44	−0.44	0.194	1116	14	196
1114	1116.05	−2.05	4.211	1116	−2	4
		$SSE =$	12.094		$SST =$	458

We will use an F-test to compare the two variances. The estimate of experimental error is: $s_E^2 = SSE/\nu_E = 12.094/(n-p) = 12.094/(6-3) = 4.031$. We could compare s_E^2 directly to the variance of the data about the average, but it is a more sensitive test of the model to partition the total sum of squares, SST, into two pieces, SSE and what is left over, which we will denote by SSR. The degrees of freedom are partitioned as well. The premise is that if the model is unnecessary, then $SST/(n-1)$ is an estimate of the pure error variance. Since $SSE/(n-p)$ is also an estimate of the pure error variance, what is left over when we subtract SSE from SST and divide by the degrees of freedom left over should likewise give us an estimate of the pure error variance. We denote the difference in the sum of squares as $SSR = SST - SSE = 458 - 12.094 = 445.906$, and denote the difference in the degrees of freedom as $s_R^2 = (n-1) - (n-p) = 5 - 3 = 2$. When we compute the F-ratio we get:

$$F_{2,3} = \frac{445.906/2}{12.094/3} = 55.31.$$

The critical $F_{2,3}$ (at 5% significance) is 9.55 from Table A.4, so the equation is quite significant. Another measure of the value of the equation is the multiple correlation coefficient, R^2. Its value is:

$$R^2 = \frac{SSR}{SST} = \frac{445.906}{458} = 0.916$$

which indicates that 91.6% of the variability in crushing strength can be explained by the relationship with the three mixing proportions X_1, X_2, and X_3.

Since we have found the prediction equation as a whole to be useful, it is now pertinent to see if any of the coefficients can be dropped from the model.

To compute the variability in the coefficients, we use Equation 8.47 as was done for response surface problems, $Var(\hat{\mathbf{B}}) = (\mathbf{X}^T\mathbf{X})^{-1}\sigma^2$. Since we do not know σ^2, we estimate it from SSE: $s^2 = SSE/(n-p) = 12.094/(6-3) = 4.031$, or $s = 2.01$. The computation of the standard errors of the regression coefficients $\hat{b}_0$, $\hat{b}_2$, $\hat{b}_3$ are shown in Table 12.10 along with their t-ratios.

TABLE 12.10
Worksheet for Calculating t-Statistics for Flare Experiment

Coefficient	Value	c_{ii} Diagonal of $(\mathbf{X}^T\mathbf{X})^{-1}$	Standard Error of Coefficient $(s_b = s\sqrt{c_{ii}})$	t-statistic
$\hat{b}_0$	1098.91	3.06	3.51	313
$\hat{b}_2$	210	196	28.1	7.48
$\hat{b}_3$	−269	395	38.9	−6.75

From the t-ratios it can be seen that both $\hat{b}_2$ and $\hat{b}_3$ are significant (since $t > t_3^*(.95\%)$ = 3.182 in Table A.3). This indicates that the difference between the response values at the vertex ($X_1 = 0$, $X_2 = 1$, $X_3 = 0$) and the vertex ($X_1 = 1$, $X_2 = 0$, $X_3 = 0$), which is what $\hat{b}_2$ measures, is significantly different from zero. Likewise, the difference between the response values at the vertex ($X_1 = 0$, $X_2 = 0$, $X_3 = 1$) and the vertex ($X_1 = 1$, $X_2 = 0$, $X_3 = 0$), which is what $\hat{b}_3$ measures, is also significantly different from zero. In other words the response plane above the simplex region, as shown in Figure 12.3 in Section 12.2, is not parallel to the simplex plane. The significance of the b_0 coefficient indicates that the response value above the vertex ($X_1 = 1$, $X_2 = 0$, $X_3 = 0$) is significantly greater than zero.

To fit the Scheffé form of the model, the X-matrix would be

$$\mathbf{X} = \begin{bmatrix} 0.90 & 0.10 & 0.00 \\ 0.90 & 0.08 & 0.02 \\ 0.87 & 0.08 & 0.05 \\ 0.80 & 0.15 & 0.05 \\ 0.85 & 0.15 & 0.00 \\ 0.85 & 0.12 & 0.03 \end{bmatrix},$$

and the Normal equations $(\mathbf{X}^T\mathbf{X})\hat{\mathbf{B}} = \mathbf{X}^T\mathbf{Y})$ would be

$$\begin{bmatrix} 4.4619 & 0.5811 & 0.127 \\ 0.5811 & 0.0822 & 0.0167 \\ 0.127 & 0.0167 & 0.0063 \end{bmatrix} \begin{bmatrix} \hat{b}_1 \\ \hat{b}_2 \\ \hat{b}_3 \end{bmatrix} = \begin{bmatrix} 5769.31 \\ 760.04 \\ 166.65 \end{bmatrix}.$$

The fitted model is summarized in Table 12.11. In this case all the coefficients ($\hat{b}_1$, $\hat{b}_2$, and $\hat{b}_3$) are all significant (i.e., t-values were > 3.182 from Table A.3). The interpretation of the Scheffé coefficients is that the response above the pure components are significantly different from zero as described in Section 12.2. Better ways of interpreting the fitted model will be explained in Section 12.4.3.

TABLE 12.11
Summary of Fitted Scheffé Coefficients and t-Statistics
for Flare Experiment

Coefficient	Value	Standard Error of Coefficient	t-statistic
$\hat{b}_1$	1098.9	3.51	313
$\hat{b}_2$	1309.1	24.9	52.6
$\hat{b}_3$	829.5	38.8	21.4

Computer programs are normally used to fit equations to mixture data. Figure 12.12 shows a portion of the results of fitting the Scheffé quadratic model,

$$y = b_1 X_1 + b_2 X_2 + b_3 X_3 + b_4 X_4$$
$$+ b_{12} X_1 X_2 + b_{13} X_1 X_3 + b_{14} X_1 X_4 + b_{23} X_2 X_3 + b_{24} X_2 X_4 + b_{34} X_3 X_4,$$

to the flare data from Table 12.8, using the MINITAB^{TM} regression procedure. The no-intercept option was used to fit this model.

The regression output illustrates one of the problems that often occurs when fitting prediction equations to data from constrained mixture designs by the least squares regression procedure. The problem is a high degree of correlation between the independent variables X_1, X_2, X_3, etc. The correlations are due to the highly irregular experimental region illustrated in Figure 12.11. When there is too much correlation between the independent variables in the $\mathbf{X}^T \mathbf{X}$ matrix, the Normal equations will be singular within machine precision, and a computer program will not be able to calculate the regression coefficients. Even in cases where the $(\mathbf{X}^T \mathbf{X})^{-1}$ matrix does exist, a high degree of correlation between the independent variables (often referred to as multicollinearity) can cause difficulties. The variance of the estimated regression coefficients is inflated by multicollinearity, and calculated values may have the wrong magnitude and/or sign. Inconsistent results are often reached when two different people try to fit a reduced model to the data by eliminating insignificant terms, or when automatic computer routines are used to select a subset of the independent variables (i.e., stepwise regression).

Some computer regression programs will calculate regression coefficients as long as the $(\mathbf{X}^T \mathbf{X})^{-1}$ matrix exists, and will not warn the user of potential problems due to multicollinearity. In this case the user should examine the standard errors of the regression coefficients or the diagonal elements of $(\mathbf{X}^T \mathbf{X})^{-1}$. Extremely large values indicate a problem. Better indicators of multicollinearity are the VIFs (variance inflation factors) that are optionally printed out by some computer regression programs including the regression procedure in MINITAB^{TM}. Some of the VIFs in Figure 12.12, particularly those involving X_4, are quite large and are strong evidence of multicollinearity. St.John [1984] describes the use of VIFs to detect multicollinearity in mixture experiments and illustrates a solution to the problem using ridge regression. Ridge regression will provide an equation that is not quite as accurate for predicting the actual experimental data within the experimental region, but the coefficients are less likely to have the wrong sign or magnitude and the prediction equation will be more stable for extrapolations slightly outside the constrained experimental region. However, if the purpose of the mixture experiment is to develop a prediction equation solely for use within the constrained experimental region rather than to interpret

```
Regression for Mixtures: Y versus X1, X2, X3, X4

Estimated Regression Coefficients for Y (component proportions)

Term    Coef   SE Coef      T      P       VIF
X1     -1557    893.1       *      *      855.5
X2     -2351    993.8       *      *      258.8
X3     -2426    993.8       *      *      258.8
X4     14358  53419.5       *      *    41042.2
X1*X2   8300   3780.9    2.20  0.080     835.1
X1*X3   8076   3780.9    2.14  0.086     835.1
X1*X4  -6609  59506.9   -0.11  0.916   13080.0
X2*X3   3214   1964.5    1.64  0.163      28.6
X2*X4 -16982  60062.5   -0.28  0.789    3051.0
X3*X4 -17111  60062.5   -0.28  0.787    3051.0

S = 59.9121      PRESS = 257796
R-Sq = 85.96%    R-Sq(pred) = 0.00%   R-Sq(adj) = 60.70%

Analysis of Variance for Y (component proportions)

Source          DF  Seq SS    Adj SS    Adj MS     F      P
Regression       9  109926  109926.1  12214.01  3.40  0.095
  Linear         3   69466    7951.2   2650.38  0.74  0.573
  Quadratic      6   40460   40460.5   6743.41  1.88  0.253
Residual Error   5   17947   17947.3   3589.46
Total           14  127873
```

FIGURE 12.12: MINITAB[TM]Multiple Regression Output Fitting a Quadratic Model to the Extreme Vertices Design for the Flare Experiment

the actual coefficients, the multiple regression equation will be an acceptable solution. To repeat this very important point—the regression equation found in Figure 12.12 will be accurate for predicting within the experimental region (shown in Figure 12.11), but it could be very inaccurate for extrapolating outside the experimental region and should not be used for that purpose.

12.4.2   A Procedure for Choosing the Correct Model

Due to synergism or antagonism of mixing components, highly nonlinear response surfaces often result in mixture experiments. If an estimate of the experimental variance is available from replicated experimental design points, Khuri and Cornell [1987] have recommended the following model fitting procedure.

1. Fit the linear Scheffé model given by Equation 12.4 and test for lack of fit. If the lack of fit is significant, go on to Step 2. Otherwise assume the linear model is adequate, and stop.

2. Fit the quadratic Scheffé model given by Equation 12.5 and test for lack of fit. If the lack of fit is significant, go on to Step 3. Otherwise assume the quadratic model is adequate, and stop.

3. Fit the special cubic Scheffé model given by Equation 12.6 and test for lack of fit. If the lack of fit is significant, go on to Step 4. Otherwise assume the special cubic model is adequate, and stop.

4. Fit the full cubic Scheffé model given by Equation 12.7.

12.4.3 Interpretation of Fitted Models

Interpretation of the coefficients in the Scheffé mixture models fit by least squares is different than the interpretation of non-mixture quadratic models, such as those fit in Chapter 5. When a linear coefficient, $\hat{b}_i$, is not significantly different from zero, as determined by the t-statistic, $t = \hat{b}_i/s_{\hat{b}_i}$, it does not mean that the component X_i has no significant effect on the response. A linear coefficient, b_i, that is zero implies that the predicted response is zero for the pure mixture $X_i = 1.0$. This is highly unlikely.

Each linear coefficient, b_i, in a non-mixture model is a direct measure of the effect, or change in the response, Y, that will be caused by a one unit change in the coded independent variable X_i while all other independent variables or factors are held constant. In the mixture model case it is impossible to change one component while holding all others constant because all components must sum to 100%. Therefore, Snee and Marquart [1976] suggested a more appropriate measure of the effect of a mixture component X_i. Their measure is called the **total effect**, and it is calculated as:

$$E_i = \hat{b}_i - \sum_{j=1(j\neq i)}^{k} b_j/(k-1).$$

This quantity measures the difference in the predicted values at the vertex, $X_i = 1.0$, and the point $X_i = 0$, $X_j = 1/(k-1)$, for $j \neq i$. This is illustrated for a three-component case in Figure 12.13.

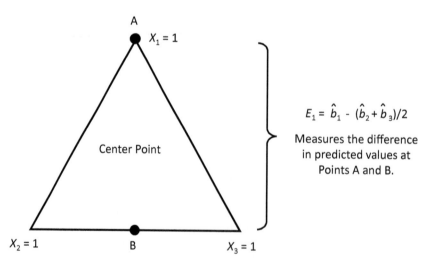

FIGURE 12.13: Total Effect of X_1, E_1

The total effect, E_i, is a good way of expressing the effects of the variables in linear models fit in unconstrained mixture regions. However, if there are constraints of the form $L_i \leq X_i \leq U_i$ or if higher order terms such as quadratic or special cubic are included in the model, Piepel [1982] has suggested a simple Effect Plot which provides a more accurate way of comparing the effects of several components. The plot is constructed by calculating the predicted values at two or more points along the line connecting the centroid of the experimental region and the vertex for the ith pseudocomponent. This is illustrated using the constrained experimental region for the coke briquette example of Section 12.3.3 in Figure 12.14.

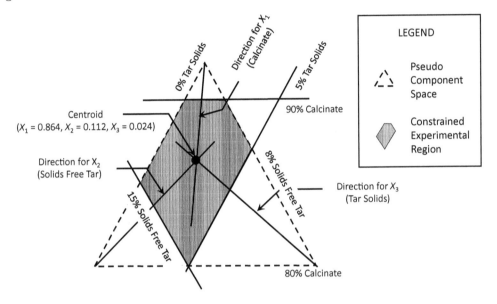

FIGURE 12.14: Directions for Effect Plot

Since the linear model ($Y = 1098.9X_1 + 1309.1X_2 + 829.5X_3$) was already fit to the briquette data, predicted values need to be calculated only at the centroid and pseudo-component vertices (represented in terms of the pure components). This is illustrated in the worksheet shown in Table 12.12. The predicted values are obtained using the equation coefficients for the Scheffé linear model. The Effect Plot is then constructed by plotting the predicted value versus the percentage change from the centroid for the ith component as shown in Table 12.12. The plot for this example is shown in Figure 12.15.

TABLE 12.12
Worksheet for Effect Plot for Briquette Experiment

	X_1	X_2	X_3	Predicted Value	\% Change from Centroid 100\% $(X_i - X_{ic})/(U_i - L_i)$		
Centroid	0.864	0.112	0.024	1115.99	X_1	X_2	X_3
X_1' vertex	0.920	0.080	0.000	1115.73	+56\%		
X_2' vertex	0.800	0.200	0.000	1140.95		+126\%	
X_3' vertex	0.800	0.080	0.120	1083.40			+192\%

The X_i in the worksheet represents the X_i component in the X_i' pseudocomponent vertex, X_{ic} represents the X_i component in the centroid, and U_i and L_i are the upper and lower constraints for component i. The percent change $100\%(X_i - X_{ic})/(U_i - L_i)$ is calculated for component 1 as $(100\%)(0.92 - 0.864)/(0.90 - 0.80) = 56\%$. In Figure 12.15, it can be seen that the effect of X_1 (Calcinate) is minor, while the effects of X_2 (Solids free tar) and X_3 (Tar Solids) are stronger with opposite signs.

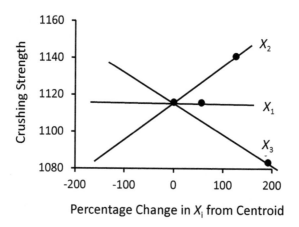

FIGURE 12.15: Effect Plot for Coke Briquette Example

If a quadratic or higher order model has been fit to the data, the predicted values should be calculated at least four points (i.e. three points plus the centroid) along the line connecting the centroid and the vertex in order to show the curvature. This is illustrated in Table 12.13 for the Herbicide example of Section 12.3.2. In the effect plot for this example, shown in Figure 12.16, it can be seen that X_2 has a strong negative and nearly linear effect, while X_1 and X_3 have quadratic effects. This indicates that there is synergism between X_1 and X_3, and that a lower percent of X_2 is desirable. It is probably worth mentioning that the Effect Plot is a visual way to get a feeling for the impact of each variable on the response. But it is NOT intended to be used as a method of seeking the optimum. That endeavor is discussed in the next section. Notice that if you picked the highest response on the Effect Plot (at $X_2 = 0$), that would not be the optimum shown in Figure 12.7 for this system.

TABLE 12.13
Worksheet for Effect Plot for Briquette Experiment

	$\mathbf{X_1}$	$\mathbf{X_2}$	$\mathbf{X_3}$	Predicted Value	% Change from Centroid $100\%\ (\mathbf{X}_i - \mathbf{X}_{ic})/(\mathbf{U}_i - \mathbf{L}_i)$		
Centroid	0.333	0.333	0.333	81.21	X_1	X_2	X_3
Point from X_1 Vertex	1.000	0.000	0.000	73.00	+67%		
Point from X_1 Vertex	0.667	0.167	0.167	80.65	+33%		
Point from X_1 Vertex	0.000	0.500	0.500	75.00	−33%		
Point from X_2 Vertex	0.000	1.000	0.000	68.00		+67%	
Point from X_2 Vertex	0.167	0.667	0.167	75.31		+33%	
Point from X_2 Vertex	0.500	0.000	0.500	86.00		−33%	
Point from X_3 Vertex	0.000	0.000	1.000	80.00			+67%
Point from X_3 Vertex	0.167	0.167	0.667	82.31			+33%
Point from X_3 Vertex	0.500	0.500	0.000	77.00			−33%

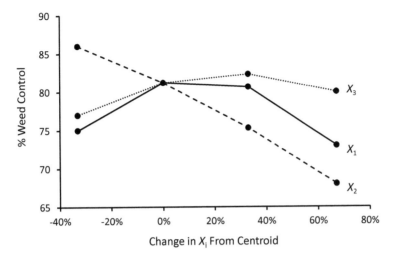

FIGURE 12.16: Effect Plot for Herbicide Example

12.4.4 Identifying Optimum Conditions

When mixture designs are used in a response surface situation, the ultimate goal is to identify the optimum mixture. This can be done in the same way that it is done in non-mixture response surfaces, namely through use of contour plots or numerical methods.

Contour plots over the simplex mixture experimental region have been illustrated previously in this chapter (e.g., Figure 12.7). Computer programs (Hare and Brown [1977], Koons and Heasly [1981], and Lawson et al. [2015]) have been published for producing these plots. When more than three mixture components are involved and constraints are placed on the experimental region, contour plots at various slices may be used to identify the optimum mixture. For example, considering the constrained experimental region for the flare mixture problem, contour plots can be made at three slices, $X_1 = 0.40$, $X_1 = 0.50$, and $X_1 = 0.60$ as shown in Figure 12.17. The resulting contour plots with the experimental region indicated are shown in Figure 12.18. From the contour plots it can be seen that the greatest illumination within the constrained experimental region is about 400 in the middle graph where $X_1 = 0.5$ and along the right most boundary where $X_4 = 0.08$.

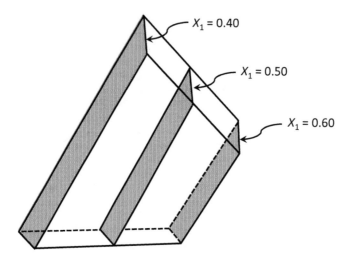

FIGURE 12.17: Slices Through Constrained Region Where Contour Plots Are Made

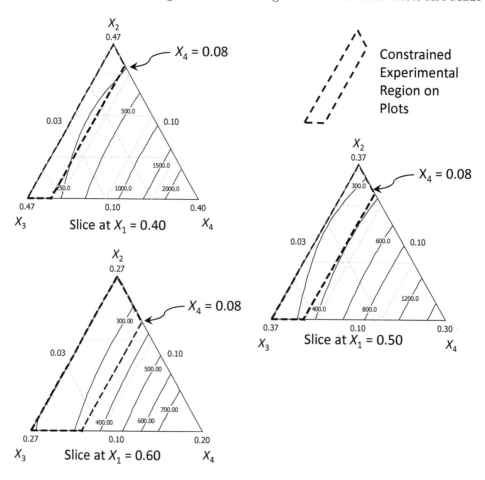

FIGURE 12.18: Contour Plots for Flare Mixture Example

12.5 Screening Experiments with Mixtures

12.5.1 Designs for Screening Experiments with Mixtures

At the beginning of any program of experimentation, the experimental environment should be diagnosed as explained in Section 1.4. How many factors or mixture components are being studied? What is the cost of experiments, and how much prior knowledge is there about the relationship of the response to mixture components? When there are a large number of mixture components under study with little understanding as to which ones are really important, it is wasteful to conduct a response surface type experiment. A screening experiment aimed at identifying the most important components is the natural first step and may reduce the amount of experimentation required in the long run. A screening experimental design should allow for fitting a linear model with as few experiments as possible.

For non-mixture problems, Plackett–Burman and Fractional Factorial designs were used in screening situations to allow fitting a linear model. In mixture problems these designs cannot be used directly because of the constraints on the experimental region. Extreme vertices designs have been described in Section 12.3.3, and without edge midpoints or face centroids, these designs can be used to fit a linear model. However, in a screening situation where there may be 10 or more mixture components, a constrained experimental region may contain 100 or more extreme vertices. Therefore a subset of the vertices from the McLean–Anderson Extreme Vertices design is normally used for a screening design (Snee and Marquart [1976]).

The subset of extreme vertices which minimizes the variances of the estimated regression coefficients, b_i's, is the subset which maximizes the determinant of the $\mathbf{X}^T\mathbf{X}$ matrix used in the regression calculations. This is the subset that should be used for a screening design.

To illustrate the principle consider the six extreme vertices for the coke briquetting example described in Section 12.3.3.2. If a four-vertex subset were to be selected for fitting the linear model, there would be five possible designs. One such design would be the first four vertices. The $\mathbf{X}$ matrix for this design would be:

$$\mathbf{X} = \begin{bmatrix} 0.90 & 0.10 & 0.00 \\ 0.90 & 0.08 & 0.02 \\ 0.87 & 0.08 & 0.05 \\ 0.80 & 0.15 & 0.05 \end{bmatrix},$$

the $\mathbf{X}^T\mathbf{X}$ matrix would be:

$$\mathbf{X}^T\mathbf{X} = \begin{bmatrix} 3.0169 & 0.3516 & 0.1015 \\ 0.3516 & 0.0453 & 0.0131 \\ 0.1015 & 0.0131 & 0.0054 \end{bmatrix}$$

and its determinant is: $\mathrm{Det}(\mathbf{X}^T\mathbf{X}) = .0000210$. The vertices and determinant of $\mathbf{X}^T\mathbf{X}$ for all five possible designs are shown in Table 12.14, and the best four-vertex design for fitting the linear model would be Design 4 since it has the maximum determinant of $\mathbf{X}^T\mathbf{X}$ = 3.70 x 10^{-5}.

In general, the computations are very time consuming to determine the subset of vertices with the maximum determinant of $\mathbf{X}^T\mathbf{X}$. However, there are computer programs available which can automate the process. Commercially available experimental design programs such as SAS^{TM}QC-proc optex and JMP^{TM}(SASInstitute [2015]) contain determinant maximizing

TABLE 12.14
Four-Vertex Designs for Coke Briquetting Example

Design	Vertices Included	Determinant of $(\mathbf{X}^T\mathbf{X})$
1	1, 2, 3, 4	2.10 x 10^{-5}
2	1, 2, 3, 5	1.20 x 10^{-5}
3	1, 2, 4, 5	2.35 x 10^{-5}
4	1, 3, 4, 5	3.70 x 10^{-5}
5	2, 3, 4, 5	3.33 x 10^{-5}

routines as well as the R-package AlgDesign (Wheeler [2014]). To use these programs, a list of all the vertices in the constrained region would be obtained using the algorithm described in Section 12.3.3.2. Next the program would be used to choose the subset which had the maximum $\mathrm{Det}(\mathbf{X}^T\mathbf{X})$.

If the above-mentioned computer programs are not available, Piepel [1990] has shown that a design developed from the non-mixture screening designs presented in Chapter 4 will be almost as good. His procedure is a slight modification of the algorithm described in Section 12.3.3.2 and can be outlined for the k component case as follows:

1. Rank the k components in order of increasing range, $U_i - L_i$. Call X_i the component with the ith largest range.

2. Set up a fractional factorial or Plackett–Burman design in the $k - 1$ components with the smallest ranges using the upper and lower constraint values as factor levels.

3. Calculate the value of the omitted component X_k (the one with the largest range) as:
$$X_k = 1.0 - \sum_{i=1}^{k-1} X_i \text{ for each of the runs in the fractional factorial design obtained in}$$
Step 2.

4. If the calculated value of X_k in Step 3 falls within its upper and lower constraints the run is a design point. Otherwise set X_k equal to the violated limit (i.e., U_k if the computed X_k from Step 3 is $> U_k$ or L_k if the computed $X_k < L_k$) and adjust (by an amount equal to the difference between the computed value of X_k and its violated limit) the first possible component (working from X_{k-1} to X_1) such that a value within the required limits is obtained.

5. If the adjustment required in Step 4 cannot be accommodated by a single component, make as much of the adjustment as possible on X_{k-1}, then as much of what remains as possible on X_{k-2}, etc. Note that each adjusted component except possibly the last one adjusted will be at it's upper or lower constraint.

6. Add a center point to the design.

This algorithm is illustrated with a gasoline blending example described by Cornell [1981]. The seven components are:
$X_1 =$ Natural Gasoline
$X_2 =$ Polymer

X_3 = Thermally cracked naphtha
X_4 = Straightrun
X_5 = Reformate
X_6 = Catalytically cracked naphtha
X_7 = Alkylate

As stated in Step 1, the variables are ranked inversely according to their ranges as shown below:

Component	Upper Limit	Lower Limit	Range
X_1	0.08	0.00	0.08
X_2	0.12	0.00	0.12
X_3	0.12	0.00	0.12
X_4	0.21	0.00	0.21
X_5	0.62	0.00	0.62
X_6	0.62	0.00	0.62
X_7	0.74	0.00	0.74

Next, according to Step 2, an appropriate screening design should be selected. Since seven components are to be screened, a reasonable design is a 12-run Plackett–Burman design as described in Chapter 4. An eight-run fractional factorial design would leave only 1 degree of freedom for error and would probably not provide enough information. A 16 run fractional factorial would be another reasonable choice. In this example, we illustrate with the 12-run Plackett–Burman design.

TABLE 12.15
Plackett–Burman Design for Gasoline Blending

Run	X_1	X_2	X_3	X_4	X_5	X_6	X_7
1	0.08	0.12	0.00	0.21	0.62	0.62	−0.65
2	0.08	0.00	0.12	0.21	0.62	0.00	−0.03
3	0.00	0.12	0.12	0.21	0.00	0.00	0.55
4	0.08	0.12	0.12	0.00	0.00	0.00	0.68
5	0.08	0.12	0.00	0.00	0.00	0.62	0.18
6	0.08	0.00	0.00	0.00	0.62	0.00	0.30
7	0.00	0.00	0.00	0.21	0.00	0.62	0.17
8	0.00	0.00	0.12	0.00	0.62	0.62	−0.36
9	0.00	0.12	0.00	0.21	0.62	0.00	0.05
10	0.08	0.00	0.12	0.21	0.00	0.62	−0.03
11	0.00	0.12	0.12	0.00	0.62	0.62	−0.48
12	0.00	0.00	0.00	0.00	0.00	0.00	1.00

The first six columns in Table 12.15 are copied from Table B.1-1 of Appendix B. The six components with the smallest ranges are copied in succession to the columns and the + and − signs in the table are replaced by the upper and lower ranges of the components, respectively. Next, according to Step 3, the value of X_7 is calculated as $1 - (X_1+X_2+X_3+X_4+X_5+X_6)$. For Runs 1, 2, 8, 10, and 11 it can be seen that the computed value for X_7 exceeds its lower constraint of 0.0 and for Run 12 it exceeds its upper constraint of 0.74. According to Step 4, the computed value of X_7, in Run 1, is replaced by its lower constraint 0.0 and the difference $(-0.65 - 0.0) = -0.65$ must be adjusted (i.e., added to the component with the next largest range (X_6). Since adding −0.65 to 0.62 would exceed the lower constraint for X_6, X_6 is set to its lower constraint 0.0 and the remainder $-0.65 + 0.62 = 0.03$ must be adjusted from the next component to the left, X_5, according to Step 5. Thus the altered component values for Run 1 would be $X_1 = 0.08$, $X_2 = 0.12$, $X_3 = 0.0$, $X_4 = 0.21$, $X_5 = 0.59$, $X_6 = 0.0$, and $X_7 = 0.0$. This procedure is repeated for Runs 2, 8, 10, and 11 where the computed value of X_7 also exceeded the lower constraint. In these runs only single adjustments to components X_5, X_6, X_6, and X_6, respectively, were required. For Run 12 where the computed value of X_7 exceeded the upper constraint, it is replaced by its upper constraint, 0.74, and in this case the positive difference $(1.0 - 0.74) = 0.26$ is again added to the component with the next largest range, X_6. Finally according to Step 6 a center point is added to the design by averaging each of the components over the 12 runs. The resulting screening design in 7 mixture components is shown in Table 12.16.

TABLE 12.16
Complete Screening Design for Gasoline Blending

Run	Nat. Gas. X_1	Polymer X_2	Therm. Cracked Naptha X_3	Straight-run X_4	Reform. X_5	Catalyt. Cracked Naptha X_6	Alkylate X_7
1	0.08	0.12	0.00	0.21	0.59	0.00	0.00
2	0.08	0.00	0.12	0.21	0.59	0.00	0.00
3	0.00	0.12	0.12	0.21	0.00	0.00	0.55
4	0.08	0.12	0.12	0.00	0.00	0.00	0.68
5	0.08	0.12	0.00	0.00	0.00	0.62	0.18
6	0.08	0.00	0.00	0.00	0.62	0.00	0.30
7	0.00	0.00	0.00	0.21	0.00	0.62	0.17
8	0.00	0.00	0.12	0.00	0.62	0.26	0.00
9	0.00	0.12	0.00	0.21	0.62	0.00	0.05
10	0.08	0.00	0.12	0.21	0.00	0.59	0.00
11	0.00	0.12	0.12	0.00	0.62	0.14	0.00
12	0.00	0.00	0.00	0.00	0.00	0.26	0.74
13	0.04	0.06	0.06	0.105	0.305	0.208	0.222

Once a screening design has been selected, the experiments are run, and the linear model is fit to the data using a regression program. The effect plots described in Sections 12.4.3 are used to determine which components are most important and should be used in further response surface type experiments. The effect plots take the place of the Normal probability plots or Pareto diagrams used to identify the important factors in non-mixture screening experiments. The mixture components having the steepest lines in the effect plots are most important and those whose lines have slopes near zero (like X_1, in Figure 12.15) can be dropped from further experimentation.

12.5.2   An Example

Snee and Marquart [1976] presented an example of a screening experiment in 8 mixture components. Economic requirements dictated that four of the components (X_1 X_2 X_5 and X_6) should be included in the mixture and the effects of the other four components were unknown. The factor ranges were as follows:

$$0.10 \leq X_1 \leq 0.45$$
$$0.05 \leq X_2 \leq 0.50$$
$$0.00 \leq X_3 \leq 0.10$$

$$
\begin{aligned}
0.00 &\leq X_4 \leq 0.10 \\
0.10 &\leq X_5 \leq 0.60\ . \\
0.05 &\leq X_6 \leq 0.20 \\
0.00 &\leq X_7 \leq 0.05 \\
0.00 &\leq X_8 \leq 0.05
\end{aligned}
$$

Using the algorithm described in Section 12.3.3.2, 188 extreme vertices were found in the constrained experimental region. It was decided to run a design consisting of 16 vertices plus four replicates of a centroid. A computer algorithm was used to select the subset of 16 vertices to maximize the determinant of the $\mathbf{X}^T\mathbf{X}$ matrix and the resulting design plus the observed responses are shown in Table 12.17.

TABLE 12.17
Design and Response for Snee and Marquart's Example

Blend	Run Order	X_1	X_2	X_3	X_4	X_5	X_6	X_7	X_8	Y
1	18	0.10	0.50	0.00	0.00	0.10	0.20	0.05	0.05	30
2	3	0.10	0.05	0.00	0.00	0.55	0.20	0.05	0.05	113
3	11	0.10	0.50	0.00	0.10	0.10	0.20	0.00	0.00	17
4	17	0.15	0.05	0.00	0.10	0.60	0.05	0.05	0.00	94
5	1	0.10	0.05	0.10	0.00	0.55	0.20	0.00	0.00	89
6	13	0.10	0.50	0.10	0.10	0.10	0.05	0.00	0.05	18
7	7	0.10	0.05	0.10	0.10	0.55	0.05	0.00	0.05	90
8	12	0.40	0.05	0.10	0.10	0.10	0.20	0.05	0.00	20
9	16	0.35	0.05	0.10	0.10	0.10	0.20	0.05	0.05	21
10	8	0.30	0.50	0.00	0.00	0.10	0.05	0.00	0.05	15
11	6	0.10	0.50	0.10	0.00	0.20	0.05	0.05	0.00	28
12	14	0.45	0.05	0.00	0.00	0.45	0.05	0.00	0.00	48
13	4	0.45	0.20	0.00	0.10	0.10	0.05	0.05	0.05	18
14	19	0.45	0.15	0.00	0.10	0.10	0.20	0.00	0.00	7
15	2	0.45	0.25	0.10	0.00	0.10	0.05	0.05	0.00	16
16	9	0.45	0.10	0.10	0.00	0.10	0.20	0.00	0.05	19
17	5	0.259	0.222	0.05	0.05	0.244	0.125	0.025	0.025	38
18	10	0.259	0.222	0.05	0.05	0.244	0.125	0.025	0.025	30
19	15	0.259	0.222	0.05	0.05	0.244	0.125	0.025	0.025	35
20	20	0.259	0.222	0.05	0.05	0.244	0.125	0.025	0.025	40

The results of a MINITAB^{TM} regression fitting the linear model,

$$
Y = b_1 X_1 + b_2 X_2 + b_3 X_3 + b_4 X_4 + b_5 X_5 + b_6 X_6 + b_7 X_7 + b_8 X_8,
$$

are shown in Figure 12.19.

Although the t-statistics for b_2, b_3, and b_4 show these coefficients are not significantly different from zero, they should not be dropped from the equation, as they would be in a non-mixture screening experiment. These coefficients only represent the predicted value at the pure component mixtures. In order to judge which coefficients should be left in the model it is necessary to examine the Effects Plot. The Effects Plot is used in screening designs for mixtures in the same way that normal or half-normal plots of effects are used in the analysis of unreplicated factorials in the non-mixture case. The Effect Plot (not shown) for this problem was constructed in the manner described for the coke briquetting and herbicide examples. From this plot it was seen that X_6 has a very minor effect, since its effect

```
Regression for Mixtures: Y versus X1, X2, X3, X4, X5, X6, X7, X8

Estimated Regression Coefficients for Y (component proportions)
```

Term	Coef	SE Coef	T	P	VIF
X1	$-$33.32	7.620	$-$4.37	0.001	3.132
X2	$-$10.26	6.438	$-$1.59	0.137	2.048
X3	$-$2.70	27.361	$-$0.10	0.923	2.081
X4	$-$19.74	27.451	$-$0.72	0.486	2.095
X5	150.40	5.978	25.16	0.000	2.040
X6	46.55	17.039	2.73	0.018	3.610
X7	165.45	55.721	2.97	0.012	2.158
X8	188.65	56.368	3.35	0.006	2.208

```
S = 5.68972      PRESS = 1196.13
R-Sq = 97.91%   R-Sq(pred) = 93.56%   R-Sq(adj) = 96.69%

Analysis of Variance for Y (component proportions)
```

Source	DF	Seq SS	Adj SS	Adj MS	F	P
Regression	7	18173.7	18173.7	2596.25	80.20	0.000
Linear	7	18173.7	18173.7	2596.25	80.20	0.000
Residual Error	12	388.5	388.5	32.37		
Lack-of-Fit	9	331.7	331.7	36.86	1.95	0.317
Pure Error	3	56.8	56.8	18.92		
Total	19	18562.2				

FIGURE 12.19: MINITAB^{TM}Equation Fit to Snee and Marquart's Example Data

line had a near zero slope. The mixture system could therefore be simplified by considering X_6 to be inert and refitting the equation in the other seven components. Before refitting, new compositions of the other seven components were computed $(X_i^* = X_i/(1-X_6)$ so that the sum of the components still added to 1. After refitting, Snee and Marquart [1976] also noticed that, within the computed standard errors of the coefficients, that the following coefficients were nearly equal: $b_1 \approx b_4$, $b_2 \approx b_3$, and $b_5 \approx b_7 \approx b_8$. Therefore the mixture component system could be further simplified to a three-component system,

$$Z_1 = (X_1 + X_4)/(1-X_6)$$
$$Z_2 = (X_2 + X_3)/(1-X_6)$$
$$Z_3 = (X_5 + X_7 + X_8)/(1-X_6),$$

and with 20 data points the quadratic equation

$$Y = -11.6Z_1 + 3.6Z_2 + 158.3Z_3 + 28.7Z_1Z_2 - 70.9Z_1Z_3 - 80.0Z_2Z_3$$

was fit to the data. Figure 12.20 shows a contour plot of this equation. The optimum mixture can be seen to be near the vertex where $Z_3 = 1$, or $(X_5+X_7+X_8)/(1-X_6) = 1$. Within the constraints of the system this implies that components X_1, X_2, X_3, X_4, and X_6

should be set to their lower constraints, and highest amounts of X_5, X_7, and X_8 should be used for the remaining part of the mixture.

This example illustrates, much like the non-mixture example shown in Section 7.4, how the result of a screening experiment can quickly lead to optimization when many factors are involved. It is always better to start a research project with more components and a screening type design than it is to optimize with respect to a few variables, and later find out that results are not optimal since some important component was left out of the study.

In summary, in mixture screening experiments the Effect Plot helps to identify important components. And equality of some regression coefficients can also lead to simplification of the fitted model.

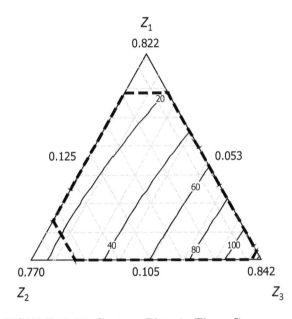

FIGURE 12.20: Contour Plots in Three Components

12.6 Mixture Experiments with Process Variables

In mixture experiments, the qualities or characteristics of the product are sometimes influenced by process variables in addition to proportions of the mixing components. For example in the production of silicon carbide ceramics for use in advanced heat engines, not only do the proportions of SiC powder and Boron influence the strength of the parts, but also process variables like the method of mixing, sintering time, and sintering temperature.

12.6.1 Designs for Mixture Experiments with Process Variables

Designs for mixture experiments with process variables are created by combining the non-mixture designs discussed in Chapters 3, 4, and 7 with the mixture designs discussed in Sections 12.3 and 12.5. For example, Figure 12.21 shows the combination of a linear simplex mixture design in three components with a 2^2 factorial design in two process variables. The experiments are listed in Table 12.18. It can be seen that the total number of runs in the experiment is equal to the product of the number of runs in the design for process variables (i.e., four) and the number of runs in the mixture design (i.e., again four). One of the most common situations in mixture experiments is when the total amount of the mixture may influence the product qualities, as well as the proportion of individual components. Of course before running a list of experiments like that shown in Table 12.18, the run order would be randomized to prevent biasing the estimates of either the process variable effects or mixture component coefficients.

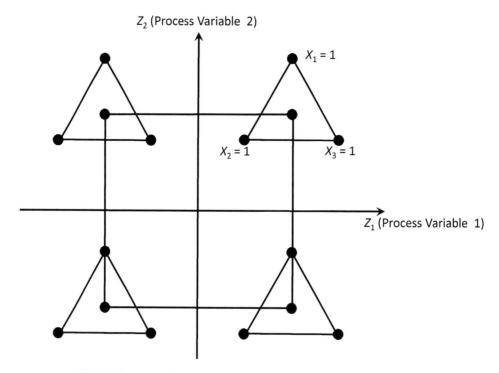

FIGURE 12.21: Combination of Simplex and Factorial Designs

In general any mixture design can be combined with any design for process variables by just repeating all the runs in the mixture design for each combination of runs in the process variable design.

12.6.2   Models for Mixture Experiments with Process Variables

The models $Y = \mathrm{f}(X_1, X_2, X_3, Z_1, Z_2)$ to be fit to the data resulting from a mixture experiment with process variables can be determined from the separate models appropriate for the specific mixture design and process variable design used. For example, the mixture design used to form Table 12.18 is a linear simplex design and the appropriate model for this design is the linear model, $Y = b_1X_1 + b_2X_2 + b_3X_3$. The process variable design used for Table 12.18 is the 2^2 factorial design, and an appropriate model for this design is the linear model with interaction, $Y = a_0 + a_1Z_1 + a_2Z_2 + a_{12}Z_1Z_2$. The model for the combined and mixture process variable experiment is determined by multiplying the two separate models together. The complete model is:

TABLE 12.18

Combination of Simplex and Factorial
Design

Run Number	Mixture Components			Process Variables	
	X_1	X_2	X_3	Z_1	Z_2
1	1	0	0	$-$	$-$
2	0	1	0	$-$	$-$
3	0	0	1	$-$	$-$
4	1/3	1/3	1/3	$-$	$-$
5	1	0	0	$+$	$-$
6	0	1	0	$+$	$-$
7	0	0	1	$+$	$-$
8	1/3	1/3	1/3	$+$	$-$
9	1	0	0	$-$	$+$
10	0	1	0	$-$	$+$
11	0	0	1	$-$	$+$
12	1/3	1/3	1/3	$-$	$+$
13	1	0	0	$+$	$+$
14	0	1	0	$+$	$+$
15	0	0	1	$+$	$+$
16	1/3	1/3	1/3	$+$	$+$

$$Y = (b_1 X_1 + b_2 X_2 + b_3 X_3)(a_0 + a_1 Z_1 + a_2 Z_2 + a_{12} Z_1 Z_2), \text{ or}$$

$$Y = b_1 a_0 X_1 + b_1 a_1 Z_1 X_1 + b_1 a_2 Z_2 X_1 + b_1 a_{12} Z_1 Z_2 X_1$$
$$+ b_2 a_0 X_2 + b_2 a_1 Z_1 X_2 + b_2 a_2 Z_2 X_2 + b_2 a_{12} Z_1 Z_2 X_2$$
$$+ b_3 a_0 X_3 + b_3 a_1 Z_1 X_3 + b_3 a_2 Z_2 X_3 + b_3 a_{12} Z_1 Z_2 X_3.$$

Letting $b_i a_j = d_j^i$, we can write the model as:

$$Y = d_0^1 X_1 + d_1^1 Z_1 X_1 + d_2^1 Z_2 X_1 + d_{12}^1 Z_1 Z_2 X_1 \qquad (12.10)$$
$$+ d_0^2 X_2 + d_1^2 Z_1 X_2 + d_2^2 Z_2 X_2 + d_{12}^2 Z_1 Z_2 X_2$$
$$+ d_0^3 X_3 + d_1^3 Z_1 X_3 + d_2^3 Z_2 X_3 + d_{12}^3 Z_1 Z_2 X_3.$$

In general this multiplication procedure can be used to determine the appropriate model for any mixture and process variable design. These models can then be fit to the data using the regression procedure outlined in Chapter 8, and optimum conditions can be found through use of contour plots or non-linear optimization techniques.

12.6.3   An Example

Cornell [1988] describes an experiment involving the mixture of three plasticizers to produce vinyl for automobile seat covers. The three plasticizers (X_1, X_2, and X_3) were subject to the following constraints:

$$
\begin{array}{ccccc}
0.47 & \leq & X_1 & \leq & 0.85 \\
0.00 & \leq & X_2 & \leq & 0.25 \\
0.15 & \leq & X_3 & \leq & 0.28
\end{array}
$$

and the extreme vertices design plus a centroid shown in Table 12.19 was used to study them. This five-point design allows fitting the linear model, $Y = b_1X_1 + b_2X_2 + b_3X_3$, which was thought appropriate for the mixture components.

TABLE 12.19
Extreme Vertices for Plasticizer Components

Blend Number	Plasticizer Proportions			
	X_1	X_2	X_3	
1	0.85	0.000	0.150	
2	0.72	0.000	0.280	
3	0.60	0.250	0.150	
4	0.47	0.250	0.280	
5	0.66	0.125	0.215	Centroid

The experiment also included two process variables: rate of extrusion (Z_1) and temperature at drying (Z_2). Both of these variables were studied at two levels in a 2^2 factorial design, which supported the model $Y = a_0 + a_1Z_1 + a_2Z_2 + a_{12}Z_1Z_2$. Two replicate experiments were performed at each of the 5×4 combinations of conditions obtained by combining the mixture and process variables designs, to obtain a total of 40 runs. The run conditions and the measured responses (scaled thickness values) are shown in Table 12.20. In the table the first 20 runs and the second 20 runs are repeat experiments at the same conditions. These were used to fit the combined model in the form of Equation 12.10. Part of the results from fitting the equation via MINITAB™ is shown in Figure 12.22.

```
Regression for Mixtures: Y versus X1, X2, X3, Z1, Z2

Estimated Regression Coefficients for Y (component proportions)

Term        Coef   SE Coef     T       P     VIF
X1         15.94    1.344      *       *   7.190
X2         14.44    2.934      *       *   2.135
X3        -16.75    4.575      *       *   9.155
X1*Z1       2.95    1.344    2.20   0.036  7.190
X2*Z1      -0.55    2.934   -0.19   0.853  2.135
X3*Z1      -2.82    4.575   -0.62   0.543  9.155
X1*Z2      -0.57    1.344   -0.42   0.675  7.190
X2*Z2      -1.57    2.934   -0.54   0.597  2.135
X3*Z2       2.31    4.575    0.51   0.617  9.155
X1*Z1*Z2    0.26    1.344    0.19   0.850  7.190
X2*Z1*Z2    1.26    2.934    0.43   0.672  2.135
X3*Z1*Z2   -0.70    4.575   -0.15   0.879  9.155
```

FIGURE 12.22: Fitted Model for Plasticizer Components

The residuals from the fitted model were plotted versus the mixture proportions X_1, X_2, and X_3. The results are shown in Figures 12.23, 12.24, and 12.25. From these figures it can be seen that there is some curvature in the mixture components that has not been accounted for in the fitted model. Since curvature in the Scheffé mixture model is accounted

TABLE 12.20
Run Conditions and the Measured Responses

Run	X_1	X_2	X_3	Z_1	Z_2	Y
1	0.85	0.000	0.150	−1.00	−1.00	8
2	0.75	0.000	0.280	−1.00	−1.00	6
3	0.60	0.250	0.150	−1.00	−1.00	10
4	0.47	0.250	0.280	−1.00	−1.00	4
5	0.66	0.125	0.215	−1.00	−1.00	11
6	0.85	0.000	0.150	1.00	−1.00	12
7	0.75	0.000	0.280	1.00	−1.00	9
8	0.60	0.250	0.150	1.00	−1.00	13
9	0.47	0.250	0.280	1.00	−1.00	6
10	0.66	0.125	0.215	1.00	−1.00	15
11	0.85	0.000	0.150	−1.00	1.00	7
12	0.75	0.000	0.280	−1.00	1.00	7
13	0.60	0.250	0.150	−1.00	1.00	9
14	0.47	0.250	0.280	−1.00	1.00	5
15	0.66	0.125	0.215	−1.00	1.00	9
16	0.85	0.000	0.150	1.00	1.00	12
17	0.75	0.000	0.280	1.00	1.00	10
18	0.60	0.250	0.150	1.00	1.00	14
19	0.47	0.250	0.280	1.00	1.00	6
20	0.66	0.125	0.215	1.00	1.00	13
21	0.85	0.000	0.150	−1.00	−1.00	7
22	0.75	0.000	0.280	−1.00	−1.00	5
23	0.60	0.250	0.150	−1.00	−1.00	11
24	0.47	0.250	0.280	−1.00	−1.00	5
25	0.66	0.125	0.215	−1.00	−1.00	10
26	0.85	0.000	0.150	1.00	−1.00	10
27	0.75	0.000	0.280	1.00	−1.00	8
28	0.60	0.250	0.150	1.00	−1.00	12
29	0.47	0.250	0.280	1.00	−1.00	3
30	0.66	0.125	0.215	1.00	−1.00	11
31	0.85	0.000	0.150	−1.00	1.00	8
32	0.75	0.000	0.280	−1.00	1.00	6
33	0.60	0.250	0.150	−1.00	1.00	10
34	0.47	0.250	0.280	−1.00	1.00	4
35	0.66	0.125	0.215	−1.00	1.00	7
36	0.85	0.000	0.150	1.00	1.00	11
37	0.75	0.000	0.280	1.00	1.00	9
38	0.60	0.250	0.150	1.00	1.00	12
39	0.47	0.250	0.280	1.00	1.00	5
40	0.66	0.125	0.215	1.00	1.00	9

for by the cross product terms X_1X_2, X_1X_3, and X_2X_3 three additional regression models were fit to the data each adding a different cross product term. It was impossible to add all three cross product terms to the regression model simultaneously because there were only five extreme vertices in the design. The model with the X_1X_2 term added fit the data best, and the MINITAB^{TM} regression report for this model is shown in Figure 12.26.

The significant positive coefficient for X_1Z_1 indicates that high levels of the process variable, Z_1 (rate of extrusion), increases Y (scaled thickness) at mixture combinations with high proportions of Plasticizer 1 (i.e., near the pure component $X_1 = 1.0$), and the

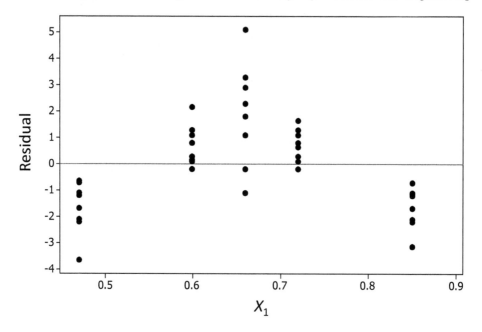

FIGURE 12.23: Residuals Versus X_1

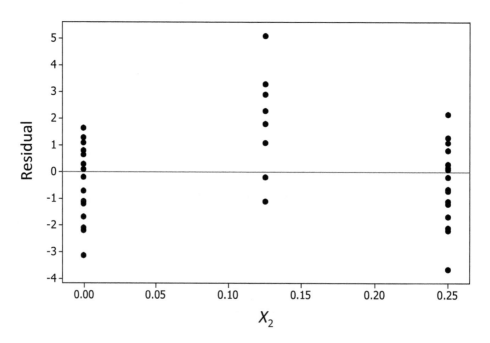

FIGURE 12.24: Residuals Versus X_2

significant coefficient for $X_1 X_2$ indicates that there is non-linear blending with mixtures of X_1 and X_2 that produces higher values of Y when mixtures of X_1 and X_2 are used.

The optimum conditions can be located by first producing four prediction equations by fixing (Z_1, Z_2) at the combinations $(-1,-1)$, $(-1,1)$, $(1,-1)$, and $(1,1)$. Next make contour

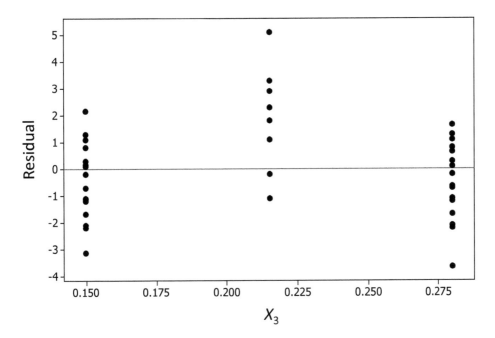

FIGURE 12.25: Residuals Versus X_3

Regression for Mixtures: Y versus X1, X2, X3, Z1, Z2

Estimated Regression Coefficients for Y (component proportions)

Term	Coef	SE Coef	T	P	VIF
X1	11.48	1.0445	*	*	11.613
X2	-69.54	12.2713	*	*	99.857
X3	-2.63	3.4633	*	*	14.028
X1*X2	148.64	21.4857	6.92	0.000	93.850
X1*Z1	2.95	0.8219	3.59	0.001	7.190
X2*Z1	-0.55	1.7945	-0.30	0.763	2.135
X3*Z1	-2.82	2.7978	-1.01	0.323	9.155
X1*Z2	-0.57	0.8219	-0.69	0.494	7.190
X2*Z2	-1.57	1.7945	-0.88	0.389	2.135
X3*Z2	2.31	2.7978	0.83	0.415	9.155
X1*Z1*Z2	0.26	0.8219	0.31	0.757	7.190
X2*Z1*Z2	1.26	1.7945	0.70	0.490	2.135
X3*Z1*Z2	-0.70	2.7978	-0.25	0.803	9.155

FIGURE 12.26: Expanded Model for Plasticizer Components

plots of Y versus X_1, X_2, and X_3 at each of the four Z_1, Z_2 combinations. Examination of the four contour plots will reveal the optimum.

12.7   Summary

In this chapter we have discussed the ideas of experimental design and data analysis in the context of mixture experiments. We have discussed the experimental designs and mathematical models appropriate for both screening experiments and optimization experiments. Below is a list of important concepts we have discussed in this chapter.

mixture component
experimental region for mixture experiments
slack variable model
Scheffé model (linear, quadratic, and special cubic)
interpretation of coefficients in Scheffé model
simplex design (linear and quadratic)
formulas for calculating model coefficients from simplex designs
constrained mixture experimental region
pseudocomponents
extreme vertices design
Snee and Marquart's algorithm for finding extreme vertices
multicollinearity
effect plot
screening designs for mixture experiments
mixture experiments with process variables (experiments and models)

12.8   Exercises

1. When there are three mixture components X_1, X_2, X_3 and no constraints,

   (a) List the experiments or mixtures that would be required to fit a special cubic model in these components: X_1, X_2, and X_3.

   (b) How would this list change if the components were subject to the following constraints:
   $0.20 \leq X_1 \leq 1.0$
   $0.25 \leq X_2 \leq 1.0$
   $0.15 \leq X_3 \leq 1.0$

   (c) List the experiments required to fit the linear model in five components: X_1, X_2, X_3, X_4, and X_5.

2. Soo et al. [1978] investigated the texture of shrimp patties made by blending proportions of isolated soy protein (X_1), sodium chloride (X_2), sodium tripolyphosphate (X_3), and Alaskan shrimp (X_4). The proportions of the ingredients were bounded by

the constraints:

$0.05 \leq X_1 \leq 0.10$
$0.01 \leq X_2 \leq 0.03$
$0.001 \leq X_3 \leq 0.005$
$0.865 \leq X_4 \leq 1.00$

The experimental design and resulting data are shown below:

Run	ISP X_1	NaCl X_2	STP X_3	Shrimp X_4	Texture Y
1	0.1	0.03	0.005	0.865	9.88
2	0.1	0.03	0.001	0.869	9.35
3	0.1	0.01	0.005	0.885	9.85
4	0.1	0.01	0.001	0.889	9.65
5	0.05	0.03	0.005	0.915	9.35
6	0.05	0.03	0.001	0.919	7.9
7	0.05	0.01	0.005	0.935	7.65
8	0.05	0.01	0.001	0.939	7.85
9	0.1	0.02	0.003	0.877	9.75
10	0.075	0.03	0.003	0.892	8.03
11	0.075	0.02	0.005	0.9	8.03
12	0.075	0.02	0.001	0.904	8.6
13	0.075	0.01	0.003	0.912	8.05
14	0.05	0.02	0.003	0.927	7.65
15	0.075	0.02	0.003	0.902	8.18
16	0.1	0.03	0.005	0.865	9.6
17	0.1	0.01	0.005	0.885	9.55
18	0.05	0.03	0.001	0.919	7.72
19	0.05	0.01	0.001	0.939	7.63
20	0.075	0.02	0.003	0.902	8.48

(a) Fit the Scheffé quadratic model to the data using regression.

(b) Using the coding and scaling:
$X_1' = (X_1 - 0.075)/0.025$, $X_2' = (X_2 - 0.02)/0.01$, and $X_3' = (X_3 - 0.003)/0.002$, fit the slack variable model to the coded and scaled proportions.

Chapter 13

Practical Suggestions for Successful Experimentation

13.1   Introduction

In this book we have presented some basic experimental designs and corresponding statistical methods of data analysis that are very effective for increasing knowledge about a phenomenon under study where: 1) factors can be deliberately manipulated; 2) responses of interest can be measured; and 3) knowledge must be gained by trial and error or experimentation, because there is no theory or known facts to rely upon. Statistical data analysis techniques lead to objective conclusions, and most of the methods we present in this book can be carried out with a simple spreadsheet program. In Chapter One we discussed some of the common pitfalls researchers often encounter when trying to draw conclusions from historical data, or by using classical approaches to experimentation. Throughout the rest of the book we present specific experimental plans, or designs, that can be used to avoid these pitfalls and increase the validity and objectivity of research.

The experimental plans we present can be copied from appendix tables or spreadsheets on the book website. There are various plans or designs that are appropriate for different situations. Different types of designs, and the appropriate statistical data analysis for each, can be used in a sequence of experiments to increase knowledge about a phenomenon from a low state to a high state. In this way sequential use of various designs forms a complete strategy for increasing knowledge.

13.2   Points to Remember

- **Do not use the classical one-at-a-time approach to experimentation.** This is the approach that has been used traditionally by holding all factors constant but one that is varied sequentially over a range. There is no set plan laid out to begin a sequence of experiments like this, and the results can be very easily biased by unknown influences such as learning curves in experimental technique, or changes in background variables. Further since there is no planned replication with this technique, there is no way to determine if any effects seen are real or just due to experimental error. One of the facts that has been illustrated in this book with many real examples is the presence of interactions. An interaction means the effect of one factor on the response may be different depending on the level at which another factor is held constant. Interactions are the key to solving many research problems, and they simply cannot be detected with one-at-a-time experimentation.

Even when interactions are not present, the number of experiments required by the one-at-a-time approach to reach the same level of precision you get with a fractional factorial design is greatly increased.

The one-at-a-time approach has also been attempted with mixture experiments, but since the factors are proportions that add up to the total amount in the mixture, it is impossible to vary one factor while holding all others constant. If one proportion in a mixture increases at least one other must decrease. This leads to correlated factors as described in Section 1.4, and it becomes impossible to sort out which factor has the effect.

- **Start with screening experiments**. In the preliminary stages of experimentation, it is always better to start with a screening experiment to identify the most important factors. In screening experiments the effects of a long list of factors can be explored in an efficient way. Brainstorming sessions with people knowledgeable about the process under study, and simple tools such as cause and effect diagrams will usually provide an adequate list. When many factors are studied simultaneously, surprises often result. For example, the largest significant effect discovered in the screening experiment described in Section 7.4.6 was not originally even planned to be studied. Just as experimenting with one factor holding all others constant is a bad idea; in preliminary stages of experimentation, experimenting with as many factors as is feasible is a good idea. In the early stages of experimentation don't be overly influenced by a minority opinion of what may and may not be important. Include as many factors as you can, and vary the factors over a wide range so that the effects can be seen above the experimental noise.

- **Save 50%–70% percent of the budget for experimentation for follow-up**. Although screening experiments are efficient for identifying the important factors in a long list of potential factors, they are not always efficient for detecting interactions or optimal factor settings. Save plenty of budget for performing foldover fractions to unconfound main effects and interactions, adding additional experiments to allow fitting quadratic prediction models if necessary, and performing confirmation experiments to check fitted model predictions.

- **Use adequate replication**. Make sure that you include enough replication to detect effects large enough to have practical or economical importance. To determine statistical significance, effects are always compared to the noise or variability among repeat experiments at the same conditions. By increasing the number of replicates, the standard error of estimated effects can be reduced, and the sensitivity of the statistical test increased.

  The formula presented in Section 4.2 is a starting point. It can be used to determine how many experiments are required in a factorial or screening experiment to detect effects of a specified size, when you have a preliminary estimate of the variability of the noise. This formula can also be used backwards to determine how large an effect can be detected when a given number of experiments (based on the budget) can be run. When the budget will not support an adequate number of experiments, searching for effects will be like looking for golf balls in a wheat field, and time may be better spent trying to study and reduce the amount of experimental error using designs like those described in Chapter 6.

- **Always analyze data statistically**. Statistical analysis of results guarantees approximate validity. The chance of false positive or false negative conclusions can be controlled by selecting the number of replicates and choosing the confidence level when statistical methods of data analysis are used.

- **Randomize the order of experiments**. Proper randomization of experiments must be performed in order to certify results of statistical analysis.

- **Represent results and conclusions in an equation or graphical form**. This is the best way to convey results to others (as well as yourself in the future). Linear and quadratic equations for presenting the results of experiments and graphical displays such as effect plots, interaction plots, and contour plots have been described in this book as methods for quantifying and visualizing models fitted to experimental data.

- **Always report prediction error limits on model predictions**. Variability is always present when experiments are repeated. By providing prediction limits with any predictions made from models fit to experimental data, the models become more valuable and predictions can be more easily verified with confirmation experiments.

Appendix A

TABLE A.1 Standard Normal Distribution Tail Areas

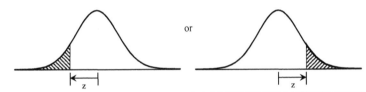

z	x.x0	x.x1	x.x2	x.x3	x.x4	x.x5	x.x6	x.x7	x.x8	x.x9
0.0	0.5000	0.4960	0.4920	0.4880	0.4840	0.4801	0.4761	0.4721	0.4681	0.4641
0.1	0.4602	0.4562	0.4522	0.4483	0.4443	0.4404	0.4364	0.4325	0.4286	0.4247
0.2	0.4207	0.4168	0.4129	0.4090	0.4052	0.4013	0.3974	0.3936	0.3897	0.3859
0.3	0.3821	0.3783	0.3745	0.3707	0.3669	0.3632	0.3594	0.3557	0.3520	0.3483
0.4	0.3446	0.3409	0.3372	0.3336	0.3300	0.3264	0.3228	0.3192	0.3156	0.3121
0.5	0.3085	0.3050	0.3015	0.2981	0.2946	0.2912	0.2877	0.2843	0.2810	0.2776
0.6	0.2743	0.2709	0.2676	0.2643	0.2611	0.2578	0.2546	0.2514	0.2483	0.2451
0.7	0.2420	0.2389	0.2358	0.2327	0.2296	0.2266	0.2236	0.2206	0.2177	0.2148
0.8	0.2119	0.2090	0.2061	0.2033	0.2005	0.1977	0.1949	0.1922	0.1894	0.1867
0.9	0.1841	0.1814	0.1788	0.1762	0.1736	0.1711	0.1685	0.1660	0.1635	0.1611
1.0	0.1587	0.1562	0.1539	0.1515	0.1492	0.1469	0.1446	0.1423	0.1401	0.1379
1.1	0.1357	0.1335	0.1314	0.1292	0.1271	0.1251	0.1230	0.1210	0.1190	0.1170
1.2	0.1151	0.1131	0.1112	0.1093	0.1075	0.1056	0.1038	0.1020	0.1003	0.0985
1.3	0.0968	0.0951	0.0934	0.0918	0.0901	0.0885	0.0869	0.0853	0.0838	0.0823
1.4	0.0808	0.0793	0.0778	0.0764	0.0749	0.0735	0.0721	0.0708	0.0694	0.0681
1.5	0.0668	0.0655	0.0643	0.0630	0.0618	0.0606	0.0594	0.0582	0.0571	0.0559
1.6	0.0548	0.0537	0.0526	0.0516	0.0505	0.0495	0.0485	0.0475	0.0465	0.0455
1.7	0.0446	0.0436	0.0427	0.0418	0.0409	0.0401	0.0392	0.0384	0.0375	0.0367
1.8	0.0359	0.0351	0.0344	0.0336	0.0329	0.0322	0.0314	0.0307	0.0301	0.0294
1.9	0.0287	0.0281	0.0274	0.0268	0.0262	0.0256	0.0250	0.0244	0.0239	0.0233
2.0	0.0228	0.0222	0.0217	0.0212	0.0207	0.0202	0.0197	0.0192	0.0188	0.0183
2.1	0.0179	0.0174	0.0170	0.0166	0.0162	0.0158	0.0154	0.0150	0.0146	0.0143
2.2	0.0139	0.0136	0.0132	0.0129	0.0125	0.0122	0.0119	0.0116	0.0113	0.0110
2.3	0.0107	0.0104	0.0102	0.0099	0.0096	0.0094	0.0091	0.0089	0.0087	0.0084
2.4	0.0082	0.0080	0.0078	0.0075	0.0073	0.0071	0.0069	0.0068	0.0066	0.0064
2.5	0.0062	0.0060	0.0059	0.0057	0.0055	0.0054	0.0052	0.0051	0.0049	0.0048
2.6	0.0047	0.0045	0.0044	0.0043	0.0041	0.0040	0.0039	0.0038	0.0037	0.0036
2.7	0.0035	0.0034	0.0033	0.0032	0.0031	0.0030	0.0029	0.0028	0.0027	0.0026
2.8	0.0026	0.0025	0.0024	0.0023	0.0023	0.0022	0.0021	0.0021	0.0020	0.0019
2.9	0.0019	0.0018	0.0018	0.0017	0.0016	0.0016	0.0015	0.0015	0.0014	0.0014
3.0	0.0013	0.0013	0.0013	0.0012	0.0012	0.0011	0.0011	0.0011	0.0010	0.0010
4.0	3.17E-05									
5.0	2.87E-07									
6.0	9.87E-10									
7.0	1.28E-12									

TABLE A.2 Right Tail Areas of the χ^2 Distribution with ν Degrees of Freedom

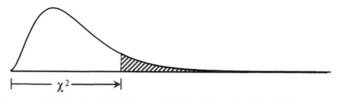

ν	0.99	0.975	0.95	0.9	0.5	0.10	0.05	0.025	0.01
1	1.57E-04	9.82E-04	3.93E-03	1.58E-02	0.4549	2.706	3.841	5.024	6.635
2	0.02010	0.05064	0.1026	0.2107	1.386	4.605	5.991	7.378	9.210
3	0.1148	0.2158	0.3518	0.5844	2.366	6.251	7.815	9.348	11.34
4	0.2971	0.4844	0.7107	1.064	3.357	7.779	9.488	11.14	13.28
5	0.5543	0.8312	1.145	1.610	4.351	9.236	11.07	12.83	15.09
6	0.8721	1.237	1.635	2.204	5.348	10.64	12.59	14.45	16.81
7	1.239	1.690	2.167	2.833	6.346	12.02	14.07	16.01	18.48
8	1.646	2.180	2.733	3.490	7.344	13.36	15.51	17.53	20.09
9	2.088	2.700	3.325	4.168	8.343	14.68	16.92	19.02	21.67
10	2.558	3.247	3.940	4.865	9.342	15.99	18.31	20.48	23.21
11	3.053	3.816	4.575	5.578	10.34	17.28	19.68	21.92	24.72
12	3.571	4.404	5.226	6.304	11.34	18.55	21.03	23.34	26.22
13	4.107	5.009	5.892	7.042	12.34	19.81	22.36	24.74	27.69
14	4.660	5.629	6.571	7.790	13.34	21.06	23.68	26.12	29.14
15	5.229	6.262	7.261	8.547	14.34	22.31	25.00	27.49	30.58
16	5.812	6.908	7.962	9.312	15.34	23.54	26.30	28.85	32.00
17	6.408	7.564	8.672	10.09	16.34	24.77	27.59	30.19	33.41
18	7.015	8.231	9.390	10.86	17.34	25.99	28.87	31.53	34.81
19	7.633	8.907	10.12	11.65	18.34	27.20	30.14	32.85	36.19
20	8.260	9.591	10.85	12.44	19.34	28.41	31.41	34.17	37.57
21	8.897	10.28	11.59	13.24	20.34	29.62	32.67	35.48	38.93
22	9.542	10.98	12.34	14.04	21.34	30.81	33.92	36.78	40.29
23	10.20	11.69	13.09	14.85	22.34	32.01	35.17	38.08	41.64
24	10.86	12.40	13.85	15.66	23.34	33.20	36.42	39.36	42.98
25	11.52	13.12	14.61	16.47	24.34	34.38	37.65	40.65	44.31
26	12.20	13.84	15.38	17.29	25.34	35.56	38.89	41.92	45.64
27	12.88	14.57	16.15	18.11	26.34	36.74	40.11	43.19	46.96
28	13.56	15.31	16.93	18.94	27.34	37.92	41.34	44.46	48.28
29	14.26	16.05	17.71	19.77	28.34	39.09	42.56	45.72	49.59
30	14.95	16.79	18.49	20.60	29.34	40.26	43.77	46.98	50.89

TABLE A.3 Two-Sided Student's t-Statistic

t*

ν	0.90	0.95	0.98	0.99
1	6.314	12.706	31.821	63.657
2	2.920	4.303	6.965	9.925
3	2.353	3.182	4.541	5.841
4	2.132	2.776	3.747	4.604
5	2.015	2.571	3.365	4.032
6	1.943	2.447	3.143	3.707
7	1.895	2.365	2.998	3.499
8	1.860	2.306	2.896	3.355
9	1.833	2.262	2.821	3.250
10	1.812	2.228	2.764	3.169
11	1.796	2.201	2.718	3.106
12	1.782	2.179	2.681	3.055
13	1.771	2.160	2.650	3.012
14	1.761	2.145	2.624	2.977
15	1.753	2.131	2.602	2.947
16	1.746	2.120	2.583	2.921
17	1.740	2.110	2.567	2.898
18	1.734	2.101	2.552	2.878
19	1.729	2.093	2.539	2.861
20	1.725	2.086	2.528	2.845
21	1.721	2.080	2.518	2.831
22	1.717	2.074	2.508	2.819
23	1.714	2.069	2.500	2.807
24	1.711	2.064	2.492	2.797
25	1.708	2.060	2.485	2.787
26	1.706	2.056	2.479	2.779
27	1.703	2.052	2.473	2.771
28	1.701	2.048	2.467	2.763
29	1.699	2.045	2.462	2.756
30	1.697	2.042	2.457	2.750
40	1.684	2.021	2.423	2.704
60	1.671	2.000	2.390	2.660
120	1.658	1.980	2.358	2.617
∞	1.645	1.960	2.326	2.576

TABLE A.4 Upper 5% Points of the F-Distribution with ν_1 and ν_2 Degrees of Freedom

ν_2	ν_1 1	2	3	4	5	6	7	8	9	10	12	15	20	25	30	40	60	120
1	161.4	199.5	215.7	224.6	230.2	234.0	236.8	238.9	240.5	241.9	243.9	245.9	248.0	249.3	250.1	251.1	252.2	253.3
2	18.51	19.00	19.16	19.25	19.30	19.33	19.35	19.37	19.38	19.40	19.41	19.43	19.45	19.46	19.46	19.47	19.48	19.49
3	10.13	9.55	9.28	9.12	9.01	8.94	8.89	8.85	8.81	8.79	8.74	8.70	8.66	8.63	8.62	8.59	8.57	8.55
4	7.71	6.94	6.59	6.39	6.26	6.16	6.09	6.04	6.00	5.96	5.91	5.86	5.80	5.77	5.75	5.72	5.69	5.66
5	6.61	5.79	5.41	5.19	5.05	4.95	4.88	4.82	4.77	4.74	4.68	4.62	4.56	4.52	4.50	4.46	4.43	4.40
6	5.99	5.14	4.76	4.53	4.39	4.28	4.21	4.15	4.10	4.06	4.00	3.94	3.87	3.83	3.81	3.77	3.74	3.70
7	5.59	4.74	4.35	4.12	3.97	3.87	3.79	3.73	3.68	3.64	3.57	3.51	3.44	3.40	3.38	3.34	3.30	3.27
8	5.32	4.46	4.07	3.84	3.69	3.58	3.50	3.44	3.39	3.35	3.28	3.22	3.15	3.11	3.08	3.04	3.01	2.97
9	5.12	4.26	3.86	3.63	3.48	3.37	3.29	3.23	3.18	3.14	3.07	3.01	2.94	2.89	2.86	2.83	2.79	2.75
10	4.96	4.10	3.71	3.48	3.33	3.22	3.14	3.07	3.02	2.98	2.91	2.85	2.77	2.73	2.70	2.66	2.62	2.58
11	4.84	3.98	3.59	3.36	3.20	3.09	3.01	2.95	2.90	2.85	2.79	2.72	2.65	2.60	2.57	2.53	2.49	2.45
12	4.75	3.89	3.49	3.26	3.11	3.00	2.91	2.85	2.80	2.75	2.69	2.62	2.54	2.50	2.47	2.43	2.38	2.34
13	4.67	3.81	3.41	3.18	3.03	2.92	2.83	2.77	2.71	2.67	2.60	2.53	2.46	2.41	2.38	2.34	2.30	2.25
14	4.60	3.74	3.34	3.11	2.96	2.85	2.76	2.70	2.65	2.60	2.53	2.46	2.39	2.34	2.31	2.27	2.22	2.18
15	4.54	3.68	3.29	3.06	2.90	2.79	2.71	2.64	2.59	2.54	2.48	2.40	2.33	2.28	2.25	2.20	2.16	2.11
16	4.49	3.63	3.24	3.01	2.85	2.74	2.66	2.59	2.54	2.49	2.42	2.35	2.28	2.23	2.19	2.15	2.11	2.06
17	4.45	3.59	3.20	2.96	2.81	2.70	2.61	2.55	2.49	2.45	2.38	2.31	2.23	2.18	2.15	2.10	2.06	2.01
18	4.41	3.55	3.16	2.93	2.77	2.66	2.58	2.51	2.46	2.41	2.34	2.27	2.19	2.14	2.11	2.06	2.02	1.97
19	4.38	3.52	3.13	2.90	2.74	2.63	2.54	2.48	2.42	2.38	2.31	2.23	2.16	2.11	2.07	2.03	1.98	1.93
20	4.35	3.49	3.10	2.87	2.71	2.60	2.51	2.45	2.39	2.35	2.28	2.20	2.12	2.07	2.04	1.99	1.95	1.90
21	4.32	3.47	3.07	2.84	2.68	2.57	2.49	2.42	2.37	2.32	2.25	2.18	2.10	2.05	2.01	1.96	1.92	1.87
22	4.30	3.44	3.05	2.82	2.66	2.55	2.46	2.40	2.34	2.30	2.23	2.15	2.07	2.02	1.98	1.94	1.89	1.84
23	4.28	3.42	3.03	2.80	2.64	2.53	2.44	2.37	2.32	2.27	2.20	2.13	2.05	2.00	1.96	1.91	1.86	1.81
24	4.26	3.40	3.01	2.78	2.62	2.51	2.42	2.36	2.30	2.25	2.18	2.11	2.03	1.97	1.94	1.89	1.84	1.79
25	4.24	3.39	2.99	2.76	2.60	2.49	2.40	2.34	2.28	2.24	2.16	2.09	2.01	1.96	1.92	1.87	1.82	1.77
26	4.23	3.37	2.98	2.74	2.59	2.47	2.39	2.32	2.27	2.22	2.15	2.07	1.99	1.94	1.90	1.85	1.80	1.75
27	4.21	3.35	2.96	2.73	2.57	2.46	2.37	2.31	2.25	2.20	2.13	2.06	1.97	1.92	1.88	1.84	1.79	1.73
28	4.20	3.34	2.95	2.71	2.56	2.45	2.36	2.29	2.24	2.19	2.12	2.04	1.96	1.91	1.87	1.82	1.77	1.71
29	4.18	3.33	2.93	2.70	2.55	2.43	2.35	2.28	2.22	2.18	2.10	2.03	1.94	1.89	1.85	1.81	1.75	1.70
30	4.17	3.32	2.92	2.69	2.53	2.42	2.33	2.27	2.21	2.16	2.09	2.01	1.93	1.88	1.84	1.79	1.74	1.68
40	4.08	3.23	2.84	2.61	2.45	2.34	2.25	2.18	2.12	2.08	2.00	1.92	1.84	1.78	1.74	1.69	1.64	1.58
60	4.00	3.15	2.76	2.53	2.37	2.25	2.17	2.10	2.04	1.99	1.92	1.84	1.75	1.69	1.65	1.59	1.53	1.47
120	3.92	3.07	2.68	2.45	2.29	2.18	2.09	2.02	1.96	1.91	1.83	1.75	1.66	1.60	1.55	1.50	1.43	1.35
∞	3.84	3.00	2.60	2.37	2.21	2.10	2.01	1.94	1.88	1.83	1.75	1.67	1.57	1.51	1.46	1.39	1.32	1.22

TABLE A.5 95% Confidence Values for the Studentized Range Statistic, q(0.05, k, ν)

Error Degrees of Freedom, ν	Number of Means Being Compared, k								
	3	**4**	**5**	**6**	**8**	**10**	**12**	**14**	**16**
1	27.0	32.8	37.2	40.5	45.4	49.1	52.0	54.3	56.3
2	8.33	9.80	10.89	11.73	13.03	13.99	14.7	15.4	15.9
3	5.91	6.83	7.51	8.04	8.85	9.46	9.95	10.3	10.7
4	5.04	5.76	6.29	6.71	7.35	7.83	8.21	8.52	8.79
5	4.60	5.22	5.67	6.03	6.58	6.99	7.32	7.60	7.83
6	4.34	4.90	5.31	5.63	6.12	6.49	6.79	7.03	7.24
7	4.16	4.68	5.06	5.36	5.80	6.15	6.43	6.66	6.85
8	4.04	4.53	4.89	5.17	5.60	5.92	6.18	6.39	6.57
9	3.95	4.42	4.76	5.02	5.43	5.74	5.98	6.19	6.36
10	3.88	4.33	4.66	4.91	5.30	5.60	5.83	6.03	6.19
11	3.82	4.26	4.58	4.82	5.20	5.49	5.71	5.90	6.06
12	3.77	4.20	4.51	4.75	5.12	5.40	5.62	5.80	5.95
13	3.73	4.15	4.46	4.69	5.05	5.32	5.53	5.71	5.86
14	3.70	4.11	4.41	4.64	4.99	5.25	5.46	5.64	5.79
15	3.67	4.08	4.37	4.59	4.94	5.20	5.40	5.57	5.72
16	3.65	4.05	4.34	4.56	4.90	5.15	5.35	5.52	5.66
17	3.62	4.02	4.31	4.52	4.86	5.11	5.31	5.47	5.61
18	3.61	4.00	4.28	4.49	4.83	5.07	5.27	5.43	5.57
19	3.59	3.98	4.26	4.47	4.79	5.04	5.23	5.39	5.53
20	3.58	3.96	4.24	4.45	4.77	5.01	5.20	5.36	5.49
24	3.53	3.90	4.17	4.37	4.68	4.92	5.10	5.25	5.38
30	3.48	3.84	4.11	4.30	4.60	4.83	5.00	5.15	5.27
40	3.44	3.79	4.04	4.23	4.52	4.74	4.90	5.04	5.16
60	3.40	3.74	3.98	4.16	4.44	4.65	4.81	4.94	5.06
120	3.36	3.68	3.92	4.10	4.47	4.56	4.71	4.84	4.95
∞	3.31	3.63	3.86	4.03	4.29	4.47	4.62	4.74	4.85

TABLE A.6 Multipliers to Obtain the Lenth ME and SME at the $\alpha = 0.05$ Level

m	for ME	for SME	m	for ME	for SME
7	2.296	4.866	19	2.120	4.080
8	2.272	4.734	23	2.094	4.011
11	2.213	4.450	26	2.081	3.901
15	2.157	4.207	27	2.077	3.973
17	2.137	4.134	31	2.066	3.925

TABLE A.7 Orthogonal Polynomials

Number of Factor Levels, I = 3

Factor Level	Linear c_{1i}	Quadratic c_{2i}
1	-1	1
2	0	-2
3	1	1

Number of Factor Levels, I = 4

Factor Level	Linear c_{1i}	Quadratic c_{2i}	Cubic c_{3i}
1	-3	1	-1
2	-1	-1	3
3	1	-1	-3
4	3	1	1

Number of Factor Levels, I = 5

Factor Level	Linear c_{1i}	Quadratic c_{2i}	Cubic c_{3i}	Quartic c_{4i}
1	-2	2	-1	1
2	-1	-1	2	-4
3	0	-2	0	6
4	1	-1	-2	-4
5	2	2	1	1

TABLE A.8 Factors ($d_2{}^*$) for Converting the Average Range ($\bar{R}$) into a Standard Deviation (σ)

n	1	2	3	4	5	8	10	∞
2	1.41	1.28	1.23	1.21	1.19	1.17	1.16	1.13
3	1.91	1.81	1.77	1.75	1.74	1.72	1.72	1.69
4	2.24	2.15	2.12	2.11	2.10	2.08	2.08	2.06
5	2.48	2.40	2.38	2.37	2.39	2.35	2.34	2.33
6	2.67	2.60	2.58	2.57	2.56	2.55	2.55	2.53
7	2.83	2.77	2.75	2.74	2.73	2.72	2.72	2.70
8	2.96	2.91	2.89	2.88	2.87	2.87	2.86	2.85
9	3.08	3.02	3.01	3.00	2.99	2.99	2.98	2.97
10	3.19	3.13	3.11	3.10	3.10	3.09	3.09	3.08

Appendix B

Appendix B.1 Screening Designs

TABLE B.1-1 12-Run Plackett-Burman Design

Run	Mean	X_1	X_2	X_3	X_4	X_5	X_6	X_7	X_8	X_9	X_{10}	X_{11}
1	+	+	+	−	+	+	+	−	−	−	+	−
2	+	+	−	+	+	+	−	−	−	+	−	+
3	+	−	+	+	+	−	−	−	+	−	+	+
4	+	+	+	+	−	−	−	+	−	+	+	−
5	+	+	+	−	−	−	+	−	+	+	−	+
6	+	+	−	−	−	+	−	+	+	−	+	+
7	+	−	−	−	+	−	+	+	−	+	+	+
8	+	−	−	+	−	+	+	−	+	+	+	−
9	+	−	+	−	+	+	−	+	+	+	−	−
10	+	+	−	+	+	−	+	+	+	−	−	−
11	+	−	+	+	−	+	+	+	−	−	−	+
12	+	−	−	−	−	−	−	−	−	−	−	−

TABLE B.1-2 20-Run Plackett-Burman Design

Run	Mean	X_1	X_2	X_3	X_4	X_5	X_6	X_7	X_8	X_9	X_{10}	X_{11}	X_{12}	X_{13}	X_{14}	X_{15}	X_{16}	X_{17}	X_{18}	X_{19}
1	+	+	+	−	−	+	+	+	+	−	+	−	+	−	−	−	−	+	+	−
2	+	+	−	−	+	+	+	+	−	+	−	+	−	−	−	−	+	+	−	+
3	+	−	−	+	+	+	+	−	+	−	+	−	−	−	−	+	+	−	+	+
4	+	−	+	+	+	+	−	+	−	+	−	−	−	−	+	+	−	+	+	−
5	+	+	+	+	+	−	+	−	+	−	−	−	−	+	+	−	+	+	−	−
6	+	+	+	+	−	+	−	+	−	−	−	−	+	+	−	+	+	−	−	+
7	+	+	+	−	+	−	+	−	−	−	−	+	+	−	+	+	−	−	+	+
8	+	+	−	+	−	+	−	−	−	−	+	+	−	+	+	−	−	+	+	+
9	+	−	+	−	+	−	−	−	−	+	+	−	+	+	−	−	+	+	+	+
10	+	+	−	+	−	−	−	−	+	+	−	+	+	−	−	+	+	+	+	−
11	+	−	+	−	−	−	−	+	+	−	+	+	−	−	+	+	+	+	−	+
12	+	+	−	−	−	−	+	+	−	+	+	−	−	+	+	+	+	−	+	−
13	+	−	−	−	−	+	+	−	+	+	−	−	+	+	+	+	−	+	−	+
14	+	−	−	−	+	+	−	+	+	−	−	+	+	+	+	−	+	−	+	−
15	+	−	−	+	+	−	+	+	−	−	+	+	+	+	−	+	−	+	−	−
16	+	−	+	+	−	+	+	−	−	+	+	+	+	−	+	−	+	−	−	−
17	+	+	+	−	+	+	−	−	+	+	+	+	−	+	−	+	−	−	−	−
18	+	+	−	+	+	−	−	+	+	+	+	−	+	−	+	−	−	−	−	+
19	+	−	+	+	−	−	+	+	+	+	−	+	−	+	−	−	−	−	+	+
20	+	−	−	−	−	−	−	−	−	−	−	−	−	−	−	−	−	−	−	−

TABLE B.1-3 24-Run Plackett-Burman Design

Run	Mean	X1	X2	X3	X4	X5	X6	X7	X8	X9	X10	X11	X12	X13	X14	X15	X16	X17	X18	X19	X20	X21	X22	X23
1	+	+	+	−	−	−	+	+	+	−	−	+	+	+	−	+	+	+	+	−	+	+	+	+
2	+	+	+	+	−	−	−	+	+	+	−	−	+	+	+	−	+	+	+	+	−	+	+	+
3	+	+	+	+	+	−	−	−	+	+	+	−	−	+	+	+	−	+	+	+	+	−	+	+
4	+	+	+	+	+	+	−	−	−	+	+	+	−	−	+	+	+	−	+	+	+	+	−	+
5	+	+	+	+	+	+	+	−	−	−	+	+	+	−	−	+	+	+	−	+	+	+	+	−
6	+	−	+	+	+	+	+	+	−	−	−	+	+	+	−	−	+	+	+	−	+	+	+	+
7	+	+	−	+	+	+	+	+	+	−	−	−	+	+	+	−	−	+	+	+	−	+	+	+
8	+	−	+	−	+	+	+	+	+	+	−	−	−	+	+	+	−	−	+	+	+	−	+	+
9	+	+	−	+	−	+	+	+	+	+	+	−	−	−	+	+	+	−	−	+	+	+	−	+
10	+	+	+	−	+	−	+	+	+	+	+	+	−	−	−	+	+	+	−	−	+	+	+	−
11	+	−	+	+	−	+	−	+	+	+	+	+	+	−	−	−	+	+	+	−	−	+	+	+
12	+	−	−	+	+	−	+	−	+	+	+	+	+	+	−	−	−	+	+	+	−	−	+	+
13	+	+	−	−	+	+	−	+	−	+	+	+	+	+	+	−	−	−	+	+	+	−	−	+
14	+	+	+	−	−	+	+	−	+	−	+	+	+	+	+	+	−	−	−	+	+	+	−	−
15	+	−	+	+	−	−	+	+	−	+	−	+	+	+	+	+	+	−	−	−	+	+	+	−
16	+	−	−	+	+	−	−	+	+	−	+	−	+	+	+	+	+	+	−	−	−	+	+	+
17	+	+	−	−	+	+	−	−	+	+	−	+	−	+	+	+	+	+	+	−	−	−	+	+
18	+	−	+	−	−	+	+	−	−	+	+	−	+	−	+	+	+	+	+	+	−	−	−	+
19	+	+	−	+	−	−	+	+	−	−	+	+	−	+	−	+	+	+	+	+	+	−	−	−
20	+	−	+	−	+	−	−	+	+	−	−	+	+	−	+	−	+	+	+	+	+	+	−	−
21	+	−	−	+	−	+	−	−	+	+	−	−	+	+	−	+	−	+	+	+	+	+	+	−
22	+	−	−	−	+	−	+	−	−	+	+	−	−	+	+	−	+	−	+	+	+	+	+	+
23	+	−	−	−	−	+	−	+	−	−	+	+	−	−	+	+	−	+	−	+	+	+	+	+
24	+	−	−	−	−	−	−	−	−	−	−	−	−	−	−	−	−	−	−	−	−	−	−	−

TABLE B.1-4 28-Run Plackett-Burman Design

Run	Mean	X_1	X_2	X_3	X_4	X_5	X_6	X_7	X_8	X_9	X_{10}	X_{11}	X_{12}	X_{13}	X_{14}	X_{15}	X_{16}	X_{17}	X_{18}	X_{19}	X_{20}	X_{21}	X_{22}	X_{23}	X_{24}	X_{25}	X_{26}	X_{27}
1	+	+	−	+	+	+	+	−	−	−	−	+	−	−	−	+	−	−	+	+	+	+	+	−	+	+	−	+
2	+	+	+	−	+	+	+	−	−	−	+	−	+	+	+	−	+	−	−	−	+	+	+	+	−	+	+	−
3	+	−	+	+	+	+	+	+	+	+	−	−	−	−	+	−	−	+	−	+	−	+	−	+	+	−	+	+
4	+	−	−	+	+	−	+	+	+	+	+	−	+	−	+	−	−	−	+	+	−	−	+	+	−	+	−	+
5	+	−	−	+	+	+	−	+	+	+	−	+	−	+	−	+	+	−	−	+	+	+	−	+	+	+	−	−
6	+	−	−	+	−	+	+	+	−	+	−	−	+	−	−	−	+	+	−	−	+	+	+	−	+	−	+	+
7	+	−	−	+	−	−	−	+	+	−	+	+	−	+	−	+	−	+	−	−	−	−	+	−	+	+	+	−
8	+	+	+	−	−	−	−	−	−	+	−	−	+	−	−	−	−	−	+	+	+	+	+	+	−	−	+	+
9	+	+	+	−	+	−	−	−	+	+	+	+	−	+	+	−	−	−	−	+	+	−	−	+	+	+	−	+
10	+	+	+	+	+	−	+	+	+	−	+	+	+	+	+	+	−	−	−	−	+	+	+	−	+	−	−	+
11	+	+	+	+	−	+	−	+	−	+	−	−	−	+	+	+	+	+	+	−	−	−	−	−	−	+	−	−
12	+	−	+	+	+	+	+	−	+	+	−	−	−	+	+	+	+	+	+	−	−	+	−	+	−	−	+	−
13	+	−	−	−	−	+	−	−	+	−	−	−	+	+	+	+	+	+	+	+	−	−	−	+	−	−	−	+
14	+	+	−	−	+	+	+	+	+	+	+	+	+	−	−	−	+	+	+	−	−	+	+	−	+	+	−	−
15	+	+	+	−	+	−	+	+	−	−	+	+	−	−	+	+	+	−	+	+	+	−	−	−	−	−	+	−
16	+	+	+	+	+	−	+	−	−	+	+	+	+	−	+	−	−	+	−	−	−	−	+	−	+	−	+	−
17	+	−	−	+	−	+	−	−	−	+	+	+	+	−	−	−	+	+	+	−	+	+	−	+	−	+	−	+
18	+	−	+	+	−	+	−	+	−	−	−	−	+	+	+	+	+	−	+	+	−	−	+	+	−	−	−	−
19	+	+	+	−	+	−	+	−	−	−	+	−	+	+	−	−	−	+	−	−	+	+	+	+	+	−	−	−
20	+	+	+	+	−	−	+	+	+	−	+	−	−	+	+	+	+	+	+	+	+	−	+	+	−	+	−	−
21	+	−	+	+	+	+	−	−	−	+	+	+	+	−	+	−	−	−	+	−	−	+	+	−	+	+	+	−
22	+	+	−	−	−	−	−	−	−	+	+	+	+	−	+	+	+	+	−	−	−	−	+	+	+	+	+	+
23	+	−	−	+	+	+	+	+	+	−	−	−	−	+	+	−	−	+	+	−	+	+	−	+	+	+	+	+
24	+	+	+	−	−	−	−	−	−	+	+	+	+	+	−	+	+	−	−	−	+	+	+	−	+	+	+	+
25	+	−	−	+	+	+	−	+	+	+	−	+	+	+	−	+	−	+	+	+	+	+	−	−	−	−	+	+
26	+	+	−	−	−	−	−	−	+	−	+	−	−	+	+	+	+	+	+	+	+	+	−	−	−	−	+	−
27	+	−	+	+	+	+	+	+	+	+	−	+	+	+	+	−	+	+	+	+	+	−	−	−	−	−	+	+
28	+	−	−	−	−	−	−	−	−	−	−	−	−	−	−	+	+	−	−	−	−	−	−	−	−	−	−	−

TABLE B.1-5 Eight-Run Fractional Factorial Design for Five to Seven Factors

Run	Mean	X_1	X_2	X_3	X_4	X_5	X_6	X_7
1	+	−	−	−	+	+	+	−
2	+	+	−	−	−	−	+	+
3	+	−	+	−	−	+	−	+
4	+	+	+	−	+	−	−	−
5	+	−	−	+	+	−	−	+
6	+	+	−	+	−	+	−	−
7	+	−	+	+	−	−	+	−
8	+	+	+	+	+	+	+	+

CONFOUNDINGS

$X_1 = 1 + 24 + 35 + 67$
$X_2 = 2 + 14 + 36 + 57$
$X_3 = 3 + 15 + 26 + 47$
$X_4 = 4 + 12 + 37 + 56$
$X_5 = 5 + 13 + 27 + 46$
$X_6 = 6 + 17 + 23 + 45$
$X_7 = 7 + 16 + 25 + 34$

TABLE B.1-6 Eight-Run Fractional Factorial Design for Four Factors

Run	Mean	X_1	X_2	X_3	X_4	E_1	E_2	E_3
1	+	−	−	−	−	+	+	+
2	+	+	−	−	+	−	−	+
3	+	−	+	−	+	−	+	−
4	+	+	+	−	−	+	−	−
5	+	−	−	+	+	+	−	−
6	+	+	−	+	−	−	+	−
7	+	−	+	+	−	−	−	+
8	+	+	+	+	+	+	+	+

CONFOUNDINGS

$E_1 = 12 + 34$
$E_2 = 13 + 24$
$E_3 = 14 + 23$

TABLE B.1-7 16-Run Fractional Factorial Design for 9–15 Factors
(Main Effects Confounded with Two-Factor Interactions)

Run	Mean	X_1	X_2	X_3	X_4	X_5	X_6	X_7	X_8	X_9	X_{10}	X_{11}	X_{12}	X_{13}	X_{14}	X_{15}
1	+	−	−	−	−	−	−	−	−	+	+	+	+	+	+	+
2	+	+	−	−	−	+	−	+	+	−	−	−	−	+	+	+
3	+	−	+	−	−	+	+	−	+	−	−	+	+	−	−	+
4	+	+	+	−	−	−	+	+	−	+	+	−	−	−	−	+
5	+	−	−	+	−	+	+	+	−	−	+	−	+	−	+	−
6	+	+	−	+	−	−	+	−	+	+	−	+	−	−	+	−
7	+	−	+	+	−	−	−	+	+	+	−	−	+	+	−	−
8	+	+	+	+	−	+	−	−	−	−	+	+	−	+	−	−
9	+	−	−	−	+	−	+	+	+	−	+	+	−	+	−	−
10	+	+	−	−	+	+	+	−	−	+	−	−	+	+	−	−
11	+	−	+	−	+	+	−	+	−	+	−	+	−	−	+	−
12	+	+	+	−	+	−	−	−	+	−	+	−	+	−	+	−
13	+	−	−	+	+	+	−	−	+	+	+	−	−	−	−	+
14	+	+	−	+	+	−	−	+	−	−	−	+	+	−	−	+
15	+	−	+	+	+	−	+	−	−	−	−	−	−	+	+	+
16	+	+	+	+	+	+	+	+	+	+	+	+	+	+	+	+

CONFOUNDINGS

$x_1 = 1 + 2,10 + 3,11 + 4,12 + 5,13 + 6,9 + 7,15 + 8,14$
$x_2 = 2 + 1,10 + 3\ 13 + 4,14 + 5,11 + 6,15 + 7,9 + 8,12$
$x_3 = 3 + 1,11 + 2,13 + 4,15 + 5,10 + 6,14 + 7,12 + 8,9$
$x_4 = 4 + 1,12 + 2,14 + 3,15 + 5,9 + 6,13 + 7,11 + 8,10$
$x_5 = 5 + 1,13 + 2,11 + 3,10 + 4,9 + 6,12 + 7,14 + 8,15$
$x_6 = 6 + 1,9 + 2,15 + 3,14 + 4,13 + 5,12 + 7,10 + 8,11$
$x_7 = 7 + 1,15 + 2,9 + 3,12 + 4,11 + 5,14 + 6,10 + 8,13$
$x_8 = 8 + 1,14 + 2,12 + 3,9 + 4,10 + 5,15 + 6,11 + 7,13$
$x_9 = 9 + 1,6 + 2,7 + 3,8 + 4,5 + 10,15 + 11,14 + 12,13$
$x_{10} = 10 + 1,2 + 3,5 + 4,8 + 6,7 + 9,15 + 11,13 + 12,14$
$x_{11} = 11 + 1,3 + 2,5 + 4,7 + 6,8 + 9,14 + 10,13 + 12,15$
$x_{12} = 12 + 1,4 + 2,8 + 3,7 + 5,6 + 9,13 + 10,14 + 11,15$
$x_{13} = 13 + 1,5 + 2,3 + 4,6 + 7,8 + 9,12 + 10,11 + 14,15$
$x_{14} = 14 + 1,8 + 2,4 + 3,6 + 5,7 + 9,11 + 10,12 + 13,15$
$x_{15} = 15 + 1,7 + 2,6 + 3,4 + 5,8 + 9,10 + 11,12 + 13,14$

$I = 1,2,3,5 = 2,3,4,6 = 1,3,4,7 = 1,2,4,8 = 1,2,3,4,9$
$= 1,2,10 = 1,3,11 = 1,4,12 = 2,3,13 = 2,4,14 = 3,4,15$

TABLE B.1-8 16-Run Fractional Factorial Design for Six to Eight Factors (Main Effects Clear of Two-Factor Interactions)

Run	Mean	X_1	X_2	X_3	X_4	X_5	X_6	X_7	X_8	E_1	E_2	E_3	E_4	E_5	E_6	E_7
1	+	+	−	−	−	−	−	−	−	+	+	+	+	+	+	+
2	+	+	+	−	−	−	+	−	+	+	−	−	−	+	−	+	+
3	+	+	−	+	−	−	+	+	−	+	−	+	+	−	−	+	−
4	+	+	+	+	−	−	−	+	+	−	+	−	−	−	+	+	−
5	+	+	−	−	+	−	+	+	+	−	+	−	+	−	−	−	+
6	+	+	+	−	+	−	−	+	−	+	−	+	−	−	+	−	+
7	+	+	−	+	+	−	−	−	+	+	−	−	+	+	+	−	−
8	+	+	+	+	+	−	+	−	−	−	+	+	−	+	−	−	−
9	+	+	−	−	−	+	−	+	+	+	+	+	−	+	−	−	−
10	+	+	+	−	−	+	+	+	−	−	−	−	+	+	+	−	−
11	+	+	−	+	−	+	+	−	+	−	−	+	−	−	+	−	+
12	+	+	+	+	−	+	−	−	−	+	+	−	+	−	−	−	+
13	+	+	−	−	+	+	+	−	−	+	+	−	−	−	+	+	−
14	+	+	+	−	+	+	−	−	+	−	−	+	+	−	−	+	−
15	+	+	−	+	+	+	−	+	−	−	−	−	−	+	−	+	+
16	+	+	+	+	+	+	+	+	+	+	+	+	+	+	+	+

CONFOUNDINGS

E_1 = 12 + 35 + 48 + 67
E_2 = 13 + 25 + 47 + 68
E_3 = 14 + 28 + 37 + 56
E_4 = 15 + 23 + 46 + 78
E_5 = 16 + 27 + 38 + 45
E_6 = 17 + 26 + 34 + 58
E_7 = 18 + 24 + 36 + 57

I = 1235 + 2346 + 1347 + 1248

TABLE B.1-9 16-Run Fractional Factorial Design for Five Factors (Main Effects and Two-Factor Interactions Clear of Each Other)

Run	Mean	X_1	X_2	X_3	X_4	X_5	X_1X_2	X_1X_3	X_1X_4	X_1X_5	X_2X_3	X_2X_4	X_2X_5	X_3X_4	X_3X_5	X_4X_5
1	+	−	−	−	−	+	+	+	+	−	+	+	−	+	−	−
2	+	+	−	−	−	−	−	−	−	−	+	+	+	+	+	+
3	+	−	+	−	−	−	−	+	+	+	−	−	−	+	+	+
4	+	+	+	−	−	+	+	−	−	+	−	−	+	+	−	−
5	+	−	−	+	−	−	+	−	+	+	−	+	+	−	−	+
6	+	+	−	+	−	+	−	+	−	+	−	+	−	−	+	−
7	+	−	+	+	−	+	−	−	+	−	+	−	+	−	+	−
8	+	+	+	+	−	−	+	+	−	−	+	−	−	−	−	+
9	+	−	−	−	+	−	+	+	−	+	+	−	+	−	+	−
10	+	+	−	−	+	+	−	−	+	+	+	−	−	−	−	+
11	+	−	+	−	+	+	−	+	−	−	−	+	+	−	−	+
12	+	+	+	−	+	−	+	−	+	−	−	+	−	−	+	−
13	+	−	−	+	+	+	+	−	−	−	−	−	−	+	+	+
14	+	+	−	+	+	−	−	+	+	−	−	−	+	+	−	−
15	+	−	+	+	+	−	−	−	−	+	+	+	−	+	−	−
16	+	+	+	+	+	+	+	+	+	+	+	+	+	+	+	+

CONFOUNDINGS

None (other than with third order or fourth order interactions)

$$I = 12345$$

TABLE B.1-10 32-Run Fractional Factorial Design for 17–31 Factors
(Main Effects Confounded with Two-Factor Interactions)

Run	Mean	X1	X2	X3	X4	X5	X6	X7	X8	X9	X10	X11	X12	X13	X14	X15	X16	X17	X18	X19	X20	X21	X22	X23	X24	X25	X26	X27	X28	X29	X30	X31
1	+	−	−	−	−	−	+	+	+	+	+	+	+	+	+	+	−	−	−	−	−	−	−	−	−	−	+	+	+	+	+	−
2	+	+	−	−	−	−	−	−	−	−	+	+	+	+	+	+	+	+	+	+	+	+	−	−	−	−	−	−	−	−	+	+
3	+	−	+	−	−	−	−	+	+	+	−	−	−	+	+	+	+	+	+	−	−	−	+	+	+	−	−	−	−	+	−	+
4	+	+	+	−	−	−	+	−	−	−	−	−	−	+	+	+	−	−	−	+	+	+	+	+	+	−	+	+	+	−	−	−
5	+	−	−	+	−	−	+	−	+	+	−	+	+	−	−	+	+	−	−	+	+	−	+	+	−	+	−	−	+	−	−	+
6	+	+	−	+	−	−	−	+	−	−	−	+	+	−	−	+	−	+	+	−	−	+	+	+	−	+	+	+	−	+	−	−
7	+	−	+	+	−	−	−	−	+	+	+	−	−	−	−	+	−	+	+	+	+	−	−	−	+	+	+	+	−	−	+	−
8	+	+	+	+	−	−	+	+	−	−	+	−	−	−	−	+	+	−	−	−	−	+	−	−	+	+	−	−	+	+	+	+
9	+	−	−	−	+	−	+	+	−	−	+	−	+	−	+	−	−	+	−	+	−	+	+	−	+	+	−	+	−	−	−	+
10	+	+	−	−	+	−	−	−	+	+	+	−	+	−	+	−	+	−	+	−	+	−	+	−	+	+	+	−	+	+	−	−
11	+	−	+	−	+	−	−	+	−	+	−	+	−	−	+	−	+	−	+	+	−	+	−	+	−	+	+	−	+	−	+	−
12	+	+	+	−	+	−	+	−	+	−	−	+	−	−	+	−	−	+	−	−	+	−	−	+	−	+	−	+	−	+	+	+
13	+	−	−	+	+	−	+	−	−	+	−	−	+	+	−	−	+	+	−	−	+	+	−	+	+	−	+	−	−	+	+	−
14	+	+	−	+	+	−	−	+	+	−	−	−	+	+	−	−	−	−	+	+	−	−	−	+	+	−	−	+	+	−	+	+
15	+	−	+	+	+	−	−	−	−	+	+	+	−	+	−	−	−	−	+	−	+	+	+	−	−	−	−	+	+	+	−	+
16	+	+	+	+	+	−	+	+	+	−	+	+	−	+	−	−	+	+	−	+	−	−	+	−	−	−	+	−	−	−	−	−
17	+	−	−	−	−	+	+	+	+	−	+	+	−	+	−	−	−	−	+	−	+	+	−	+	+	+	+	−	−	−	−	+
18	+	+	−	−	−	+	−	−	−	+	+	+	−	+	−	−	+	+	−	+	−	−	−	+	+	+	−	+	+	+	−	−
19	+	−	+	−	−	+	−	+	+	−	−	−	+	+	−	−	+	+	−	−	+	+	+	−	−	+	−	+	+	−	+	−
20	+	+	+	−	−	+	+	−	−	+	−	−	+	+	−	−	−	−	+	+	−	−	+	−	−	+	+	−	−	+	+	+
21	+	−	−	+	−	+	+	−	+	−	−	+	−	−	+	−	+	−	+	+	−	+	+	−	+	−	−	+	−	+	+	−
22	+	+	−	+	−	+	−	+	−	+	−	+	−	−	+	−	−	+	−	−	+	−	+	−	+	−	+	−	+	−	+	+
23	+	−	+	+	−	+	−	−	+	−	+	−	+	−	+	−	−	+	−	+	−	+	−	+	−	−	+	−	+	+	−	+
24	+	+	+	+	−	+	+	+	−	+	+	−	+	−	+	−	+	−	+	−	+	−	−	+	−	−	−	+	−	−	−	−
25	+	−	−	−	+	+	+	+	−	−	+	−	−	−	−	+	−	+	+	+	+	−	+	+	−	−	−	−	+	+	+	−
26	+	+	−	−	+	+	−	−	+	+	+	−	−	−	−	+	+	−	−	−	−	+	+	+	−	−	+	+	−	−	+	+
27	+	−	+	−	+	+	−	+	−	−	−	+	+	−	−	+	+	−	−	+	+	−	−	−	+	−	+	+	−	+	−	+
28	+	+	+	−	+	+	+	−	+	+	−	+	+	−	−	+	−	+	+	−	−	+	−	−	+	−	−	−	+	−	−	−
29	+	−	−	+	+	+	+	−	−	−	−	−	−	+	+	+	+	+	+	−	−	−	−	−	−	+	+	+	+	−	−	+
30	+	+	−	+	+	+	−	+	+	+	−	−	−	+	+	+	−	−	−	+	+	+	−	−	−	+	−	−	−	+	−	−
31	+	−	+	+	+	+	−	−	−	−	+	+	+	+	+	+	−	−	−	−	−	−	+	+	+	+	−	−	−	−	+	−
32	+	+	+	+	+	+	+	+	+	+	+	+	+	+	+	+	+	+	+	+	+	+	+	+	+	+	+	+	+	+	+	+

GENERATORS: 6=1,2 7=1,3 8=1,4 9=1,5 10=2,3 11=2,4 12=2,5 13=3,4 14=3,5 15=4,5 16=1,2,3 17=1,2,4 18=1,2,5 19=1,3,4 20=1,3,5 21=1,4,5 22=2,3,4 23=2,3,5 24=2,4,5 25=3,4,5 26=1,2,3,4 27=1,2,3,5 28=1,2,4,5 29=1,3,4,5 30=2,3,4,5 31=1,2,3,4,5

TABLE B.1-11 32-Run Fractional Factorial Design for 7–16 Factors (Main Effects Clear of Two-Factor Interactions)

Run	Mean	X_1	X_2	X_3	X_4	X_5	X_6	X_7	X_8	X_9	X_{10}	X_{11}	X_{12}	X_{13}	X_{14}	X_{15}	E_1	E_2	E_3	E_4	E_5	E_6	E_7	E_8	E_9	E_{10}	E_{11}	E_{12}	E_{13}	E_{14}	E_{15}	E_{16}
1	+	−	−	−	−	−	+	−	+	+	−	+	+	+	+	+	−	−	−	+	+	−	−	−	−	−	+	+	+	+	+	−
2	+	+	+	−	−	−	−	+	+	+	+	+	+	+	+	+	+	+	+	−	−	+	+	+	+	+	+	−	−	+	+	+
3	+	−	+	+	−	−	−	+	−	+	−	+	−	+	+	+	+	+	+	+	+	+	+	+	+	+	+	+	+	−	−	+
4	+	+	−	+	−	−	+	−	−	+	+	+	−	+	+	+	−	−	−	−	−	−	−	−	−	−	+	−	−	−	−	−
5	+	−	−	+	+	−	+	+	+	−	−	−	+	+	+	+	+	+	−	+	+	−	−	+	−	−	−	+	+	−	−	+
6	+	+	+	+	+	−	−	−	+	−	+	−	+	+	+	+	−	−	+	−	−	+	+	−	+	+	−	−	−	+	+	−
7	+	−	+	−	+	−	−	−	−	−	−	−	−	+	+	+	−	−	+	+	+	+	+	−	+	+	−	+	+	+	+	−
8	+	+	−	−	+	−	+	+	−	−	+	−	−	+	+	+	+	+	−	−	−	−	−	+	−	−	−	−	−	−	−	+
9	+	−	−	−	−	+	+	−	+	+	−	+	+	−	−	−	−	−	+	+	+	+	+	+	+	−	−	+	−	+	+	−
10	+	+	+	−	−	+	−	+	+	+	+	+	+	−	−	−	+	+	−	−	−	−	−	−	−	+	−	−	+	+	+	+
11	+	−	+	+	−	+	−	+	−	+	−	+	−	−	−	−	+	+	−	+	+	−	−	−	−	+	−	+	−	−	−	+
12	+	+	−	+	−	+	+	−	−	+	+	+	−	−	−	−	−	−	+	−	−	+	+	+	+	−	−	−	+	−	−	−
13	+	−	−	+	+	+	+	+	+	−	−	−	+	−	−	−	+	+	+	+	+	+	+	−	+	−	+	+	−	−	−	+
14	+	+	+	+	+	+	−	−	+	−	+	−	+	−	−	−	−	−	−	−	−	−	−	+	−	+	+	−	+	−	−	−
15	+	−	+	−	+	+	−	−	−	−	−	−	−	−	−	−	−	−	−	+	+	+	+	+	+	+	+	+	−	+	+	−
16	+	+	−	−	+	+	+	+	−	−	+	−	−	−	−	−	+	+	+	−	−	−	−	−	−	−	+	−	+	+	+	+
17	+	−	−	−	−	+	+	−	+	+	+	−	−	−	−	−	+	+	+	+	+	+	+	+	+	−	−	−	+	−	−	+
18	+	+	+	−	−	+	−	+	+	+	−	−	−	−	−	−	−	−	−	−	−	−	−	−	−	+	−	+	−	−	−	−
19	+	−	+	+	−	+	−	+	−	+	+	−	+	−	−	−	−	−	−	+	+	−	−	−	−	+	−	−	+	+	+	−
20	+	+	−	+	−	+	+	−	−	+	−	−	+	−	−	−	+	+	+	−	−	+	+	+	+	−	−	+	−	+	+	+
21	+	−	−	+	+	+	+	+	+	−	+	+	−	−	−	−	−	−	+	+	+	+	+	−	+	−	+	−	+	+	+	−
22	+	+	+	+	+	+	−	−	+	−	−	+	−	−	−	−	+	+	−	−	−	−	−	+	−	+	+	+	−	−	−	+
23	+	−	+	−	+	+	−	−	−	−	+	+	+	−	−	−	+	+	−	+	+	+	+	+	+	+	+	−	+	−	−	+
24	+	+	−	−	+	+	+	+	−	−	−	+	+	−	−	−	−	−	+	−	−	−	−	−	−	−	+	+	−	+	+	−
25	+	−	−	−	−	−	+	−	+	+	+	−	−	+	+	+	+	+	−	+	+	+	+	+	+	−	−	+	−	+	+	−
26	+	+	+	−	−	−	−	+	+	+	−	−	−	+	+	+	−	−	+	−	−	−	−	−	−	+	−	−	+	+	+	+
27	+	−	+	+	−	−	−	+	−	+	+	−	+	+	+	+	−	−	+	+	+	−	−	−	−	+	−	+	−	−	−	+
28	+	+	−	+	−	−	+	−	−	+	−	−	+	+	+	+	+	+	−	−	−	+	+	+	+	−	−	−	+	−	−	−
29	+	−	−	+	+	−	+	+	+	−	+	+	−	+	+	+	−	−	−	+	+	+	+	−	+	−	+	+	−	−	−	+
30	+	+	+	+	+	−	−	−	+	−	−	+	−	+	+	+	+	+	+	−	−	−	−	+	−	+	+	−	+	−	−	−
31	+	−	+	−	+	−	−	−	−	−	+	+	+	+	+	+	+	+	+	+	+	+	+	+	+	+	+	+	−	+	+	−
32	+	+	−	−	+	−	+	+	−	−	−	+	+	+	+	+	−	−	−	−	−	−	−	−	−	−	+	−	+	+	+	+

TABLE B.1-12 Confoundings for 32-Run Fractional Factorial Design for 7–16 Factors (Main Effects Clear of Two-Factor Interactions)

E_1	=	1,2	+	3,6	+	4,7	+	5,8	+	9,12	+	10,13	+	11,14	+	15,16
E_2	=	1,3	+	2,6	+	4,9	+	5,10	+	7,12	+	8,13	+	11,15	+	14,16
E_3	=	1,4	+	2,7	+	3,9	+	5,11	+	6,12	+	8,14	+	10,15	+	13,16
E_4	=	1,5	+	2,8	+	3,10	+	4,11	+	6,13	+	7,14	+	9,15	+	12,16
E_5	=	1,6	+	2,3	+	4,12	+	5,13	+	7,9	+	8,10	+	11,16	+	14,15
E_6	=	1,7	+	2,4	+	3,12	+	5,14	+	6,9	+	8,11	+	10,16	+	13,15
E_7	=	1,8	+	2,5	+	3,13	+	4,14	+	6,10	+	7,11	+	9,16	+	12,15
E_8	=	1,9	+	2,12	+	3,4	+	5,15	+	6,7	+	8,16	+	10,11	+	13,14
E_9	=	1,10	+	2,13	+	3,5	+	4,15	+	6,8	+	7,16	+	9,11	+	12,14
E_{10}	=	1,11	+	2,14	+	3,15	+	4,5	+	6,16	+	7,8	+	9,10	+	12,13
E_{11}	=	1,12	+	2,9	+	3,7	+	4,6	+	5,16	+	8,15	+	10,14	+	11,13
E_{12}	=	1,13	+	2,10	+	3,8	+	4,16	+	5,6	+	7,15	+	9,14	+	11,12
E_{13}	=	1,14	+	2,11	+	3,16	+	4,8	+	5,7	+	6,15	+	9,13	+	10,12
E_{14}	=	1,15	+	2,16	+	3,11	+	4,10	+	5,9	+	6,14	+	7,13	+	8,12
E_{15}	=	1,16	+	2,15	+	3,14	+	4,13	+	5,12	+	6,11	+	7,10	+	8,9

GENERATORS

6 = 1,2,3	7 = 1,2,4	8 = 1,2,5	9 = 1,3,4
10 = 1,3,5	11 = 1,4,5	12 = 2,3,4	13 = 2,3,5
14 = 2,4,5	15 = 3,4,5	16 = 1,2,3,4,5	

TABLE B.1-13 32-Run Fractional Factorial Design for Six Factors
(Main Effects and Two-Factor Interactions Clear of Each Other)

Run	Mean	X_1	X_2	X_3	X_4	X_5	X_6	X_1X_2	X_1X_3	X_1X_4	X_1X_5	X_1X_6	X_2X_3	X_2X_4	X_2X_5	X_2X_6	X_3X_4	X_3X_5	X_3X_6	X_4X_5	X_4X_6	X_5X_6	E_1	E_2	E_3	E_4	E_5	E_6	E_7	E_8	E_9	E_{10}
1	+	−	−	−	−	−	−	+	+	+	+	+	+	+	+	+	+	+	+	+	+	+	−	−	−	−	−	−	−	−	−	−
2	+	+	−	−	−	−	+	−	−	−	−	+	+	+	+	−	+	+	−	+	−	−	+	+	+	−	+	+	−	+	−	−
3	+	−	+	−	−	−	+	−	+	+	+	−	−	−	−	+	+	+	−	+	−	−	+	+	+	−	−	−	+	−	+	+
4	+	+	+	−	−	−	−	+	−	−	−	−	−	−	−	−	+	+	+	+	+	+	−	−	−	−	+	+	+	+	+	+
5	+	−	−	+	−	−	+	+	−	+	+	−	−	+	+	−	−	−	+	+	−	−	+	−	−	+	+	+	−	−	+	+
6	+	+	−	+	−	−	−	−	+	−	−	+	+	+	+	+	−	−	−	+	+	+	−	+	+	+	−	−	−	+	+	+
7	+	−	+	+	−	−	−	−	−	+	+	+	+	−	−	−	−	−	−	+	+	+	−	+	+	+	+	+	+	−	−	−
8	+	+	+	+	−	−	+	+	+	−	−	+	+	−	−	+	−	−	+	+	−	−	+	−	−	+	−	−	+	+	−	−
9	+	−	−	−	+	−	+	+	+	−	+	−	+	−	+	−	−	+	−	−	+	−	−	+	−	+	+	−	+	+	−	+
10	+	+	−	−	+	−	−	−	−	+	−	−	+	−	+	+	−	+	+	−	−	+	+	−	+	+	−	+	+	−	−	+
11	+	−	+	−	+	−	−	−	+	−	+	+	−	+	−	−	−	+	+	−	−	+	+	−	+	+	+	−	−	+	+	−
12	+	+	+	−	+	−	+	+	−	+	−	+	−	+	−	+	−	+	−	−	+	−	−	+	−	+	−	+	−	−	+	−
13	+	−	−	+	+	−	−	+	−	−	+	+	−	−	+	+	+	−	+	−	−	+	+	+	−	−	−	+	+	+	+	−
14	+	+	−	+	+	−	+	−	+	+	−	+	−	−	+	−	+	−	+	−	+	−	−	−	+	−	+	−	+	−	+	−
15	+	−	+	+	+	−	+	−	−	−	+	−	+	+	−	+	+	−	−	−	+	−	−	−	+	−	−	+	−	+	−	+
16	+	+	+	+	+	−	−	+	+	+	−	−	+	+	−	−	+	−	−	−	−	+	+	+	−	−	+	−	−	−	−	+
17	+	−	−	−	−	+	+	+	+	+	−	−	+	+	−	−	+	−	−	−	−	+	−	−	+	+	−	+	+	+	+	−
18	+	+	−	−	−	+	−	−	−	−	+	−	+	+	−	+	+	−	+	−	+	−	+	+	−	+	+	−	+	−	+	−
19	+	−	+	−	−	+	−	−	+	+	−	+	−	−	+	−	+	−	+	−	+	−	+	+	−	+	−	+	−	+	−	+
20	+	+	+	−	−	+	+	+	−	−	+	+	−	−	+	+	+	−	−	−	−	+	−	−	+	+	+	−	−	−	−	+
21	+	−	−	+	−	+	−	+	−	+	−	+	−	+	−	+	−	+	+	−	+	−	+	−	+	−	−	−	+	+	−	+
22	+	+	−	+	−	+	+	−	+	−	+	+	−	+	−	−	−	+	+	−	−	+	−	+	−	−	+	+	+	−	−	+
23	+	−	+	+	−	+	+	−	−	+	−	−	+	−	+	+	−	+	−	−	−	+	−	+	−	−	+	−	−	+	+	+
24	+	+	+	+	−	+	−	+	+	−	+	−	+	−	+	−	−	+	−	−	+	−	+	−	+	−	−	+	−	−	+	−
25	+	−	−	−	+	+	−	+	+	−	−	+	+	−	−	+	−	−	+	+	−	−	−	+	+	−	+	+	−	−	+	+
26	+	+	−	−	+	+	+	−	−	+	+	+	+	−	−	−	−	−	−	+	+	+	+	−	−	−	−	−	−	+	+	+
27	+	−	+	−	+	+	+	−	+	−	−	−	−	+	+	+	−	−	−	+	+	+	+	−	−	−	+	+	+	−	−	−
28	+	+	+	−	+	+	−	+	−	+	+	−	−	+	+	−	−	−	+	+	−	−	−	+	+	−	−	−	+	+	−	−
29	+	−	−	+	+	+	+	+	−	−	−	−	−	−	−	−	+	+	+	+	+	+	+	−	+	+	+	+	−	−	−	−
30	+	+	−	+	+	+	−	−	+	+	+	−	−	−	−	+	+	+	−	+	−	−	−	+	−	+	−	−	−	+	−	−
31	+	−	+	+	+	+	−	−	−	−	−	+	+	+	+	−	+	+	−	+	−	−	−	+	−	+	+	+	+	−	+	+
32	+	+	+	+	+	+	+	+	+	+	+	+	+	+	+	+	+	+	+	+	+	+	+	+	+	+	+	+	+	+	+	+

TABLE B.1-14 Confoundings for 32-Run Fractional Factorial Design with Six Factors

$\mathbf{E}_1$	=	123 + 456
$\mathbf{E}_2$	=	124 + 356
$\mathbf{E}_3$	=	125 + 346
$\mathbf{E}_4$	=	126 + 345
$\mathbf{E}_5$	=	134 + 256
$\mathbf{E}_6$	=	135 + 246
$\mathbf{E}_7$	=	136 + 245
$\mathbf{E}_8$	=	145 + 236
$\mathbf{E}_9$	=	146 + 235
$\mathbf{E}_{10}$	=	156 + 234

GENERATOR:
6 = 12345

DEFINING RELATION:
I = 123456

TABLE B.1-15 L_{18} Orthogonal Array Design — 2×3^7

Run	1	2	3	4	5	6	7	8
1	1	1	1	1	1	1	1	1
2	1	1	2	2	2	2	2	2
3	1	1	3	3	3	3	3	3
4	1	2	1	1	2	2	3	3
5	1	2	2	2	3	3	1	1
6	1	2	3	3	1	1	2	2
7	1	3	1	2	1	3	2	3
8	1	3	2	3	2	1	3	1
9	1	3	3	1	3	2	1	2
10	2	1	1	3	3	2	2	1
11	2	1	2	1	1	3	3	2
12	2	1	3	2	2	1	1	3
13	2	2	1	2	3	1	3	2
14	2	2	2	3	1	2	1	3
15	2	2	3	1	2	3	2	1
16	2	3	1	3	2	3	1	2
17	2	3	2	1	3	1	2	3
18	2	3	3	2	1	2	3	1

Appendix B.2 Full Factorial Designs

TABLE B.2-1 The Factorial Pattern of Experiments

	Run	X_1	X_2	X_3	X_4	X_5
k = 1	1	−	−	−	−	−
	2	+	−	−	−	−
k = 2	3	−	+	−	−	−
	4	+	+	−	−	−
k = 3	5	−	−	+	−	−
	6	+	−	+	−	−
	7	−	+	+	−	−
	8	+	+	+	−	−
	9	−	−	−	+	−
	10	+	−	−	+	−
	11	−	+	−	+	−
	12	+	+	−	+	−
	13	−	−	+	+	−
	14	+	−	+	+	−
k = 4	15	−	+	+	+	−
	16	+	+	+	+	−
	17	−	−	−	−	+
	18	+	−	−	−	+
	19	−	+	−	−	+
	20	+	+	−	−	+
	21	−	−	+	−	+
	22	+	−	+	−	+
	23	−	+	+	−	+
	24	+	+	+	−	+
	25	−	−	−	+	+
	26	+	−	−	+	+
	27	−	+	−	+	+
	28	+	+	−	+	+
	29	−	−	+	+	+
	30	+	−	+	+	+
k = 5	31	−	+	+	+	+
	32	+	+	+	+	+

TABLE B.2-2 Computation Table for 2^2 Design

| | Computation Table | | |
| | Design Table | | |
Run	Mean	X_1	X_2	X_1X_2
1	+	−	−	+
2	+	+	−	−
3	+	−	+	−
4	+	+	+	+

TABLE B.2-3 Computation Table for 2^3 Design

| | Computation Table | | | | | | | |
| | Design Table | | | | | | | |
Run	Mean	X_1	X_2	X_3	X_1X_2	X_1X_3	X_2X_3	$X_1X_2X_3$
1	+	−	−	−	+	+	+	−
2	+	+	−	−	−	−	+	+
3	+	−	+	−	−	+	−	+
4	+	+	+	−	+	−	−	−
5	+	−	−	+	+	−	−	+
6	+	+	−	+	−	+	−	−
7	+	−	+	+	−	−	+	−
8	+	+	+	+	+	+	+	+

TABLE B.2-4 Computation Table for 2^4 Design

Run	Design Table					Computation Table										
	Mean	X_1	X_2	X_3	X_4	X_1X_2	X_1X_3	X_1X_4	X_2X_3	X_2X_4	X_3X_4	$X_1X_2X_3$	$X_1X_2X_4$	$X_1X_3X_4$	$X_2X_3X_4$	$X_1X_2X_3X_4$
1	+	−	−	−	−	+	+	+	+	+	+	−	−	−	−	+
2	+	+	−	−	−	−	−	−	+	+	+	+	+	+	−	−
3	+	−	+	−	−	−	+	+	−	−	+	+	+	−	+	−
4	+	+	+	−	−	+	−	−	−	−	+	−	−	+	+	+
5	+	−	−	+	−	+	−	+	−	+	−	+	−	+	+	−
6	+	+	−	+	−	−	+	−	−	+	−	−	+	−	+	+
7	+	−	+	+	−	−	−	+	+	−	−	−	+	+	−	+
8	+	+	+	+	−	+	+	−	+	−	−	+	−	−	−	−
9	+	−	−	−	+	+	+	−	+	−	−	−	+	+	+	−
10	+	+	−	−	+	−	−	+	+	−	−	+	−	−	+	+
11	+	−	+	−	+	−	+	−	−	+	−	+	−	+	−	+
12	+	+	+	−	+	+	−	+	−	+	−	−	+	−	−	−
13	+	−	−	+	+	+	−	−	−	−	+	+	+	−	−	+
14	+	+	−	+	+	−	+	+	−	−	+	−	−	+	−	−
15	+	−	+	+	+	−	−	−	+	+	+	−	−	−	+	−
16	+	+	+	+	+	+	+	+	+	+	+	+	+	+	+	+

TABLE B.2-5 Computation Table for 2^5 Design

Run	Mean	X_1	X_2	X_3	X_4	X_5	X_1X_2	X_1X_3	X_1X_4	X_1X_5	X_2X_3	X_2X_4	X_2X_5	X_3X_4	X_3X_5	X_4X_5	$X_1X_2X_3$	$X_1X_2X_4$	$X_1X_2X_5$	$X_1X_3X_4$	$X_1X_3X_5$	$X_1X_4X_5$	$X_2X_3X_4$	$X_2X_3X_5$	$X_2X_4X_5$	$X_3X_4X_5$	$X_1X_2X_3X_4$	$X_1X_2X_3X_5$	$X_1X_2X_4X_5$	$X_1X_3X_4X_5$	$X_2X_3X_4X_5$	$X_1X_2X_3X_4X_5$
1	+	−	−	−	−	−	+	+	+	+	+	+	+	+	+	+	−	−	−	−	−	−	−	−	−	−	+	+	+	+	+	−
2	+	+	−	−	−	−	−	−	−	−	+	+	+	+	+	+	+	+	+	+	+	+	−	−	−	−	−	−	−	−	+	+
3	+	−	+	−	−	−	−	+	+	+	−	−	−	+	+	+	+	+	+	−	−	−	+	+	+	−	−	−	−	+	−	+
4	+	+	+	−	−	−	+	−	−	−	−	−	−	+	+	+	−	−	−	+	+	+	+	+	+	−	+	+	+	−	−	−
5	+	−	−	+	−	−	+	−	+	+	−	+	+	−	−	+	+	−	−	+	+	−	+	+	−	+	−	−	+	−	−	+
6	+	+	−	+	−	−	−	+	−	−	−	+	+	−	−	+	−	+	+	−	−	+	+	+	−	+	+	+	−	+	−	−
7	+	−	+	+	−	−	−	−	+	+	+	−	−	−	−	+	−	+	+	+	+	−	−	−	+	+	+	+	−	−	+	−
8	+	+	+	+	−	−	+	+	−	−	+	−	−	−	−	+	+	−	−	−	−	+	−	−	+	+	−	−	+	+	+	+
9	+	−	−	−	+	−	+	+	−	+	+	−	+	−	+	−	−	+	−	+	−	+	+	−	+	+	−	+	−	−	−	+
10	+	+	−	−	+	−	−	−	+	−	+	−	+	−	+	−	+	−	+	−	+	−	+	−	+	+	+	−	+	+	−	−
11	+	−	+	−	+	−	−	+	−	+	−	+	−	−	+	−	+	−	+	+	−	+	−	+	−	+	+	−	+	−	+	−
12	+	+	+	−	+	−	+	−	+	−	−	+	−	−	+	−	−	+	−	−	+	−	−	+	−	+	−	+	−	+	+	+
13	+	−	−	+	+	−	+	−	−	+	−	−	+	+	−	−	+	+	−	−	+	+	−	+	+	−	+	−	−	+	+	−
14	+	+	−	+	+	−	−	+	+	−	−	−	+	+	−	−	−	−	+	+	−	−	−	+	+	−	−	+	+	−	+	+
15	+	−	+	+	+	−	−	−	−	+	+	+	−	+	−	−	−	−	+	−	+	+	+	−	−	−	−	+	+	+	−	+
16	+	+	+	+	+	−	+	+	+	−	+	+	−	+	−	−	+	+	−	+	−	−	+	−	−	−	+	−	−	−	−	−
17	+	−	−	−	−	+	+	+	+	−	+	+	−	+	−	−	−	−	+	−	+	+	−	+	+	+	+	−	−	−	−	+
18	+	+	−	−	−	+	−	−	−	+	+	+	−	+	−	−	+	+	−	+	−	−	−	+	+	+	−	+	+	+	−	−
19	+	−	+	−	−	+	−	+	+	−	−	−	+	+	−	−	+	+	−	−	+	+	+	−	−	+	−	+	+	−	+	−
20	+	+	+	−	−	+	+	−	−	+	−	−	+	+	−	−	−	−	+	+	−	−	+	−	−	+	+	−	−	+	+	+
21	+	−	−	+	−	+	+	−	+	−	−	+	−	−	+	−	+	−	+	+	−	+	+	−	+	−	−	+	−	+	+	−
22	+	+	−	+	−	+	−	+	−	+	−	+	−	−	+	−	−	+	−	−	+	−	+	−	+	−	+	−	+	−	+	+
23	+	−	+	+	−	+	−	−	+	−	+	−	+	−	+	−	−	+	−	+	−	+	−	+	−	−	+	−	+	+	−	+
24	+	+	+	+	−	+	+	+	−	+	+	−	+	−	+	−	+	−	+	−	+	−	−	+	−	−	−	+	−	−	−	−
25	+	−	−	−	+	+	+	+	−	−	+	−	−	−	−	+	−	+	+	+	+	−	+	+	−	−	−	−	+	+	+	−
26	+	+	−	−	+	+	−	−	+	+	+	−	−	−	−	+	+	−	−	−	−	+	+	+	−	−	+	+	−	−	+	+
27	+	−	+	−	+	+	−	+	−	−	−	+	+	−	−	+	+	−	−	+	+	−	−	−	+	−	+	+	−	+	−	+
28	+	+	+	−	+	+	+	−	+	+	−	+	+	−	−	+	−	+	+	−	−	+	−	−	+	−	−	−	+	−	−	−
29	+	−	−	+	+	+	+	−	−	−	−	−	−	+	+	+	+	+	+	−	−	−	−	−	−	+	+	+	+	−	−	+
30	+	+	−	+	+	+	−	+	+	+	−	−	−	+	+	+	−	−	−	+	+	+	−	−	−	+	−	−	−	+	−	−
31	+	−	+	+	+	+	−	−	−	−	+	+	+	+	+	+	−	−	−	−	−	−	+	+	+	+	−	−	−	−	+	−
32	+	+	+	+	+	+	+	+	+	+	+	+	+	+	+	+	+	+	+	+	+	+	+	+	+	+	+	+	+	+	+	+

Appendix B.3 Central Composite Designs

TABLE B.3-1 Central Composite Design for Two Factors

Run	X_1	X_2
1	-1	-1
2	1	-1
3	-1	1
4	1	1
5	$-\alpha$	0
6	α	0
7	0	$-\alpha$
8	0	α
9	0	0
10	0	0
11	0	0
12	0	0
13	0	0

Note: $\alpha = \sqrt{2} = 1.41$

TABLE B.3-2 Central Composite Design for Three Factors

Run	X_1	X_2	X_3	
1	-1	-1	-1	
2	1	-1	-1	
3	-1	1	-1	
4	1	1	-1	
5	-1	-1	1	
6	1	-1	1	Block 1
7	-1	1	1	
8	1	1	1	
9	0	0	0	
10	0	0	0	
11	0	0	0	
12	$-\alpha$	0	0	
13	α	0	0	
14	0	$-\alpha$	0	
15	0	α	0	
16	0	0	$-\alpha$	Block 2
17	0	0	α	
18	0	0	0	
19	0	0	0	
20	0	0	0	

Note: $\alpha = \sqrt{3} = 1.72$

TABLE B.3-3 Central Composite Design for Four Factors

Run	X_1	X_2	X_3	X_4	Block
1	1	-1	-1	-1	
2	-1	1	-1	-1	
3	-1	-1	1	-1	
4	1	1	1	-1	
5	-1	-1	-1	1	
6	1	1	-1	1	Block 1
7	1	-1	1	1	
8	-1	1	1	1	
9	0	0	0	0	
10	0	0	0	0	
11	0	0	0	0	
12	-1	-1	-1	-1	
13	1	1	-1	-1	
14	1	-1	1	-1	
15	-1	1	1	-1	
16	1	-1	-1	1	
17	-1	1	-1	1	Block 2
18	-1	-1	1	1	
19	1	1	1	1	
20	0	0	0	0	
21	0	0	0	0	
22	0	0	0	0	
23	$-\alpha$	0	0	0	
24	α	0	0	0	
25	0	$-\alpha$	0	0	
26	0	α	0	0	
27	0	0	$-\alpha$	0	Block 3
28	0	0	α	0	
29	0	0	0	α	
30	0	0	0	$-\alpha$	
31	0	0	0	0	
32	0	0	0	0	
33	0	0	0	0	

Notes:
(1) $\alpha = \sqrt{4} = 2.00$
(2) When run in only one block,
    only Seven center points are needed.

TABLE B.3-4 Central Composite Design for Five Factors

Run	X_1	X_2	X_3	X_4	X_5	Block
1	-1	-1	-1	-1	1	
2	1	-1	-1	-1	-1	
3	-1	1	-1	-1	-1	
4	1	1	-1	-1	1	
5	-1	-1	1	-1	-1	
6	1	-1	1	-1	1	
7	-1	1	1	-1	1	
8	1	1	1	-1	-1	
9	-1	-1	-1	1	-1	Block 1
10	1	-1	-1	1	1	
11	-1	1	-1	1	1	
12	1	1	-1	1	-1	
13	-1	-1	1	1	1	
14	1	-1	1	1	-1	
15	-1	1	1	1	-1	
16	1	1	1	1	1	
17	0	0	0	0	0	
18	0	0	0	0	0	
19	0	0	0	0	0	
20	$-\alpha$	0	0	0	0	
21	α	0	0	0	0	
22	0	$-\alpha$	0	0	0	
23	0	α	0	0	0	
24	0	0	$-\alpha$	0	0	Block 2
25	0	0	α	0	0	
26	0	0	0	α	0	
27	0	0	0	$-\alpha$	0	
28	0	0	0	0	α	
29	0	0	0	0	$-\alpha$	
30	0	0	0	0	0	
31	0	0	0	0	0	
32	0	0	0	0	0	

Note: $\alpha = \sqrt{4} = 2.00$

Appendix B.4 Box-Behnken Designs

TABLE B.4-1 Box-Behnken Design for Three Factors

Run	X_1	X_2	X_3
1	-1	-1	0
2	1	-1	0
3	-1	1	0
4	1	1	0
5	-1	0	-1
6	1	0	-1
7	-1	0	1
8	1	0	1
9	0	-1	-1
10	0	1	-1
11	0	-1	1
12	0	1	1
13	0	0	0
14	0	0	0
15	0	0	0

TABLE B.4-2 Box-Behnken Design for Four Factors

Run	X_1	X_2	X_3	X_4	Block
1	-1	-1	0	0	
2	1	-1	0	0	
3	-1	1	0	0	
4	1	1	0	0	
5	0	0	-1	-1	
6	0	0	1	-1	Block 1
7	0	0	-1	1	
8	0	0	1	1	
9	0	0	0	0	
10	-1	0	-1	0	
11	1	0	-1	0	
12	-1	0	1	0	
13	1	0	1	0	
14	0	-1	0	-1	
15	0	1	0	-1	Block 2
16	0	-1	0	1	
17	0	1	0	1	
18	0	0	0	0	
19	-1	0	0	-1	
20	1	0	0	-1	
21	-1	0	0	1	
22	1	0	0	1	
23	0	-1	-1	0	
24	0	1	-1	0	Block 3
25	0	-1	1	0	
26	0	1	1	0	
27	0	0	0	0	

TABLE B.4-3 Box-Behnken Design for Five Factors

	Block 1						Block 2				
Run	X_1	X_2	X_3	X_4	X_5	Run	X_1	X_2	X_3	X_4	X_5
1	-1	-1	0	0	0	24	0	-1	-1	0	0
2	1	-1	0	0	0	25	0	1	-1	0	0
3	-1	1	0	0	0	26	0	-1	1	0	0
4	1	1	0	0	0	27	0	1	1	0	0
5	0	0	-1	-1	0	28	-1	0	0	-1	0
6	0	0	1	-1	0	29	1	0	0	-1	0
7	0	0	-1	1	0	30	-1	0	0	1	0
8	0	0	1	1	0	31	1	0	0	1	0
9	0	-1	0	0	-1	32	0	0	-1	0	-1
10	0	1	0	0	-1	33	0	0	1	0	-1
11	0	-1	0	0	1	34	0	0	-1	0	1
12	0	1	0	0	1	35	0	0	1	0	1
13	-1	0	-1	0	0	36	-1	0	0	0	-1
14	1	0	-1	0	0	37	1	0	0	0	-1
15	-1	0	1	0	0	38	-1	0	0	0	1
16	1	0	1	0	0	39	1	0	0	0	1
17	0	0	0	-1	-1	40	0	-1	0	-1	0
18	0	0	0	1	-1	41	0	1	0	-1	0
19	0	0	0	-1	1	42	0	-1	0	1	0
20	0	0	0	1	1	43	0	1	0	1	0
21	0	0	0	0	0	44	0	0	0	0	0
22	0	0	0	0	0	45	0	0	0	0	0
23	0	0	0	0	0	46	0	0	0	0	0

Appendix B.5 Small Composite Designs

TABLE B.5-1 Small Composite Design for Two Factors

Run	X_1	X_2
1	1	-1
2	1	1
3	$-\alpha$	0
4	α	0
5	0	$-\alpha$
6	0	α

Notes:

$\alpha = 1.19$

$n_0 =$ Two center points can be added for error estimate

TABLE B.5-2 Small Composite Design for Three Factors

Run	X_1	X_2	X_3
1	-1	-1	1
2	1	-1	-1
3	-1	1	-1
4	1	1	1
5	$-\alpha$	0	0
6	α	0	0
7	0	$-\alpha$	0
8	0	α	0
9	0	0	$-\alpha$
10	0	0	α

Notes:

$\alpha = 1.41$

$n_0 =$ Two center points can be added for error estimate

TABLE B.5-3 Small Composite Design for Four Factors

Run	X_1	X_2	X_3	X_4
1	1	-1	-1	-1
2	-1	1	-1	-1
3	-1	-1	1	-1
4	1	1	1	-1
5	1	-1	-1	1
6	-1	1	-1	1
7	-1	-1	1	1
8	1	1	1	1
9	$-\alpha$	0	0	0
10	α	0	0	0
11	0	$-\alpha$	0	0
12	0	α	0	0
13	0	0	$-\alpha$	0
14	0	0	α	0
15	0	0	0	α
16	0	0	0	$-\alpha$

Notes:
$\alpha = 1.68$
$n_0 =$ Two center points can be added for error estimate

TABLE B.5-4 Small Composite Design for Five Factors

Run	X_1	X_2	X_3	X_4	X_5
1	-1	-1	-1	1	1
2	1	-1	-1	1	-1
3	-1	1	-1	-1	1
4	1	1	-1	-1	-1
5	-1	-1	1	-1	-1
6	1	-1	1	-1	1
7	-1	1	1	1	-1
8	1	1	1	1	1
9	$-\alpha$	0	0	0	0
10	α	0	0	0	0
11	0	$-\alpha$	0	0	0
12	0	α	0	0	0
13	0	0	$-\alpha$	0	0
14	0	0	α	0	0
15	0	0	0	α	0
16	0	0	0	$-\alpha$	0
17	0	0	0	0	α
18	0	0	0	0	$-\alpha$

Notes:
$\alpha = 1.82$
$n_0 =$ Two center points can be added for error estimate

Appendix B.6 Simplex Designs for Mixtures

TABLE B.6-1 Simplex Design for Three Mixture Components

Run	X_1	X_2	X_3	Symbolic Response	Coefficient Estimates for Special Cubic Scheffé Model
1	1	0	0	Y_1	$\beta_1 = Y_1$
2	0	1	0	Y_2	$\beta_2 = Y_2$
3	0	0	1	Y_3	$\beta_3 = Y_3$
4	1/2	1/2	0	Y_{12}	$\beta_{12} = 4Y_{12} - 2(Y_1 + Y_2)$
5	1/2	0	1/2	Y_{13}	$\beta_{13} = 4Y_{13} - 2(Y_1 + Y_3)$
6	0	1/2	1/2	Y_{23}	$\beta_{23} = 4Y_{23} - 2(Y_2 + Y_3)$
7	1/3	1/3	1/3	$Y_{123(1)}$	
8	1/3	1/3	1/3	$Y_{123(2)}$	$\beta_{123} = 27\bar{Y}_{123} - 12(Y_{12} + Y_{13} + Y_{23})$
9	1/3	1/3	1/3	$Y_{123(3)}$	$\qquad + 3(Y_1 + Y_2 + Y_3)$

Runs 1-3 are the linear design.

Adding Runs 4–6 creates a quadratic design.

Adding Runs 7–9 creates a special cubic design with a replicated center point.

TABLE B.6-2(a) Simplex Design for Four Mixture Components

Run	X_1	X_2	X_3	X_4	Symbolic Response
1	1	0	0	0	Y_1
2	0	1	0	0	Y_2
3	0	0	1	0	Y_3
4	0	0	0	1	Y_4
5	1/2	1/2	0	0	Y_{12}
6	1/2	0	1/2	0	Y_{13}
7	1/2	0	0	1/2	Y_{14}
8	0	1/2	1/2	0	Y_{23}
9	0	1/2	0	1/2	Y_{24}
10	0	0	1/2	1/2	Y_{34}
11	1/3	1/3	1/3	0	Y_{123}
12	1/3	1/3	0	1/3	Y_{124}
13	1/3	0	1/3	1/3	Y_{134}
14	0	1/3	1/3	1/3	Y_{234}
15	1/4	1/4	1/4	1/4	$Y_{1234(1)}$
16	1/4	1/4	1/4	1/4	$Y_{1234(2)}$
17	1/4	1/4	1/4	1/4	$Y_{1234(3)}$

Runs 1–4 are the linear design.
Adding Runs 5–10 creates a quadratic design.
Adding Runs 11–14 creates a special cubic design.
Runs 15–17 provide a replicated center point
for checking goodness of fit.

TABLE B.6-2(b) Coefficient Estimates for Four-Component Mixture Design

Coefficient Estimates for Special Cubic Scheffé Model
$\beta_1 = Y_1$
$\beta_2 = Y_2$
$\beta_3 = Y_3$
$\beta_4 = Y_4$
$\beta_{12} = 4Y_{12} - 2(Y_1+Y_2)$
$\beta_{13} = 4Y_{13} - 2(Y_1+Y_3)$
$\beta_{14} = 4Y_{14} - 2(Y_1+Y_4)$
$\beta_{23} = 4Y_{23} - 2(Y_2+Y_3)$
$\beta_{24} = 4Y_{24} - 2(Y_2+Y_4)$
$\beta_{34} = 4Y_{34} - 2(Y_3+Y_4)$
$\beta_{123} = 27Y_{123} - 12(Y_{12}+Y_{13}+Y_{23}) + 3(Y_1+Y_2+Y_3)$
$\beta_{124} = 27Y_{124} - 12(Y_{12}+Y_{14}+Y_{24}) + 3(Y_1+Y_2+Y_4)$
$\beta_{134} = 27Y_{134} - 12(Y_{13}+Y_{14}+Y_{34}) + 3(Y_1+Y_3+Y_4)$
$\beta_{234} = 27Y_{234} - 12(Y_{23}+Y_{24}+Y_{34}) + 3(Y_2+Y_3+Y_4)$

TABLE B.6-3(a) Simplex Design for Five Mixture Components

Run	X_1	X_2	X_3	X_4	X_5	Symbolic Response
1	1	0	0	0	0	Y_1
2	0	1	0	0	0	Y_2
3	0	0	1	0	0	Y_3
4	0	0	0	1	0	Y_4
5	0	0	0	0	1	Y_5
6	1/2	1/2	0	0	0	Y_{12}
7	1/2	0	1/2	0	0	Y_{13}
8	1/2	0	0	1/2	0	Y_{14}
9	1/2	0	0	0	1/2	Y_{15}
10	0	1/2	1/2	0	0	Y_{23}
11	0	1/2	0	1/2	0	Y_{24}
12	0	1/2	0	0	1/2	Y_{25}
13	0	0	1/2	1/2	0	Y_{34}
14	0	0	1/2	0	1/2	Y_{35}
15	0	0	0	1/2	1/2	Y_{45}
16	1/3	1/3	1/3	0	0	Y_{123}
17	1/3	1/3	0	1/3	0	Y_{124}
18	1/3	1/3	0	0	1/3	Y_{125}
19	1/3	0	1/3	1/3	0	Y_{134}
20	1/3	0	1/3	0	1/3	Y_{135}
21	1/3	0	0	1/3	1/3	Y_{145}
22	0	1/3	1/3	1/3	0	Y_{234}
23	0	1/3	1/3	0	1/3	Y_{235}
24	0	1/3	0	1/3	1/3	Y_{245}
25	0	0	1/3	1/3	1/3	Y_{345}
26	1/5	1/5	1/5	1/5	1/5	$Y_{12345(1)}$
27	1/5	1/5	1/5	1/5	1/5	$Y_{12345(2)}$
28	1/5	1/5	1/5	1/5	1/5	$Y_{12345(3)}$

Runs 1–5 are the linear design.
Adding Runs 6–15 creates a quadratic design.
Adding Runs 16–25 creates a special cubic design.
Runs 26–28 provide a replicated center point
for checking goodness of fit.

TABLE B.6-3(b) Coefficient Estimates for Five-Component Mixture Design

Coefficient Estimates for Special Cubic Scheffé Model
$\beta_1 = Y_1$
$\beta_2 = Y_2$
$\beta_3 = Y_3$
$\beta_4 = Y_4$
$\beta_5 = Y_5$
$\beta_{12} = 4Y_{12} - 2(Y_1+Y_2)$
$\beta_{13} = 4Y_{13} - 2(Y_1+Y_3)$
$\beta_{14} = 4Y_{14} - 2(Y_1+Y_4)$
$\beta_{15} = 4Y_{15} - 2(Y_1+Y_5)$
$\beta_{23} = 4Y_{23} - 2(Y_2+Y_3)$
$\beta_{24} = 4Y_{24} - 2(Y_2+Y_4)$
$\beta_{25} = 4Y_{25} - 2(Y_2+Y_5)$
$\beta_{34} = 4Y_{34} - 2(Y_3+Y_4)$
$\beta_{35} = 4Y_{35} - 2(Y_3+Y_5)$
$\beta_{45} = 4Y_{45} - 2(Y_4+Y_5)$
$\beta_{123} = 27Y_{123} - 12(Y_{12}+Y_{13}+Y_{23}) + 3(Y_1+Y_2+Y_3)$
$\beta_{124} = 27Y_{124} - 12(Y_{12}+Y_{14}+Y_{24}) + 3(Y_1+Y_2+Y_4)$
$\beta_{125} = 27Y_{125} - 12(Y_{12}+Y_{15}+Y_{25}) + 3(Y_1+Y_2+Y_5)$
$\beta_{134} = 27Y_{134} - 12(Y_{13}+Y_{14}+Y_{34}) + 3(Y_1+Y_3+Y_4)$
$\beta_{135} = 27Y_{135} - 12(Y_{13}+Y_{15}+Y_{35}) + 3(Y_1+Y_3+Y_5)$
$\beta_{145} = 27Y_{145} - 12(Y_{14}+Y_{15}+Y_{45}) + 3(Y_1+Y_4+Y_5)$
$\beta_{234} = 27Y_{234} - 12(Y_{23}+Y_{24}+Y_{34}) + 3(Y_2+Y_3+Y_4)$
$\beta_{235} = 27Y_{235} - 12(Y_{23}+Y_{25}+Y_{35}) + 3(Y_2+Y_3+Y_5)$
$\beta_{245} = 27Y_{245} - 12(Y_{24}+Y_{25}+Y_{45}) + 3(Y_2+Y_4+Y_5)$
$\beta_{345} = 27Y_{345} - 12(Y_{34}+Y_{35}+Y_{45}) + 3(Y_3+Y_4+Y_5)$

Appendix B.7 Designs for Components of Variance

TABLE B.7.1-1 Nested Design for Two Components of Variance

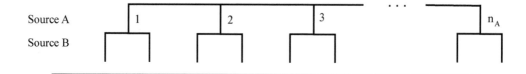

ANOVA Table for Two Components of Variance

Source	SS	ν	MS	Expected Mean Square (MS)
A	SS_A	$\nu_A = n_A - 1$	SS_A/ν_A	$\sigma_B^2 + 2\sigma_A^2$
B	SS_B	$\nu_B = n_A$	SS_B/ν_B	σ_B^2

$$T_i = \sum_j Y_{ij} \qquad SS_B = \sum_i \left[\sum_j Y_{ij}^2 - T_i^2/2 \right]$$

$$T = \sum_i T_i \qquad SS_A = \sum_i T_i^2 - T^2/2n_A$$

where Y_{ij} = the observation for the ith value of A ($i = 1, 2, \ldots, n_A$) and the jth value of B ($j = 1, 2$).

TABLE B.7.1-2 Nested Design for Three Components of Variance

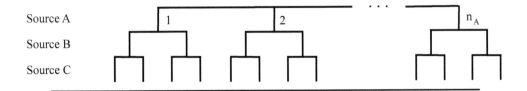

ANOVA Table for Three Components of Variance

Source	SS	ν	MS	Expected Mean Square (MS)
A	SS_A	$\nu_A = n_A - 1$	SS_A/ν_A	$\sigma_C^2 + 2\sigma_B^2 + 4\sigma_A^2$
B	SS_B	$\nu_B = n_A$	SS_B/ν_B	$\sigma_C^2 + 2\sigma_B^2$
C	SS_C	$\nu_C = 2n_A$	SS_C/ν_C	σ_C^2

$$T_{ij} = \sum_k Y_{ijk} \qquad SS_C = \sum_i \sum_j \left[\sum_k Y_{ijk}^2 - T_{ij}^2/2 \right]$$

$$T_i = \sum_j T_{ij} \qquad SS_B = \sum_i \left[\sum_j T_{ij}^2/2 - T_i^2/4 \right]$$

$$T = \sum_i T_i \qquad SS_A = \sum_i T_i^2/4 - T^2/4n_A$$

where Y_{ijk} = the observation for the ith value of A ($i = 1, 2, \ldots, n_A$), the jth value of B ($j = 1,2$), and the kth value of C ($k = 1,2$)

TABLE B.7.1-3 Nested Design for Four Components of Variance

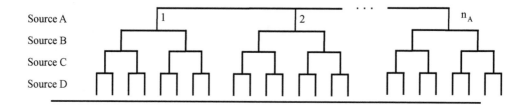

ANOVA Table for Four Components of Variance

Source	SS	ν	MS	Expected Mean Square (MS)
A	SS_A	$\nu_A = n_A - 1$	SS_A/ν_A	$\sigma_D^2 + 2\sigma_C^2 + 4\sigma_B^2 + 8\sigma_A^2$
B	SS_B	$\nu_B = n_A$	SS_B/ν_B	$\sigma_D^2 + 2\sigma_C^2 + 4\sigma_B^2$
C	SS_C	$\nu_C = 2n_A$	SS_C/ν_C	$\sigma_D^2 + 2\sigma_C^2$
D	SS_D	$\nu_D = 4n_A$	SS_D/ν_D	σ_D^2

$$T_{ijk} = \sum_l Y_{ijkl} \qquad SS_D = \sum_i \sum_j \sum_k \left[\sum_l Y_{ijkl}^2 - T_{ijk}^2/2 \right]$$

$$T_{ij} = \sum_k T_{ijk} \qquad SS_C = \sum_i \sum_j \left[\sum_k T_{ijk}^2/2 - T_{ij}^2/4 \right]$$

$$T_i = \sum_j T_{ij} \qquad SS_B = \sum_i \left[\sum_j T_{ij}^2/4 - T_i^2/8 \right]$$

$$T = \sum_i T_i \qquad SS_A = \sum_i T_i^2/8 - T^2/8n_A$$

where Y_{ijkl} = the observation for the ith value of A ($i = 1, 2, \ldots, n_A$), the jth value of B ($j = 1,2$), the kth value of C ($k = 1,2$), and the lth value of D ($l = 1,2$).

TABLE B.7.1-4 Nested Design for Five Components of Variance

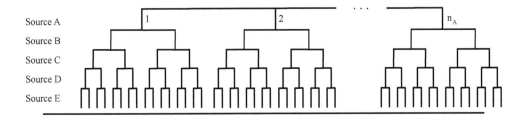

ANOVA Table for Five Components of Variance

Source	SS	ν	MS	Expected Mean Square (MS)
A	SS_A	$\nu_A = n_A - 1$	SS_A/ν_A	$\sigma_E^2 + 2\sigma_D^2 + 4\sigma_C^2 + 8\sigma_B^2 + 16\sigma_A^2$
B	SS_B	$\nu_B = n_A$	SS_B/ν_B	$\sigma_E^2 + 2\sigma_D^2 + 4\sigma_C^2 + 8\sigma_B^2$
C	SS_C	$\nu_C = 2n_A$	SS_C/ν_C	$\sigma_E^2 + 2\sigma_D^2 + 4\sigma_C^2$
D	SS_D	$\nu_D = 4n_A$	SS_D/ν_D	$\sigma_E^2 + 2\sigma_D^2$
E	SS_E	$\nu_E = 8n_A$	SS_E/ν_E	σ_E^2

$$T_{ijkl} = \sum_m Y_{ijklm} \qquad SS_D = \sum_i \sum_j \sum_k \sum_l \left[\sum_m Y_{ijklm}^2 - T_{ijkl}^2/2 \right]$$

$$T_{ijk} = \sum_l T_{ijkl} \qquad SS_D = \sum_i \sum_j \sum_k \left[\sum_l T_{ijkl}^2/2 - T_{ijk}^2/4 \right]$$

$$T_{ij} = \sum_k T_{ijk} \qquad SS_C = \sum_i \sum_j \left[\sum_k T_{ijk}^2/4 - T_{ij}^2/8 \right]$$

$$T_i = \sum_j T_{ij} \qquad SS_B = \sum_i \left[\sum_j T_{ij}^2/8 - T_i^2/16 \right]$$

$$T = \sum_i T_i \qquad SS_A = \sum_i T_i^2/16 - T^2/16n_A$$

where Y_{ijklm} = the observation for the ith value of A $(i = 1, 2, \ldots, n_A)$, the jth value of B $(j = 1,2)$, the kth value of C $(k = 1,2)$, the lth value of D $(l = 1,2)$, and the mth value of E $(m = 1,2)$.

TABLE B.7.2-1 Staggered Nested Design for Three Components of Variance

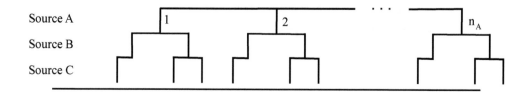

ANOVA Table for Three Components of Variance

Source	SS	ν	MS	Expected Mean Square (MS)
A	SS_A	$\nu_A = n_A - 1$	SS_A/ν_A	$\sigma_C^2 + (5/3)\sigma_B^2 + 3\sigma_A^2$
B	SS_B	$\nu_B = n_A$	SS_B/ν_B	$\sigma_C^2 + (4/3)\sigma_B^2$
C	SS_C	$\nu_C = n_A$	SS_C/ν_C	σ_C^2

$$T_{ij} = \sum_k Y_{ijk} \quad SS_C = \sum_i \sum_j \left[\sum_k Y_{ijk}^2 - T_{ij}^2/n_{ij} \right]$$

$$T_i = \sum_j T_{ij} \quad SS_B = \sum_i \left[\sum_j T_{ij}^2/n_{ij} - T_i^2/3 \right]$$

$$T = \sum_i T_i \quad SS_A = \sum_i T_i^2/3 - T^2/3n_A$$

where Y_{ijk} = the observation for the ith value of A ($i = 1, 2, \ldots, n_A$), the jth value of B ($j = 1, 2$), and the kth value of C ($k = 1, 2$)

TABLE B.7.2-2 Staggered Nested Design for Four Components of Variance

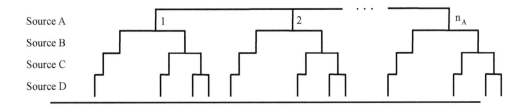

ANOVA Table for Four Components of Variance

Source	SS	ν	MS	Expected Mean Square (MS)
A	SS_A	$\nu_A = n_A - 1$	SS_A/ν_A	$\sigma_D^2 + (3/2)\sigma_C^2 + (5/2)\sigma_B^2 + 4\sigma_A^2$
B	SS_B	$\nu_B = n_A$	SS_B/ν_B	$\sigma_D^2 + (7/6)\sigma_C^2 + (3/2)\sigma_B^2$
C	SS_C	$\nu_C = n_A$	SS_C/ν_C	$\sigma_D^2 + (4/3)\sigma_C^2$
D	SS_D	$\nu_D = n_A$	SS_D/ν_D	σ_D^2

$$T_{ijk} = \sum_l Y_{ijkl} \quad SS_D = \sum_i \sum_j \sum_k \left[\sum_l Y_{ijkl}^2 - T_{ijk}^2/n_{ijk} \right]$$

$$T_{ij} = \sum_k T_{ijk} \quad SS_C = \sum_i \sum_j \left[\sum_k T_{ijk}^2/n_{ijk} - T_{ij}^2/n_{ij} \right]$$

$$T_i = \sum_j T_{ij} \quad SS_B = \sum_i \left[\sum_j T_{ij}^2/nij - T_i^2/4 \right]$$

$$T = \sum_i T_i \quad SS_A = \sum_i T_i^2/4 - T^2/4n_A$$

where Y_{ijkl} = the observation for the ith value of A ($i = 1, 2, \ldots, n_A$),
the jth value of B ($j = 1,2$), the kth value of C ($k = 1,2$),
and the lth value of D ($l = 1,2$).

TABLE B.7.2-3 Staggered Nested Design for Five Components of Variance

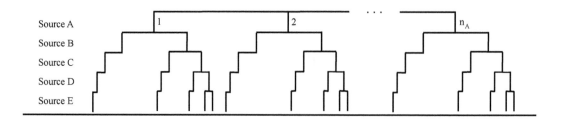

ANOVA Table for Five Components of Variance

Source	SS	ν	MS	Expected Mean Square (MS)
A	SS_A	$\nu_A = n_A - 1$	SS_A/ν_A	$\sigma_E^2 + (7/5)\sigma_D^2 + (11/5)\sigma_C^2 + (17/5)\sigma_B^2 + 5\sigma_A^2$
B	SS_B	$\nu_B = n_A$	SS_B/ν_B	$\sigma_E^2 + (11/10)\sigma_D^2 + (13/10)\sigma_C^2 + (8/5)\sigma_B^2$
C	SS_C	$\nu_C = n_A$	SS_C/ν_C	$\sigma_E^2 + (7/6)\sigma_D^2 + (3/2)\sigma_C^2$
D	SS_D	$\nu_D = n_A$	SS_D/ν_D	$\sigma_E^2 + (4/3)\sigma_D^2$
E	SS_E	$\nu_E = n_A$	SS_E/ν_E	σ_E^2

$$T_{ijkl} = \sum_m Y_{ijklm} \quad SS_D = \sum_i \sum_j \sum_k \sum_l \left[\sum_m Y_{ijklm}^2 - T_{ijkl}^2/n_{ijkl} \right]$$

$$T_{ijk} = \sum_l T_{ijkl} \quad SS_D = \sum_i \sum_j \sum_k \left[\sum_l T_{ijkl}^2/n_{ijkl} - T_{ijk}^2/n_{ijk} \right]$$

$$T_{ij} = \sum_k T_{ijk} \quad SS_C = \sum_i \sum_j \left[\sum_k T_{ijk}^2/n_{ijk} - T_{ij}^2/n_{ij} \right]$$

$$T_i = \sum_j T_{ij} \quad SS_B = \sum_i \left[\sum_j T_{ij}^2/n_{ij} - T_i^2/5 \right]$$

$$T = \sum_i T_i \quad SS_A = \sum_i T_i^2/5 - T^2/5n_A$$

where Y_{ijklm} = the observation for the ith value of A ($i = 1, 2, \ldots, n_A$), the jth value of B ($j = 1,2$), the kth value of C ($k = 1,2$), the lth value of D ($l = 1,2$), and the mth value of E ($m = 1,2$).

Appendix B.8 Definitive Screening Designs

TABLE B.8-1 Definitive Screening Design for 4 Factors

Run	X_1	X_2	X_3	X_4
1	0	1	-1	-1
2	0	-1	1	1
3	-1	0	-1	1
4	1	0	1	-1
5	-1	-1	0	-1
6	1	1	0	1
7	-1	1	1	0
8	1	-1	-1	0
9	0	0	0	0

TABLE B.8-2 Definitive Screening Design for 5 Factors

Run	X_1	X_2	X_3	X_4	X_5
1	0	1	1	-1	-1
2	0	-1	-1	1	1
3	1	0	-1	-1	1
4	-1	0	1	1	-1
5	1	-1	0	1	-1
6	-1	1	0	-1	1
7	1	-1	1	0	1
8	-1	1	-1	0	-1
9	1	1	1	1	0
10	-1	-1	-1	-1	0
11	0	0	0	0	0

TABLE B.8-3 Definitive Screening Design for 6 Factors

Run	X_1	X_2	X_3	X_4	X_5	X_6
1	0	1	-1	-1	-1	-1
2	0	-1	1	1	1	1
3	1	0	-1	1	1	-1
4	-1	0	1	-1	-1	1
5	-1	-1	0	1	-1	-1
6	1	1	0	-1	1	1
7	-1	1	1	0	1	-1
8	1	-1	-1	0	-1	1
9	1	-1	1	-1	0	-1
10	-1	1	-1	1	0	1
11	1	1	1	1	-1	0
12	-1	-1	-1	-1	1	0
13	0	0	0	0	0	0

TABLE B.8-4 Definitive Screening Design for 7 Factors

Run	X_1	X_2	X_3	X_4	X_5	X_6	X_7
1	0	1	-1	1	-1	1	-1
2	0	-1	1	-1	1	-1	1
3	-1	0	1	-1	1	1	-1
4	1	0	-1	1	-1	-1	1
5	1	-1	0	1	1	1	1
6	-1	1	0	-1	-1	-1	-1
7	1	-1	-1	0	1	-1	-1
8	-1	1	1	0	-1	1	1
9	-1	-1	1	1	0	-1	-1
10	1	1	-1	-1	0	1	1
11	-1	1	-1	1	1	0	1
12	1	-1	1	-1	-1	0	-1
13	1	1	1	1	1	-1	0
14	-1	-1	-1	-1	-1	1	0
15	0	0	0	0	0	0	0

TABLE B.8-5 Definitive Screening Design for 8 Factors

Run	X_1	X_2	X_3	X_4	X_5	X_6	X_7	X_8
1	0	-1	1	1	-1	1	1	1
2	0	1	-1	-1	1	-1	-1	-1
3	-1	0	-1	1	1	1	1	-1
4	1	0	1	-1	-1	-1	-1	1
5	-1	-1	0	1	1	-1	-1	1
6	1	1	0	-1	-1	1	1	-1
7	1	-1	1	0	1	1	-1	-1
8	-1	1	-1	0	-1	-1	1	1
9	-1	-1	1	-1	0	-1	1	-1
10	1	1	-1	1	0	1	-1	1
11	1	-1	-1	-1	1	0	1	1
12	-1	1	1	1	-1	0	-1	-1
13	-1	1	1	-1	1	1	0	1
14	1	-1	-1	1	-1	-1	0	-1
15	1	1	1	1	1	-1	1	0
16	-1	-1	-1	-1	-1	1	-1	0
17	0	0	0	0	0	0	0	0

TABLE B.8-6 Definitive Screening Design for 9 Factors

Run	X_1	X_2	X_3	X_4	X_5	X_6	X_7	X_8	X_9
1	0	1	1	1	1	1	1	1	1
2	0	-1	-1	-1	-1	-1	-1	-1	-1
3	1	0	1	-1	1	-1	-1	1	-1
4	-1	0	-1	1	-1	1	1	-1	1
5	-1	1	0	-1	1	-1	1	-1	-1
6	1	-1	0	1	-1	1	-1	1	1
7	-1	-1	1	0	1	-1	-1	-1	1
8	1	1	-1	0	-1	1	1	1	-1
9	1	-1	1	-1	0	1	1	-1	-1
10	-1	1	-1	1	0	-1	-1	1	1
11	-1	-1	-1	-1	1	0	1	1	1
12	1	1	1	1	-1	0	-1	-1	-1
13	1	1	-1	-1	1	1	0	-1	1
14	-1	-1	1	1	-1	-1	0	1	-1
15	-1	-1	-1	1	1	1	-1	0	-1
16	1	1	1	-1	-1	-1	1	0	1
17	-1	1	1	-1	-1	1	-1	1	0
18	1	-1	-1	1	1	-1	1	-1	0
19	0	0	0	0	0	0	0	0	0

TABLE B.8-7 Definitive Screening Design for 10 Factors

Run	X_1	X_2	X_3	X_4	X_5	X_6	X_7	X_8	X_9	X_{10}
1	0	1	1	-1	1	1	1	1	-1	1
2	0	-1	-1	1	-1	-1	-1	-1	1	-1
3	1	0	-1	1	1	-1	1	1	-1	-1
4	-1	0	1	-1	-1	1	-1	-1	1	1
5	-1	1	0	-1	-1	-1	1	-1	-1	-1
6	1	-1	0	1	1	1	-1	1	1	1
7	-1	1	1	0	1	-1	-1	1	1	-1
8	1	-1	-1	0	-1	1	1	-1	-1	1
9	-1	-1	-1	-1	0	1	1	1	1	-1
10	1	1	1	1	0	-1	-1	-1	-1	1
11	-1	1	-1	1	1	0	1	-1	1	1
12	1	-1	1	-1	-1	0	-1	1	-1	-1
13	1	1	-1	-1	-1	-1	0	1	1	1
14	-1	-1	1	1	1	1	0	-1	-1	-1
15	1	1	1	1	-1	1	1	0	1	-1
16	-1	-1	-1	-1	1	-1	-1	0	-1	1
17	1	1	-1	-1	1	1	-1	-1	0	-1
18	-1	-1	1	1	-1	-1	1	1	0	1
19	1	-1	1	-1	1	-1	1	-1	1	0
20	-1	1	-1	1	-1	1	-1	1	-1	0
21	0	0	0	0	0	0	0	0	0	0

TABLE B.8-8 Definitive Screening Design for 11 Factors

Run	X_1	X_2	X_3	X_4	X_5	X_6	X_7	X_8	X_9	X_{10}	X_{11}
1	0	-1	1	-1	-1	-1	-1	-1	1	-1	1
2	0	1	-1	1	1	1	1	1	-1	1	-1
3	-1	0	-1	-1	1	-1	-1	-1	-1	1	1
4	1	0	1	1	-1	1	1	1	1	-1	-1
5	-1	-1	0	1	1	1	1	-1	-1	-1	1
6	1	1	0	-1	-1	-1	-1	1	1	1	-1
7	-1	-1	-1	0	-1	1	1	-1	1	1	-1
8	1	1	1	0	1	-1	-1	1	-1	-1	1
9	1	-1	-1	1	0	1	-1	1	1	1	1
10	-1	1	1	-1	0	-1	1	-1	-1	-1	-1
11	-1	-1	1	1	-1	0	-1	1	-1	1	-1
12	1	1	-1	-1	1	0	1	-1	1	-1	1
13	-1	-1	-1	1	1	-1	0	1	1	-1	-1
14	1	1	1	-1	-1	1	0	-1	-1	1	1
15	-1	1	1	1	-1	-1	1	0	1	1	1
16	1	-1	-1	-1	1	1	-1	0	-1	-1	-1
17	-1	1	-1	-1	-1	1	-1	1	0	-1	1
18	1	-1	1	1	1	-1	1	-1	0	1	-1
19	1	-1	-1	-1	-1	-1	1	1	-1	0	1
20	-1	1	1	1	1	1	-1	-1	1	0	-1
21	1	1	-1	1	-1	-1	-1	-1	-1	-1	0
22	-1	-1	1	-1	1	1	1	1	1	1	0
23	0	0	0	0	0	0	0	0	0	0	0

TABLE B.8-9 Definitive Screening Design for 12 Factors

Run	X_1	X_2	X_3	X_4	X_5	X_6	X_7	X_8	X_9	X_{10}	X_{11}	X_{12}
1	0	-1	-1	1	-1	1	-1	1	1	1	-1	1
2	0	1	1	-1	1	-1	1	-1	-1	-1	1	-1
3	-1	0	1	1	1	1	1	1	1	-1	-1	-1
4	1	0	-1	-1	-1	-1	-1	-1	-1	1	1	1
5	1	1	0	-1	1	1	-1	1	1	-1	1	1
6	-1	-1	0	1	-1	-1	1	-1	-1	1	-1	-1
7	1	-1	-1	0	1	-1	1	-1	1	-1	-1	1
8	-1	1	1	0	-1	1	-1	1	-1	1	1	-1
9	1	1	1	1	0	-1	1	1	1	1	1	1
10	-1	-1	-1	-1	0	1	-1	-1	-1	-1	-1	-1
11	1	-1	1	-1	1	0	1	1	-1	1	-1	1
12	-1	1	-1	1	-1	0	-1	-1	1	-1	1	-1
13	1	1	1	1	-1	1	0	-1	-1	-1	-1	1
14	-1	-1	-1	-1	1	-1	0	1	1	1	1	-1
15	-1	-1	1	1	1	-1	-1	0	-1	-1	1	1
16	1	1	-1	-1	-1	1	1	0	1	1	-1	-1
17	1	-1	1	1	1	1	-1	-1	0	1	1	-1
18	-1	1	-1	-1	-1	-1	1	1	0	-1	-1	1
19	1	1	-1	1	1	-1	-1	1	-1	0	-1	-1
20	-1	-1	1	-1	-1	1	1	-1	1	0	1	1
21	-1	1	-1	1	1	1	1	-1	-1	1	0	1
22	1	-1	1	-1	-1	-1	-1	1	1	-1	0	-1
23	1	-1	-1	1	-1	1	1	1	-1	-1	1	0
24	-1	1	1	-1	1	-1	-1	-1	1	1	-1	0
25	0	0	0	0	0	0	0	0	0	0	0	0

Check Figures for Selected Exercises

Exercise 2.1
(a)-(b) Mean and variance for Batch 1 are $\bar{y}=58.193$, $s=0.5419$. Data from Batch 2 looks normal, but there is a possible outlier on the high side of the normal plot for Batch 1.
(c) Pooled standard deviation is $s_P=0.758$.
(d) $t_{18}=3.57$.
(e) The critical value can be looked up in Appendix Table A.3 or by using the spreadsheet formula =T.INV.2T

Exercise 2.2
(a) MS Within Groups = 2.9933333E-011, can be obtained using the spreadsheet menu.
(c) $F_{3,5,0.05} = 5.409$ can be obtained from Appendix Table A.4 or with the spreadsheet formula =FINV.
(d) There is a significant difference.

Exercise 3.1
(a) and (c) Equal variance assumption appears justified.
(f) Effect of X_1X_2 interaction effect = -0.050, $t_8 = -2.14$ (not significant)
(g) The effect of X_3 (Temperature) is different depending on the level of X_1 (Pressure) and on the level of X_2 (Width).

Exercise 4.1
(c) The curvature effect is 0.233
(d) The standard of error of the curvature is $s_C = 2.051$ with 21 degrees of freedom.

Exercise 4.2
(a) $s_E = 0.915$ with 24 degrees of freedom when rat litters are treated as replicates.
(b) $s_E = 0.442$ with 21 degrees of freedom when rat litters are treated as blocks.

Exercise 5.1
(b) F-test for Temperature $F_{4,8} = 1003.68$ is significant.
(c) Temperature × Time interaction is significant. Maximum tannase production at Temperature = 35, Time = 96.

Exercise 6.1
(b) 98.3%

Exercise 6.2
(a) ANOVA $MS_{error} = 1.0625$
(b) $\sigma^2_{thermometer} = 0.322915$

Exercise 6.3
(b) $\hat{\sigma}^2_{Sample} = 405.333$, $\hat{\sigma}^2_{Formula} = 899.193$

Exercise 7.1
(a) Effect of $X_1 = $ -2.75
(b) X_1, X_2 and X_6 appear to be significant.
(c) $X_1 X_2 = -X_6$ etc.

Exercise 8.1
(b) $Y = 1.160 + 0.508X$
(c) Both coefficients are significant.

Exercise 9.1
$-\alpha$ for Temperature $= 116$
$+\alpha$ for Temperature $= 184$.

Exercise 10.1
Lack of fit $F_{3,5} = 31.11$ which is significant.

Exercise 10.2
(a) None of the model terms involving X_2 nor the blocking term were significant, and they should be dropped from the model.
(c) % Impurity minimized at $X_1 = 1.3$, $X_2 = -0.7$.

Exercise 11.1
(c) Model for $CFW = 87.03 - \frac{7.66}{2}X_1 + \frac{7.67}{2}X_3 - \frac{4.43}{2}X_4$

Model for Wash water usage$= 0.125 - \frac{0.108}{2}X_1 + \frac{0.108}{2}X_3 - \frac{0.050}{2}X_1X_3$

(d) To get maximum CFW with minimal wash water usage you must compromise and use the low level of X_1 and the low level of X_3

Exercise 11.2
(a)-(b) Effect of $X_6 = 0.915$ is the largest effect, but it only appears to be slightly off the line of effects in the normal plot.
(c) Maximum predicted value is FL=6.188 when $X_6 = +$.
(d)-(e) $X_6 X_7$ interaction had the highest correlation with FL, but in the model fit including X_6, X_7 and $X_6 X_7$, X_7 was not significant.
(f) Excluding X_7 from the model the maximum predicted value is FL=6.65 at $X_6 = +$ and $X_7 = -$.

Exercise 11.3
(b) After 6 steps of forward regression preserving effect hierarchy, the model included all main effects, $X_2 X_3$, $X_3 X_4$, and $X_3 X_5$. This model had no significant lack of fit and appears to fit as well as the model proposed by the authors of the article.

Exercise 12.1
(a) Similar to mixture proportions in Table 12.4
(b) Use Equation (12.9) to convert pseudo components in Table 12.4 to pure components.

Bibliography

H. Aboubakretal, M. E-Sahn, and A. El-Banna. Some factors affecting tannase production by *Aspergillus niger* van tieghem. *Brazilian Journal of MicroBiology*, 44:559–567, 2013.

V. L. Anderson and R. A. Mclean. *Design of Experiments a Realistic Approach*. Marcel Dekker, New York, 1976.

D. M. Bates and D. G. Watts. *Nonlinear Regression Analysis and its Applications*. John Wiley & Sons, New York, 2007.

C. A. Bennett. Effect of measurement error on chemical process control. *Industrial Quality Control*, 11:17–20, 1954.

C. A. Bennett and N. L. Franklin. *Statistical Analysis in Chemistry and the Chemical Industry*. John Wiley & Sons, New York, 1954.

A. H. Bowker and G. J. Lieberman. *Handbook of Industrial Statistics*. Prentice Hall, Englewood Cliffs, NJ, 1955.

G.E.P. Box. Use and abuse of regression. *Technometrics*, 8:625–629, 1966.

G.E.P. Box and J.S. Hunter. The 2^{k-p} fractional factorial designs part I. *Technometrics*, 3:311–351, 1961.

G.E.P. Box, W.G. Hunter, and J.S. Hunter. *Statistics for Experimenters*. John Wiley & Sons, New York, 1978.

J. P. Chen, S.L. Kim, and Y.P. Ting. Optimization of membrane physical and chemical cleaning by a statistically designed approach. *Journal of Membrane Science*, 219:27–45, 2003.

J. A. Cornell. *Experiments with Mixtures*. John Wiley & Sons, New York, 1981.

J. A. Cornell. Analyzing mixture experiments containing process variables: A split plot approach. *Journal of Quality Technology*, 20:2–25, 1988.

N. R. Draper and H. Smith. *Applied Regression Analysis*. John Wiley & Sons, New York, 1966.

A. Duncan. *Quality Control and Industrial Statistics*. Richard D. Irwin Inc., Homewood, IL, 1967.

A. Erler, N. deMas, P. Ramsey, and Grant Henderson. Efficient biological process characterization by definitive-screening designs: The formaldehyde treatment of a theraputic protein as a case study. *Biotechnology Letters*, 35:323–329, 2013.

R. A. Fisher. *Statistical Methods for Research Workers*. Oliver and Boyd, Edinburg, 1925.

J. Fox. The R commander: A basic-statistics graphical user interface to R. *Journal of Statistical Software*, 14:1–42, 2005.

M. F. Franklin and R. A. Bailey. Selection of defining contrasts and confounded effects in two-level experiments. *Applied Statistics*, 26, 1977.

F. A. Graybill. *Theory and Application of the Linear Model.* Duxbury Press, North Scituate, MA, 1976.

I. Guttman, S. S. Wilks, and J. S. Hunter. *Introductory Engineering Statistics, 2nd Ed.* John Wiley & Sons, New York, 1965.

L. B. Hare and P. L. Brown. Plotting response surface contours on a three component mixture space. *Journal of Quality Technology*, 9:193–197, 1977.

G.B. Hunter, F.S. Hodi, and T.W. Eager. High-cycle fatigue of weld repaired cast t-6a1-4v. *Metallurgical Transactions*, 13A:1589–1594, 1982.

J. S. Hunter. Statistical design applied to product design. *Journal of Quality Technology*, 17:210, 1985.

N. L. Johnson and F. C. Leone. *Statistics and Experimental Design*, volume II. John Wiley & Sons, New York, 1964a.

N.L. Johnson and F. C. Leone. *Statistics and Experimental Design in Egineering and the Physical Sciences.* John Wiley & Sons, New York, 1964b.

B. Jones and D. C. Montgomery. Alternatives to resolution iv screening designs in 16 runs. *Int. J. Experimental Design and Process Optimisation*, 1:285–295, 2010.

B. Jones and C. Nachtsheim. A class of three-level designs for definitive screening in the presence of second-order effects. *Journal of Quality Technology*, 43:1–15, 2011.

B. Jones and C. Nachtsheim. Definitive screening designs with added two-level factors. *Journal of Quality Technology*, 45:121–129, 2013.

R. N. Kackar and A. Shoemaker. Robust design: A cost effective method for improving manufacturing processes. *AT&T Techical Journal*, 1986.

R. N. Kackar and K. L. Tsui. Interaction graphs: Graphical aids for planning experiments. *Journal of Quality Technology*, 22:1, 1990.

A. I. Khuri and J. A. Cornell. *Response Surfaces.* Marcel Dekker, New York, 1987.

J.R. Kittrel and J. Erjavec. Response surface methods in heterogeneous kinetic modeling. *Ind. Eng. Chem. Proc. Des. Dev.*, 7:321–327, 1968.

G. F. Koons and R. H. Heasly. Response surface contour plots for mixture problems. *Journal of Quality Technology*, 13:207–214, 1981.

I. S. Kurotori. Experiments with mixtures of components having lower bounds. *Industrial Quality Control*, 8:447–454, 1966.

J. Lawson. *daewr: Data frames and functions for Design and Analysis of Experiments with R*, 2015. URL http://CRAN.R-project.org/package=daewr. R package version 1.1-4.

J. Lawson, C. Willden, and G. Piepel. *mixexp: Design and Analysis of Mixture Experiments*, 2015.

R. V. Lenth. Lenth's method for the analysis of unreplicated experiments. URL `http://www.wiley.com/legacy/wileychi/eqr/docs/sample_1.pdf`. Accessed: 2015-12-10.

R. V. Lenth. Quick and easy analysis of factorials. *Technometrics*, 31:496–473, 1989.

W. Li and C. Nachtsheim. Model robust factorial designs. *Technometrics*, 42:345–352, 2000.

W. Libbrecht, F. Deruyck, H. Poelman, A. Verberckmoes, J. Thybaut, J. DeClercq, and P. VanDerVoort. Optimization of soft templated mesoporous carbon synthesis using definitive screening design. *Chemical Engineering Journal*, 259:126–134, 2015.

J.M. Lucas. Discussion of off-line quality control by kackar. *Journal of Quality Technology*, 17:195–197, 1985.

T. Lumley. *leaps: Regression subset selection*, 2009. URL `http://CRAN.R-project.org/package=leaps`. R package version 2.9.

R. Mclean and V. Andersoni. Extreme vertices designs of mixture experiments. *Technometrics*, 22:592–596, 1966.

MINITAB. *Minitab 17.0.1*. Minitab Inc., State College, PA, 2015. URL `http://www.minitab.com/en-us/`.

L. S. Nelson. Extreme screening designs. *Journal of Quality Technology*, 14:99–100, 1982.

E. R. Ott and E. G. Schilling. *Process Quality Control: Troubleshooting and Interpretation of Data*. McGraw-Hill Publishing Company, New York, 1990.

D.G. Peters, J.M. Hayes, and G.M. Hieftje. *A Brief Introduction to Modern Chemical Analysis*. W. B. Saunders Company, Philadelphia, 1976.

M. S. Phadke, R. N. Kackar, D. V. Speeney, and M. J. Grieco. Off-line quality control in integrated circuit fabrication using experimental design. *The Bell System Technical Journal*, 61:1273–1309, 1983.

G. F. Piepel. Measuring component effects in constrained mixture experiments. *Technometrics*, 24:29–39, 1982.

G. F. Piepel. Programs for generating extreme vertices and centroids of linearly constrained experimental regions. *Journal of Quality Technology*, 20:125–139, 1988.

G. F. Piepel. Screening designs for constrained mixture experiments derived from classical screening designs. *Journal of Quality Technology*, 22:22–33, 1990.

R. L. Plackett and J. P. Burman. The design of optimum multi-factor experiments. *Biometrika*, 33:305, 1946.

R Development Core Team. R: A language and environment for statistical computing. R Foundation for statistical computing, 2003. URL `http://www.R-project.org`.

RealStat. Real statistics resource pack, 2015. URL `www.real-statistics.com`.

SASInstitute. *JMP 11.0*. SAS Institute Inc., 2015. URL `https://www.jmp.com/en_us/software.html`.

H. Scheffé. Experiments with mixtures. *Journal of the Royal Statistical Societyl B*, 22: 344–360, 1958.

J. R. Smith and J. M. Beverly. The use and analysis of staggered nested factorial designs. *Journal of Quality Technology*, 13:166–173, 1981.

R. D. Snee. Graphical analysis of process variation studies. *Journal of Quality Technology*, 15:76–88, 1983.

R. D. Snee and D. W. Marquart. Extreme vertices designs for linear mixture models. *Technometrics*, 16:399–408, 1974.

R. D. Snee and D. W. Marquart. Screening concepts and designs for experiments with mixtures. *Technometrics*, 18:19–29, 1976.

H. M. Soo, E. H. Sander, and D. W. Kress. Definition of a prediction model for determination of the effect of processing and compositional parameters on the textural characteristics of fabricated shrimp. *Journal of Food Science*, 43:1165–1171, 1978.

L. Stahle and S. Wold. Analysis of variance (anova). *Chemometrics and Intelligent Laboratory Systems*, 6:259–272, 1989.

R. C. St.John. Experiments with mixtures, ill-conditioning, and ridge regression. *Journal of Quality Technology*, 16, 1984.

Student. The probable error of the mean. *Biometrika*, 5:1–25, 1908.

J. W. Tukey. Comparing individual means in the analysis of variance. *Biometrics*, 5: 232–242, 1949.

G. F. Weber and J. Erjavec. Simultaneous so_x/no_x, control, 1987.

G. Wernimont. Quality control in the chemical industry II statistical quality control in the chemical laboratory. *Industrial Quality Control*, 1947.

B. Wheeler. *AlgDesign: Algorithmic Experimental Design*, 2014.

R.E. Wheeler. Portable power. *Technometrics*, 16:193–201, 1974.

S. Yan, J. Li, X. Chen, J. Wu, P. Wang, and J Yao. Enzymatical hydrolysis of food waste and ethanol production from the hydrolysate. *Renewable Energy*, 36:1259–1265, 2011.

Index